PLANET MARS RESEARCH FOCUS

PLANET MARS RESEARCH FOCUS

LORENZO A. COSTAS
EDITOR

Nova Science Publishers, Inc.
New York

For permission to use material from this book please contact us:
Telephone 631-231-7269; Fax 631-231-8175
Web Site: http://www.novapublishers.com

LIBRARY OF CONGRESS CATALOGING-IN-PUBLICATION DATA

Planet Mars research focus / Lorenzo A. Costas, editor.
p. cm.
Includes index.
ISBN-13: 978-1-60021-826-2 (hardcover)
ISBN-10: 1-60021-826-1 (hardcover)
1. Mars (Planet) I. Costas, Lorenzo A.
QB641.P62 2006
523.43--dc22 2007031239

Published by Nova Science Publishers, Inc. ✦ New York

CONTENTS

PREFACE

This new book presents leading-edge research on Mars which is the fourth planet from the Sun in the Solar System. The planet is named after Mars, the Roman god of war. It is also referred to as the "Red Planet" because of its reddish appearance as seen from Earth.A terrestrial planet, Mars has a thin atmosphere and surface features reminiscent both of the impact craters of the Moon and the volcanoes, valleys, deserts and polar ice caps of Earth. It is the site of Olympus Mons, the highest known mountain in the solar system, and of Valles Marineris, the largest canyon. In addition to its geographical features, Mars' rotational period and seasonal cycles are likewise similar to those of the Earth Mars has two moons, Phobos and Deimos, which are small and irregularly shaped. These may be captured asteroids, similar to 5261 Eureka, a Martian Trojan asteroid. Mars can be seen from Earth with the naked eye. Its apparent magnitude reaches −2.9, a brightness surpassed only by Venus, the Moon, and the Sun, though for much of the year Jupiter may appear brighter to the naked eye than Mars.

Expert Commentary A - Mars at the present time is the second most studied planet of the Solar System after our own Earth. Its atmosphere, surface, and sub-surface have been the subjects of several successful spacecraft and robotic missions as well as theoretical studies carried out with numerical models of increasing degree of sophistication and complexity. The interest in investigating its remote past, both from the observational and the modeling point of view, has also grown.

At present, Mars is cold, dry, bombarded by radiation and covered by oxidants; a very hostile environment for life. According to recent observations from Mars Odyssey, Mars Express and the Mars Exploration Rovers, though, between 4.5 and 3.5 billion years ago it might have been a very different planet: stable liquid water on its surface might have created a warm and moist environment, prompting the emergence of life.

What then is the future of Mars? Is the concept of "climate change" relevant for its future?

Expert Commentary B - This work presents the calculation of the mass attenuation coefficient (μ) of Martian rocks and soils, in function of the energy, based on the knowledge of their chemical compositions. The samples were analyzed by *Viking 1*, *Viking 2* and *Mars Pathfinder* missions, all of them sent to different places at the planet Mars, in different times. The WinXCOM software was employed to determine the μ values for the samples in the range from 1 keV to 100 GeV. The obtained values were practically the same for energies larger than 100 keV, but marked differences among the samples were observed for energies below 25 keV, which is the energy range of interest for the XRF system used in space probes.

These μ data calculated for several energies of Martian soils and rocks will also be useful in calculations or modeling related to new missions sent to Mars.

Chapter 1 - Robotic reconnaissance operations are called for in extreme environments such as space (including planetary atmospheres, surfaces, and subsurfaces), and hazardous or inaccessible areas on Earth. A fundamentally new planetary exploration mission concept, *Tier-Scalable Reconnaissance*, originated by Wolfgang Fink at Caltech and his collaborators, replaces engineering and safety constrained mission designs that perform local ground-level reconnaissance with rovers and landers, or global mapping with an orbiter. The tier-scalable paradigm integrates multi-tier (orbit ⇔ atmosphere ⇔ surface/subsurface) and multi-agent (orbiter(s) ⇔ blimps ⇔ surface/subsurface agents or sensors) hierarchical mission architectures, and enables reconnaissance in real time (e.g., monitoring of transient events) on global, regional, and local scales for high science return and fully autonomous robotic missions, providing independence from human intervention, yet permitting manual override at any level. A human operator may communicate to the orbiter(s), as well as command the airborne agents via the orbiters (and thereby command the ground-tier reconnaissance agents via the airborne agents). Or, a highly automated inter- and intra-tier operation mode enables autonomous reconnaissance missions, e.g., when the communication time lag prohibits meaningful teleoperation, or during times of communication blackouts. To support such tier-scalable reconnaissance missions, a high degree of operation autonomy is required, such as: (1) automatic mapping of an operational area from different vantage points; (2) automatic feature extraction and target/region-of-interest identification within the mapped operational area; (3) automatic target prioritization for follow-up examination; and (4) subsequent autonomous navigation and relocation of entire tiers or agents within tiers. The airborne and ground-tier agents can be inexpensive enough for the deployment of numerous, expendable agents that can address specific science-driven questions within the operational areas with a complementary suite of instruments. Tier-scalable reconnaissance would overcome the inherent challenge of geologic planetary surface exploration: Space/airborne agents possess overhead perspectives at different length scales and resolutions. They would map out areas of interest and acquire terrain data, processed through automated feature-extraction, feature data analysis, and science prioritization algorithms, to choose potential targets for in-situ examination by sensors or ground-agents and for determining safe passages to these targets. The sensors and ground-agents would gather in-situ data that complement the remote sensing data obtained by the space/airborne agents. Tier-scalable reconnaissance not only introduces mission redundancy and safety, but enables distributed, science-driven, and less constrained reconnaissance (both spatially and temporally) of prime locations on Mars, Titan, Venus, and elsewhere.

Chapter 2 - The determination of the nature and the spreading of the organic compounds in the universe are of particular importance for revealing the chemistry that led to the emergence of life on Earth. In spite of the great knowledge we acquired from spectroscopic observations, the extraterrestrial organic molecules detected remain rather simple and in a limited amount. However analyses of meteorites composition, as well as laboratory and theoretical works, let assume that more complex molecules, including species relevant to life, could be present in extraterrestrial environments and more particularly in the solar system. It thus appears that in situ analyses of solar system bodies of interest for astrobiology are necessary to evidence extraterrestrial prebiotic chemistry or life forms. In this case, any strategy for investigating extraterrestrial environments should: (i) focus on compounds of astrobiological interest; (ii) avoid contamination of the samples; (iii) cross various information to improve the interpretation of the analytical results.

Obviously, the main molecules of interest are those which are readily synthesised under plausible prebiotic conditions, resistant to diagenesis, and which play a major role in biochemistry. Among them, amino acids, the building blocks of proteins in all the living organisms, are currently the most studied even if they have not been detected so far and remain to be found. In the future in situ organic analyses, discriminating molecules produced abiotically from biotic ones will be added to the previous goals of identification and quantification of organic compounds. Two main ways seem currently suited for this objective: (i) the search for homochirality, and more particularly in amino acids; (ii) the detection of chemical biomarkers, by chemical and isotopic analysis.

The search for so many different organics in extraterrestrial environments is a challenging task regarding the difficulties faced to analyze low concentrated and small size samples without any contamination from the sample to the analysts and conversely. After focusing on the choice of the better targeted organic compounds for the in situ space exploration, the authors propose to review the past and future in situ analyses of organics, with a more particular interest for the future Mars Science Laboratory and Exomars missions to Mars. Indeed, these missions will carry Gas Chromatograph-Mass Spectrometer instrumentation, partly developed by the authors, which will be coupled with a chemical reactor. This association should allow the in situ analysis of complex organics which cannot have been detected by former missions, and also the chiral analysis of theses species. Hence, it could reveal for the first time traces of extraterrestrial prebiotic chemistry or even trace of life.

Chapter 3 - In the solar system and elsewhere, zones of habitability suitable for life are likely to be sparse, given the particularity of such life as science has encountered to date. Before any direct observation of present or past life on Mars is available, a forecast of the prospects for finding such a specimen must overcome many weaknesses in the existing body of knowledge. Can we be sure even whether natural or NASA-type transportation mechanisms could have already introduced terrestrial species to Mars, casting a "discovery" of the future in a notably less dramatic light? It can be shown that terrestrial extremophiles are capable of surviving across a range of Mars conditions. In this chapter, we will undertake a conceptual and ultimately statistical analysis of the probability for tracking down a rare but postulated existing lower-order life form on Mars. The Ward-Brownlee hypothesis, that microbial life forms are common in the universe but higher-order life forms are rare, is accepted as credible.

Chapter 4 - Biological activity can produce complex patterned structures in accretionary carbonate spheres (pisoliths) from hot springs. Pisoliths from a few millimeters to 50 mm in diameter were collected from a geyser at the high-salinity Arima Hot Springs near Kobe, Japan. High-resolution electron microscope images show that microorganisms produced concentric laminar layers of aragonite alternating with Si- and Fe-rich layers. Thin Fe-bearing layers are included in well-ordered aragonite layers. This Fe-bearing layer is identified as amorphous iron oxides by the diffuse ring pattern of both XRD and electron diffraction TEM. STEM elemental distribution maps show that the cementation of aragonite is associated with NaCl-rich bacterial cells and elevated phosphorous and sulfur concentrations. The filamentous microbes build the concentric framework for mineral laminas with a web-like network of microbial strands. The intricate patterns of mineralogical and bacterial variation in the pisoliths correlate with the change of water chemistry in the hot spring. These patterns could improve our understanding of nano-scale biomineralization. In addition, the terrestrial hot spring pisoliths might be a very counterpart of the tiny spherules recently found on Mars. Liquid water on Mars may persist underground and occasionally burst to the surface, forming pisolith-like spherules in the dry river canyons and craters after volcanic activity wanes. In Eagle Crater's outcrops, Opportunity found millimeter-size spherical grains (blue berries) that may have precipitated out of iron- or salt-bearing water as it evaporated. A geyser at the high-salinity and iron-rich Arima Hot Springs may have had a Mars-like climate. Recent results from Mars rovers and orbiters show that warm, wet conditions may have prevailed on the planet for long periods during its early history. If the eras of Earth-like conditions were frequent and long, life would have had a better chance of evolving on Mars. Terrestrial hot spring pisoliths might be a counterpart of the blue berries on Mars.

Chapter 5 - The analytical application of Raman spectroscopy to the identification of the survival strategies of terrestrial extremophiles through the characterisation of key biomolecular and biological signatures has provided novel information about the roles of protective chemicals synthesised by the organisms under environmentally stressed conditions. The non-destructive sampling requirement and the ability to discriminate between organic biomolecular spectral signatures and those from the inorganic host matrix in the geological record are essential requirements for the assessment of the survival strategies being employed against low wavelength ultraviolet radiation insolation, desiccation and extremes of temperatures found in terrestrial habitats termed Mars analogue sites (Table 1).

The recent announcement of the inclusion of a miniaturised Raman spectrometer as part of a combined Raman /LIBS instrument for the molecular and elemental analytical characterisation of specimens in the Martian surface and subsurface as a component of the European Space Agency ExoMars life-detection suite has focussed attention on the Raman spectra of key biomarkers of life signatures. To search for indicators of extinct or extant life signatures on Mars serves to highlight the importance of Raman spectroscopic techniques in this astrobiological mission. As a forerunner to the ExoMars mission it is imperative that the Raman spectroscopic data from extremophilic colonies from as wide a possible range of terrestrial sites of relevance to Mars is undertaken. This chapter will address the type of information that is required and which is accessible from these terrestrial studies to adequately prepare the way for the interpretation of analytical scientific data that will emerge from the Mars exploration programme, commencing with the launch of ExoMars in 2013.

Chapter 6 - Manned missions to Mars will result in a diminution of immune status among the crew and cause profound changes to the composition of the bacterial flora in the intestinal tract, nasal passages and upper airways, leading to an increased risk of infection. Experiments conducted during short-term flight and aboard orbiting space stations suggest that growth in microgravity leads to increases in bacterial antibiotic resistance and to cell wall changes. Growth under low-shear modelled microgravity (LSMMG) has convincingly shown that a reduced gravitational field acts as an environmental signal for expression of enhanced bacterial virulence in Gram-negative pathogens such as *Salmonella enterica*. Recent microbiological surveillance of the International Space Station has identified members of the bacterial genus *Staphylococcus* as major contaminants of working and living quarters. We therefore examined the effect of LSMMG on parameters of antibiotic susceptibility and virulence in clinical isolates of *Staphylococcus aureus*. This bacterial species is likely to be the most problematical staphylococcal opportunist encountered in transit. Three strains were grown under LSMMG in a High Aspect Ratio Vessel and compared with cells grown under normal gravity (NG) in the same vessel. There were no significant differences in the antibiotic susceptibility, growth rate or morphology of staphylococci grown under LSMMG compared to NG. LSMMG-induced reductions in the synthesis of the pigment staphyloxanthin were noted. Strains secreted less protein under LSMMG and reductions in hemolysin secretion were found. We noted reduced expression of hemolysins under LSMMG conditions; these virulence determinants are key components of the staphylococcal armamentarium, enabling the pathogen to cause damage to host cells and tissues. Thus, in contrast to published data on Gram-negative pathogens, modelled microgravity reduces the expression of key virulence determinants of *S. aureus*. The relevance of these findings to the risk of infection is discussed in relation to other factors that will impinge on the success of Martian expeditions.

Chapter 7 - Today we have compelling geomorphologic and geochemical evidence for the presence in the past of vast quantities of liquid water on Mars. The numerous and diverse erosive, depositional and chemical features that characterize the Martian surface include ocean-related landforms, massive layered outcrops, valley networks, dendritic gullies, anastomosing and meandering rivers, deltas and evaporitic paleolake deposits. These features are the remnant testimony of an era when the atmosphere, surface and subsurface of Mars were linked by a complex hydrological cycle, in which water melting, freezing, evaporation and sublimation processes were largely controlled by the chemical and physical characteristics of the Martian atmosphere and surface.

Traditionally, the presence of liquid water at any given time on Mars has been explained by invoking two controversial and mutually exclusive models: the "hot and wet" Mars, where an atmosphere rich in greenhouse-gases enabled surface temperatures close or above the freezing point of water, and the "cold and dry" Mars, with a hydrological cycle mainly controlled by thermal energy sources that locally and episodically enabled the presence of liquid water on the planet's surface. Both models have difficulties to explain the timing, diversity and number of erosive and depositional features that characterize the Martian atmosphere, and have been contradicted by geochemical data and atmospheric models of Mars.

Based on the observation that weathering fluids loaded with high abundances of diverse salts can remain in the liquid state for temperatures well below 0 °C in the presence of a thick CO_2 atmosphere, we envision an alternative model that accounts for the origin and fate of water throughout the entire history of Mars. During the Noachian (>3.7 Gyr. ago) Mars was a cold and wet planet with a planetary hydrological cycle that witnessed the formation of glaciers, lakes, rivers and oceans. In this "cold and wet" Mars, massive glaciers and ice-rich regolith accumulated vast amounts of ice water in the southern highlands. The seasonal partial melting of the glaciers and regolith at lower latitudes, resulted in fluvial-lacustrine systems in which meltwater was transported through large river valleys and accumulated in crustal aquifers, confined lakes and ponds. This process originated the mature, terrestrial-like, dendritic valley networks observed in orbital explorations. The chemical reequilibration of glacier meltwater with iron- and sulfur-enriched soils produced high ionic level of superloaded acidic waters, with elevated concentrations of Fe, SO_4, Br, Na, Cl, Ca, and other ions. Such solutions would not require above-freezing temperatures to flow. Additionally, water filled the Martian lowlands forming extended and long-lasting acidic oceans and seas, partially covered by sea-ice at higher latitudes. During the local summer, periods of maximum volcanic activity and/or variations in planetary orbital parameters, ponded water would evaporate to the atmosphere, and rock and water glaciers would undergo sublimation and wasting. Water accumulated in the atmosphere would then precipitate almost exclusively in the form of snowfall, because of the low temperatures. The net accumulation of snow and its metamorphosis to ice form glaciers in the highlands again would close the hydrogeological cycle.

The proposed hydrological cycle was active only during episodes of dense atmosphere in the Noachian. During the Hesperian epoch (3.7-3.1 Gyr ago), Mars underwent a series of dramatic changes that resulted in the partial loss of its atmosphere, accompanied by a loss of vast quantities of its water reservoir and by progressive crustal assimilation of H_2O. The thinning of the atmosphere in a water depleted planet likely triggered the partial freezing of the ocean and other bodies of standing waters, resulting in an ice-covered ocean on the northern plains, and ice covered lakes and ponds in the southern highlands. This ultimately led to the interruption of the hydrological cycle and represents the onset of the colder and dryer conditions that characterize Mars today. The Hesperian is also characterized by the occurrence of catastrophic flow events.

The most recent Martian epoch, the Amazonian (3.1-present), has witnessed brief and intriguing wet episodes, in which the climate was relatively warm and moist, until geologically very recent times. Multiple glacial landscapes formed and phases of tropical mountain glaciation occurred even within the past few tens of millions of years, including water runoff during the past decade.

Chapter 8 - The ability of Raman spectroscopy for the detection of bio- and geosignatures, with little or no sample manipulation have led this technique to be considered by NASA and ESA for planetary exploration in future Mars missions.

Hostile environments, such as those with cold, hot, dry, acid or salty conditions, have been proposed as Martian analogues on Earth. These hazardous habitats have not been an obstacle for the adaptation and even proliferation of many different types of micro-organisms. The survival adaptations of micro-organisms in these environments extend to both inorganic and organic strategies, especially involving the use of a wide range of radiation protective pigments

The study by Raman spectroscopy of the strategies adopted by micro-organisms in extreme terrestrial environments is a pre-requisite for the search for life in other planets. Those strategies include the use of rocks as protective habitats, mineralogical changes, mobilization of minerals, as well as, the production of specific bio-molecules. Frequently, an admixture of organic and inorganic strategies has been detected.

In this work, a Raman spectroscopic characterization of geo- and bio-strategies developed by extremophile organisms in hostile environments is presented, such cold and hot deserts, acid environments or saline deposits;

The capability of Raman spectroscopy for the detection of these biological strategies is related to different spectroscopic parameters, such as laser wavelength, spectral resolution and laser power. The influence of these parameters on the observed Raman spectra could compromise the molecular assignments and therefore the compound identification, for what the effect of different parameters in the Raman spectrum have been evaluated, analysing the requirements for a miniaturized Raman spectrometer for planetary exploration and predicting the limits for unambiguous molecular identification from the observed spectra.

Chapter 9 - In this chapter, we contend that Mars has had global-scale plate tectonics.

Consequences can be seen today in the form of traces of continental rifting with well developed shear zones that are only compatible with successive, differently oriented, horizontal stress fields.

Plate-tectonic rifting also explains the banded magnetic anomalies and their observed local offsetting.

As on Earth, plate tectonics could have had a majour role in the development and sustainability of life on Mars. This would happen by providing carbon and water recycling and magnetic sheltering from space radiation.

In: Planet Mars Research Focus
Editor: L. A. Costas, pp. 1-5
ISBN: 978-1-60021-826-2

Expert Commentary A

FUTURE CLIMATE CHANGES ON MARS: SCIENCE FICTION OR POSSIBLE REALITY?

Luca Montabone*
The Open University, Department of Physics & Astronomy, Walton Hall,
Milton Keynes MK7 6AA (United Kingdom)
University of Oxford, Atmospheric, Oceanic and Planetary Physics,
Dept. of Physics, Parks Road, Oxford OX1 3PU (United Kingdom)

Mars at the present time is the second most studied planet of the Solar System after our own Earth. Its atmosphere, surface, and sub-surface have been the subjects of several successful spacecraft and robotic missions as well as theoretical studies carried out with numerical models of increasing degree of sophistication and complexity. The interest in investigating its remote past, both from the observational and the modeling point of view, has also grown.

At present, Mars is cold, dry, bombarded by radiation and covered by oxidants; a very hostile environment for life. According to recent observations from Mars Odyssey, Mars Express and the Mars Exploration Rovers, though, between 4.5 and 3.5 billion years ago it might have been a very different planet: stable liquid water on its surface might have created a warm and moist environment, prompting the emergence of life [2].

What then is the future of Mars? Is the concept of "climate change" relevant for its future?

We are learning from the case of the Earth that there are two causes of climate change on a planet: a naturally induced climate change, which is usually a slow process due to changes in orbital parameters, internal instabilities of the climate system or the occurrence of life, and an artificially induced climate change, due to the action of intelligent beings (the human race in the case of the Earth).

The future of Mars can certainly include the former: the planet is likely to undergo dramatic changes in the tilting of its rotational axis, although on a time scale of million of years, as it already happened in the past [8]. On a much shorter time scale (years), the

* Email: l.montabone@open.ac.uk or montabone@atm.ox.ac.uk

observed shrinking of the deposits of frozen carbon dioxide near Mars' south pole and the expansion of the pits formed in the planet's perennial southern polar ice cap suggest a climate change in progress [10], although this conclusion is still controversial, as such phenomena might just be some evidence of inter-annual variability. What about the second possible cause? Could an artificially induced climate change be foreseen on Mars?

NASA, ESA and other Space Agencies worldwide advocate the human exploration of the Red Planet on a time scale of about 30 years from now, with a remote possibility of future human settlements in the long term. NASA, for instance, strengthened the concept of human exploration of Mars in its "Vision for the Space Exploration" programme after President Bush's speech on January 14th, 2004, and ESA replied with a specific programme of exploration of the Solar System (the "Aurora Programme") which also includes the future human involvement on the planet. If this scenario is realized, the future of Mars might hold the certain presence of life on its surface, in the form of human beings, for short or not-so-short periods. What could happen next?

If the human exploration of Mars will only be a one-off event, there is little if any reason to think about any possible artificial change of the Martian climate. Still, natural climate changes on the planet would be of the highest scientific interest, not to mention the possible ramifications for the study of climate changes on the Earth. Nevertheless, the human race might realize in the future that Mars could be or should be a possible second home planet, for whatever reason. In this case, the demand of improving its habitability will become an extremely important issue.

This issue started to be scientifically addressed in the early 70's [1, 3, 12], then had a further boost in the 90's (see [5], which includes an exhaustive bibliography). Currently, it is a concept for which several "Mars societies" worldwide (http://www.marssociety.org) are trying to lobby, after the publication of the popular book by Robert Zubrin, "The Case for Mars", in 1996 [14]. Laboratory experiments to study the feasibility of increasing the habitability on Mars are currently being carried out, mostly from the engineering and biological points of view. See [4], for instance, as an example of review article on nanotechnologies applied to the idea of "terraforming".

Terraforming a planet is therefore a concept which is no longer exclusive to science fiction. Terraforming ultimately means artificially inducing changes on an extraterrestrial planetary environment in order to make it similar to our Earth, for the sake of improving the habitability to the advantage of life. I write "life" and not specifically "the human race" because in my view human beings are not the only beings worth of having a chance on another planet.

In a certain sense, we (as human beings) have been transforming our own planet for centuries, and this transformation process has assumed the dramatic connotations of "climate change" and "global warming" in the most recent years, with an exponential increase in the scale of the impact on the Earth's atmosphere and climate. Therefore, the concept of terraforming an extraterrestrial planet on a local, regional or global scale can be considered as a challenging and cutting-edge engineering task ("planetary engineering") which finds its pragmatic bases and motivations in our direct experience on the Earth.

I will not enter any discussion here about the advantages or disadvantages of artificially inducing changes in the Martian environment, nor any ethical consideration about the responsibility of the terraforming concept. The interest of my commentary is merely scientific, and is aimed at suggesting a possible direction to find answers to three specific questions which could arise after having assumed that humans will decide one day that improving the habitability of Mars is a useful, if not necessary, task.

- How will the climate on Mars be affected by any artificially induced change?
- What should the magnitude of such a forcing be in order to obtain different degrees of habitability?
- What is the time scale of such climate changes and how stable can they be?

In other words, the interest is to study if, how, how much, when and for how long the habitability of Mars can be improved by introducing changes in the Martian climate system and environment provided by possible planetary engineering ideas. It is my view that the only way to find exhaustive answers to these questions is through numerical simulations of the future Martian climate.

Currently, the state-of-the-art of the numerical models of the atmosphere of Mars is extremely advanced. There are few global circulation models (GCMs) and mesoscale (or regional scale) models available worldwide, all having a completeness, accuracy and reliability comparable to those of Earth's climate models (see [6, 7, 11, 13, 15] and reference therein). The results on the present-day Martian climate which are provided by such models are continuously compared and validated against observations, and improvements in physical parameterizations are endlessly carried out, along with the increase of our knowledge of the Martian environment by means of ground-based, robotic and remote-sensed exploration.

Recently, some world-leading numerical models of the global Martian atmosphere have been re-formulated to simulate the remote past of Mars, stimulated by the new observations which seem to point towards an early epoch on Mars when aqueous processes occurred (see [9] for a review on the topic). The challenge of adapting these GCMs to the conditions which might have been present on Mars billions of years ago is remarkable, because of the important role played by water in all its states on a planet which might have been much warmer, with a much denser atmosphere than the one it is left with at the present time. The experience of the numerical modelers with the atmosphere of the Earth helped in successfully achieving this task.

Given the degree of sophistication and reliability that the numerical models of the Martian atmosphere have achieved in studying the present Martian climate and simulating its remote past, what should prevent their use to investigate the future of the climate of the Red Planet? This investigation then can be aimed either toward the natural evolution of the Martian climate or the artificially induced evolution suggested by the planetary engineering. Conversely, it might also stimulate the planetary engineering itself.

Using these numerical models to simulate the future will be an absolutely non-trivial task: many physical processes have to be taken into account, parameterized and analyzed. Some of these processes are obviously neglected for present climate studies, and only partially resolved for the past climate simulations performed to date. Examples of new processes that will have to be introduced in the models are the activation of the complete water cycle in all its complexity (hydrology) and the increased interaction between water vapour and dust (cloud formation and scavenging). Examples of physical processes that will have to be analyzed in detail are the positive (or negative) feedbacks produced by an increased carbon dioxide greenhouse effect (the possible "runaway greenhouse effect"). The parameterization of all these processes in the models will probably have to be adapted step-by-step to the evolving future climate scenario.

Nevertheless, the experience gathered for the Earth will be invaluable to adapt the present-day Martian GCMs and mesoscale models to the requirements for simulating the future, and very valuable feedbacks to the scientific community of Earth's modelers can be foreseen.

It seems beyond doubt to me that, at the present time, all research into issues such as artificially induced climate changes, terraforming, improving habitability of an extraterrestrial planet on a short or long time scale, whether applied to Mars or anywhere else, has an intrinsic motivation in defining the boundaries of the possible. Numerical simulations will obviously not eliminate the need for further, detailed, step-by-step exploration of Mars, but at the advanced stage of the current models of the Martian atmosphere and climate, they can:

- scientifically establish the feasibility of the tasks of planetary engineering;
- define pragmatically the boundaries of the possible on a long time scale, in the same way as they are helping to define the boundaries of the global warming on the Earth;
- help to foresee valuable routes into the future.

Last but not least, we will learn an even greater amount of things about our own planet, its climate change and the way climate systems respond to external forcing.

Numerical simulations of future climate changes on Mars (both natural and artificially induced) could be an invaluable way to use Mars right now as a planetary laboratory for the sake of the Earth, the future of the human race and of life in general, without having an immediate need to deal with any ethical, political, economical or technological issues.

References

[1] Averner, M.M. & MacElroy, R.D. (1976). *On the habitability of Mars: An Approach to Planetary Ecosynthesis*, NASA SP-414.

[2] Bibring, J.P. et al. (2006). Global Mineralogical and Aqueous Mars History Derived from OMEGA/Mars Express Data, *Science* 312, 400-404.

[3] Burns, J.A. & Harwit, M. (1973). Towards a More Habitable Mars-Or-The Coming Martian Spring, *Icarus,* 19, 126.

[4] Coppinger, R.J. (1996). The Drexlerian Terraforming of Mars: A New Ark for Humanity, in Newsletter of the *Molecular Manufacturing Shortcut Group*, v4, N.1, First Quarter.

[5] Fogg, M.J. Terraforming Mars: A Review of Research, *http://www.users.globalnet.co.uk/~mfogg/paper1.htm*

[6] Forget, F., et al. (1999). Improved general circulation models of the Martian atmosphere from the surface to above 80 km, *Journal of Geophys. Res.* 104, 24155-24176.

[7] Haberle, R.M., et al. (2003). The NASA/AMES general circulation model: model improvements and comparison with observations, *First workshop on Mars atmosphere modelling and observations*, Granada, Spain.

[8] Laskar, J., et al. (2004). Long term evolution and chaotic diffusion of the insolation quantities of Mars, *Astronomy & Astrophysics,* 428, 261-285.

[9] Lewis, S.R. & Read, P.L. (2006). Evidence for Climate Changes on Mars, in "*Solar System Update,*" Springer London.

[10] Malin, M.C. Caplinger, M.A. & Davies, S.D. (2001). Observational Evidence for an Active Surface Reservoir of Solid Carbon Dioxide on Mars, *Science* 294, 2146-2148.

[11] Mischna, M.A., et al. (2006). Development of a new global, scalable and generic general circulation model for studies of the Martian atmosphere, *Second workshop on Mars atmosphere modelling and observations*, Granada, Spain.

[12] Sagan, C. (1973). Planetary Engineering on Mars, *Icarus* 20, 513.

[13] Takahashi, Y.O., et al. (2006). High resolution simulations of the Martian atmosphere with a general circulation model, *Second workshop on Mars atmosphere modelling and observations*, Granada, Spain.

[14] Zubrin, R. (1996). The Case For Mars: The Plan to Settle the Red Planet and Why We Must, *Free Press.*

[15] Wilson, R.J. & Hamilton, K.P. (1996). Comprehensive model simulations of thermal tides in the Martian atmosphere, *Journal of Atmos. Sci.* 53, 1290-1326.

In: Planet Mars Research Focus
Editor: L. A. Costas, pp. 7-15

ISBN: 978-1-60021-826-2

Expert Commentary B

MASS ATTENUATION COEFFICIENT OF MARS SAMPLES OVER 1 KEV - 100 GEV ENERGY RANGE

Anderson Camargo Moreira[1,*] and Carlos Roberto Appoloni[2]
Applied Nuclear Physics Laboratory (LFNA), Physics Department, State University of Londrina, Paraná, Brazil

ABSTRACT

This work presents the calculation of the mass attenuation coefficient (μ) of Martian rocks and soils, in function of the energy, based on the knowledge of their chemical compositions. The samples were analyzed by *Viking 1*, *Viking 2* and *Mars Pathfinder* missions, all of them sent to different places at the planet Mars, in different times. The WinXCOM software was employed to determine the μ values for the samples in the range from 1 keV to 100 GeV. The obtained values were practically the same for energies larger than 100 keV, but marked differences among the samples were observed for energies below 25 keV, which is the energy range of interest for the XRF system used in space probes. These μ data calculated for several energies of Martian soils and rocks will also be useful in calculations or modeling related to new missions sent to Mars.

Keywords: Mars, rocks, soils, mass attenuation coefficient.

INTRODUCTION

More and more space missions are using equipments for *in situ* measurements of physical and chemical properties of rocks and soils of spatial objects surfaces, like natural satellites, asteroids and planets, for example. In 1967 and 1968, the first North American unmanned

* Applied Nuclear Physics Laboratory (LFNA), Physics Department, State University of Londrina, Paraná, Brazil, P.O. Box 6001, Zip Code 86051-970 / Londrina – PR, Brazil. Tel.: +55-043-33714736; fax+55-043-33714166; [1]sandersoncamargo@uel.br; [2] appoloni@uel.br

missions, forwarded to the Moon (Turkevich et al., 1969; Franzgrote et al., 1970) an equipment with alpha particles backscattering technique was employed to determine the chemical composition of samples of the lunar surface. The same kind of equipment was also used in the missions sent to Mars in 1976, called *Viking 1* and *Viking 2* missions, which analyzed surface samples in two different localities of the planet, *Chryse Planitia* (or *The Plains of Gold*) and *Utopia Planitia*, respectively (Massambani and Mantovani, 1997). Such technique had its efficiency proven in studies with terrestrial rocks of well-known chemical compositions (Economou et al., 1970).

Later, a new apparatus, the Alpha Proton X-ray Spectrometer (APXS), with a more sophisticated technique for the determination of chemical compositions, was created. The APXS can detect almost all the elements in the periodic table, with the exception of hydrogen and helium. It uses the X-ray fluorescence and the alpha and proton particles backscattering together. The equipment was sent to Mars in the year of 1997, in a mission called *Mars Pathfinder*, and obtained new data on the chemical composition of soils and rocks samples of the Martian surface in *Ares Vallis* (Economou, 2001).

Considering that the information of the mass attenuation coefficients (μ) of samples is important in a wide range of energy for the considered nuclear methodologies, this work aimed the calculation of this coefficient with the WinXCOM software, based on the knowledge of the chemical compositions of Martian soils and rocks, for energies in the range from 1 keV to 100 GeV. These μ data calculated for several energies of Martian soils and rocks will also be useful in calculations or modeling related to new missions sent to Mars. The evolution of the chemical composition determination techniques, alpha backscattering and APXS, is mentioned in this work too.

THEORY

The mass attenuation coefficient μ, whose unit is cm^2/g, is defined by the equation of the exponential absorption that characterizes the passage of a parallel beam of electromagnetic radiation through the matter, known as the Law of Lambert–Beer given by:

$$I = I_0 e^{-\mu\rho x} \tag{1}$$

where x represents the material thickness used as sample, μ the of mass attenuation coefficient, ρ the physical density of the sample, I_0 the incident beam intensity in the sample and I the emergent beam intensity of the sample.

The dependence of m with the atomic properties of the samples is given by the following relationship:

$$\mu = \sigma_{tot}\left(ZN_A/A\right) \tag{2}$$

where Z is the atomic number, N_A the Avogadro's number, A the atomic mass of the sample and σ_{tot} the sum of the cross–sections' contributions of the radiation photon interactions with the matter.

The sum of the cross-sections of the principal processes is given by (Hubbell, 1982)

$$\sigma_{tot} = \sigma_P + \sigma_R + \sigma_C + \sigma_{P_P} \tag{3}$$

where the indexes P, R, C and P_p designate the photoelectric effect, Rayleigh and Compton scattering, pair production, respectively.

CALCULATION

Calculations of the mass attenuation coefficients of several samples were carried out by the WinXCOM program (Gerward et al., 2001). The software can generate cross-sections and attenuation coefficients for elements, compounds or mixtures in the energy range between 1 keV and 100 GeV, in the form of total cross-sections and attenuation coefficients as well as partial cross-sections of the following processes: incoherent scattering, coherent scattering, photoelectric absorption, and pair production in the field of the atomic nucleus and in the field of the atomic electrons. For compounds, the quantities tabulated are the partial and total mass interaction coefficients.

The program possesses a comprehensive database for all elements over a wide range of energies, constructed through the combination of incoherent and coherent scattering cross-sections from Hubbell et al. (1977) and Hubbell et al. (1980), photoelectric absorption from Scofield (1973), and pair production cross-sections from Hubbell et al. (1980). The same cross-sections were used as in other recent tabulations of Hubbell (1977), Hubbell et al. (1980), Hubbell (1982) and Hubbell and Seltzer (1995).

The mass attenuation coefficients of 16 Mars samples i.e., three samples analyzed by the *Viking 1* mission, one sample by the *Viking 2* mission (Massambani and Mantovani, 1997) and 12 samples by the *Mars Pathfinder* mission (Economou, 2001) were calculated by the Win XCOM code.

EXPERIMENTAL

Table 1 presents the chemical compositions of samples analyzed by the *Viking 1* and *Viking 2* missions.

The alpha particle scattering technique used by the *Viking 1* and *Viking 2* missions, employed a 242Cm source of 6.11 MeV, silicon detectors and appropriate electronics, as described in Paterson et al. (1965). APXS used in the *Mars Pathfinder* mission employed a 244Cm source, which emits a monochromatic beam of 5.8MeV alpha particles. The principles of the APXS technique can be found in full detail in Economou et al. (1970).

Table 2 presents the chemical compositions of samples analyzed by the *Mars Pathfinder* mission, respectively. The data are expressed in the form of oxides percentage.

Table 1. Chemical compositions of four samples analyzed by the *Viking 1* and *Viking 2* missions (Massambani and Mantovani, 1997)

Chem. Comp.	*Viking 1-A*	*Viking 1-B*	*Viking 1-C*	*Viking 2*
Mg	5.0	5.2		
Al	3.0		2.9	
Si	20.9	20.8	20.5	30.0
S	3.1	3.8	3.8	2.6
Cl	0.7	0.8	0.9	0.6
K	<0.25			
Ca	4.0	3.8		3.6
Ti	0.5			0.61
Fe	12.7	12.6	13.1	14.2
O	50	49.7		50.4

Table 2. Chemical compositions (in oxides), of samples analyzed by the *Mars Pathfinder* mission (Economou, 2001)

Chem. Comp.	Soil A-2	Soil A-4	Soil A-5	Soil A-8	Soil A-9	Soil A-10	Soil A-15
Na_2O	3.1	3.1	3.8	2.4	2.3	2.5	2.9
MgO	10.3	9.5	8.4	7.6	7.7	8.8	8.0
Al_2O_3	10.0	10.2	10.0	10.2	9.9	9.4	9.6
SiO_2	40.9	40.9	40.5	45.4	41.9	41.0	43.1
P_2O_5	0.9	1.1	0.5	0.5	0.7	0.6	0.5
SO_3	6.0	7.0	5.7	5.6	6.7	6.4	5.3
Cl	0.7	0.8	0.8	0.9	1.3	0.9	0.8
K_2O	0.5	0.5	0.5	0.9	0.7	0.4	0.7
CaO	6.1	5.6	6.1	7.1	6.5	6.0	5.6
TiO_2	0.8	1.3	0.7	0.9	1.1	1.0	1.0
Cr_2O_3	0.3	0.4	0.5	0.1	0.2	0.3	0.3
MnO	0.5	0.4	0.2	0.3	0.1	0.4	0.3
FeO	20.0	19.2	22.2	18.2	21.0	22.4	21.9

RESULTS

Figure 1 presents the behavior of the mass attenuation coefficient (cm^2/g) versus energy (keV) showing the partial contributions of the effects Compton, coherent, photoelectric and pair production for the soil sample A-5 of Mars. The Compton effect is dominant from E*4*120 keV to approximately 9 MeV, showing that, in this energy range the Compton scattering, which is directly proportional to *Z*, can be used for the direct measuring of the density of the material or of the average atomic number.

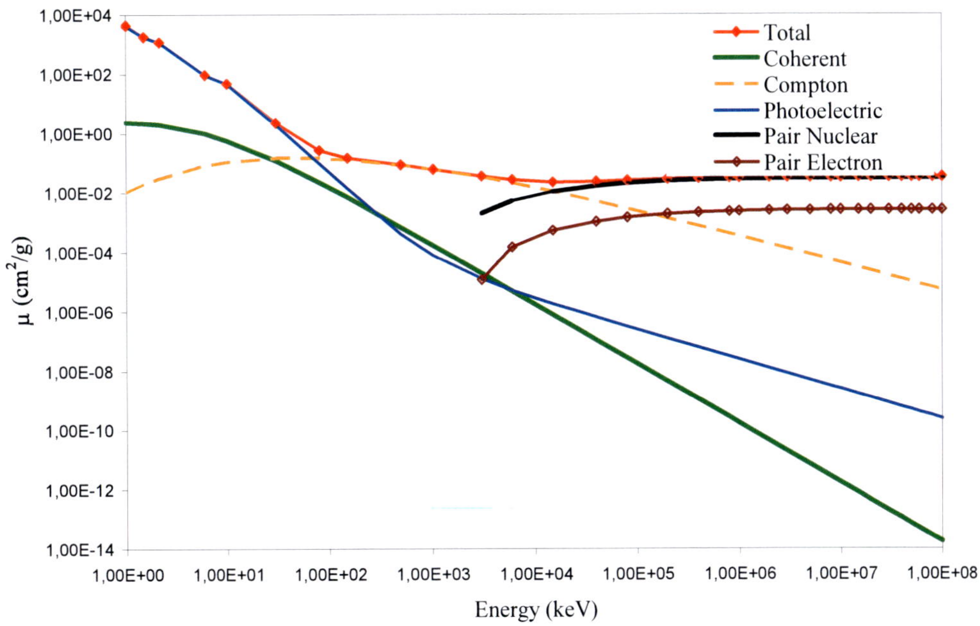

Figure 1. Total mass attenuation coefficient total and the individual contributions of the processes of radiation interaction with the matter versus energy, for Mars Pat. A-5 sample.

Table 3 gives the values of μ for the energies 1, 10, 100 keV, 1, 100MeV, 1 and 100GeV. Figure 2 presents the graph of μ (cm^2/g) versus energy (keV) for four samples of the *Viking 1* and *Viking 2* missions, in the energy interval between 10 and 50 keV.

Table 3. Chemical compositions (in oxides), of samples analyzed by the *Mars Pathfinder* mission (Economou, 2001) (continued)

Chem. Comp.	Rock A-3	Rock A-7	Rock A-16	Rock A-17	Rock A-18
Na_2O	3.4	3.6	4.3	2.5	4.3
MgO	2.5	6.2	4.8	4.6	4.0
Al_2O_3	12.2	10.9	11.0	10.3	11.7
SiO_2	53.5	47.2	47.5	53.8	49.2
P_2O_5	0.7	0.5	0.5	0.4	0.5
SO_3	2.0	4.5	3.0	1.7	3.0
Cl	0.5	0.9	0.7	0.5	0.7
K_2O	1.2	0.8	0.8	0.9	1.0
CaO	5.7	6.6	6.9	7.7	5.9
TiO_2	0.7	0.9	0.9	0.7	0.9
Cr_2O_3	0.1	0.1	0.0	0.1	0.1
MnO	0.4	0.4	0.3	0.4	0.4
FeO	17.3	17.8	19.3	16.5	18.5

Figure 3 shows the mass attenuation coefficients of four samples analyzed by the *Pathfinder* mission in the energy interval between 10 and 50 keV, in a graph of μ (cm^2/g)

versus energy (keV). In this graphic, only curves for the larger and smaller μ samples values and two intermediaries are presented to facilitate visualization.

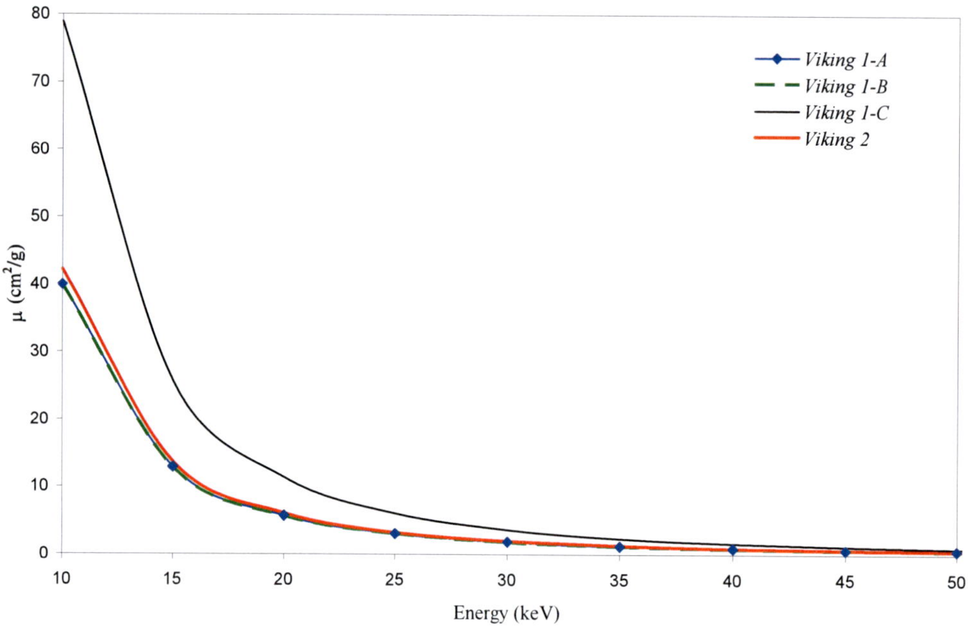

Figure 2. Mass attenuation coefficient versus energy for four samples analyzed by the *Viking* 1 and 2 missions, at 10 to 50 keV energy range.

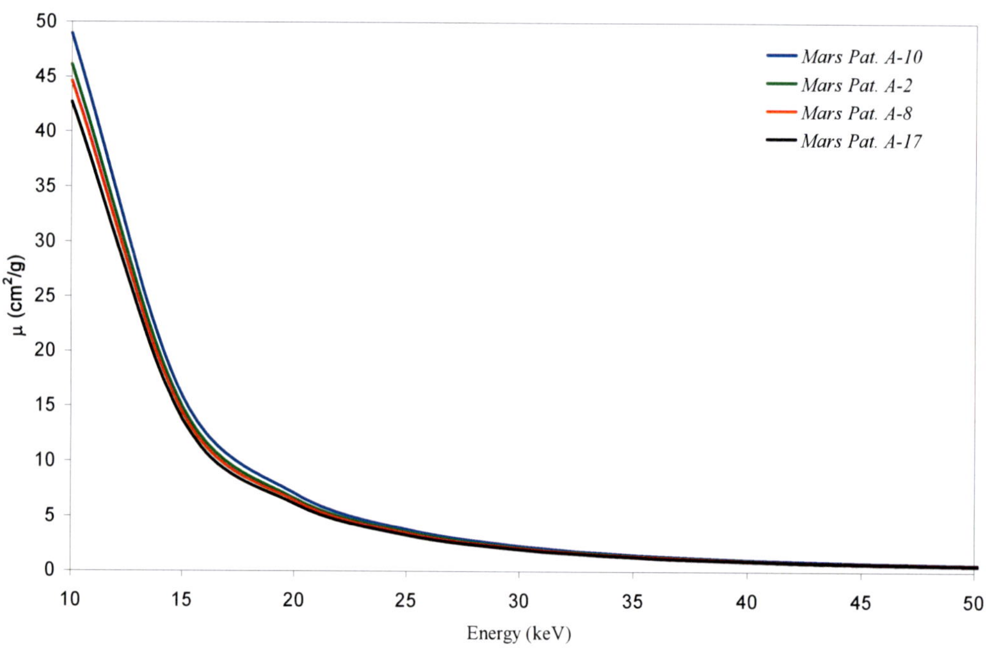

Figure 3. Mass attenuation coefficient versus energy for four samples analyzed by the *Mars Pathfinder* mission, at 10 to 50 keV energy range.

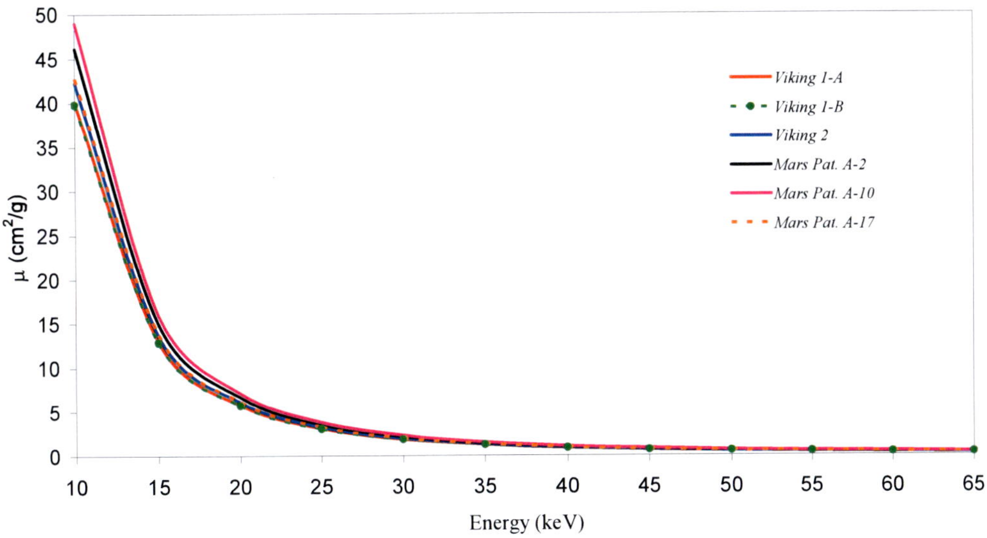

Figure 4. Mass attenuation coefficient versus energy for the Mars surfaces, at 10 to 65 keV energy range, analyzed by the three missions.

Finally, Figure 4 shows data for the mass attenuation coefficients of samples versus energy for 6 samples analyzed by *Viking 1* and *Viking 2* and *Mars Pathfinder* missions (and two techniques, alpha particles backscattering and APXS), in the range from 10 to 65 keV. In Figure 5 the same graph can be visualized, in the range from 10 to 25 keV, where it is possible to have a better differentiation between the samples.

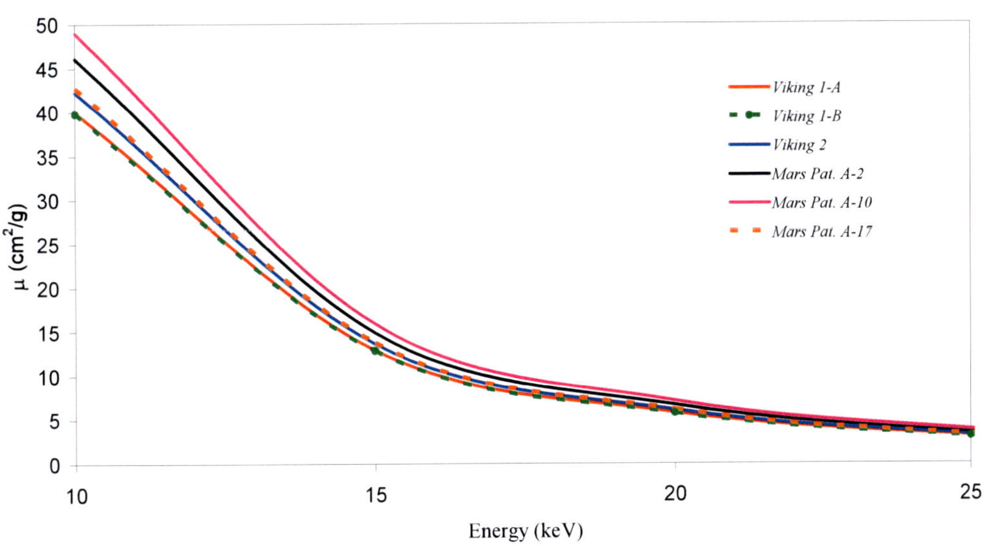

Figure 5. Mass attenuation coefficient versus energy for the Mars surfaces, at 10 to 25 keV energy range.

Calculations were accomplished taking into account the experimental deviations in the chemical compositions presented by the studies of the literature (Turkevich et al., 1969; Franzgrote et al., 1970; Paterson et al., 1969). However, these deviations did not influence significantly the values of μ, once the spread of μ deviations were typically in the interval between 0.2% and 0.9%.

The mass attenuation coefficients calculated were highly similar to each other comparing the *Vikings* mission results. However, Table 3 shows that this similarity does not prevail when data of μ for the *Viking 1* and *Viking 2* missions are compared with the data for the *Mars Pathfinder* mission. The differences found can be explained by the improvement in the technique for the determination of the chemical composition of samples used by the *Mars Pathfinder* mission (APXS), in relation to the alpha particles scattering technique used by the *Viking 1* and *Viking 2* missions, or, yet, by real differences in the chemical compositions of the soils analyzed by the three missions.

Table 4. Values of μ for the energies of 10 keV, 100 keV, 1 MeV, 100 MeV, 1 GeV and 100 GeV.

Samples	μ (cm^2/g)						
	1 keV	10 keV	100 keV	1 MeV	100 MeV	1 GeV	100 GeV
Viking 1-A	4181.495	39.95645	0.19632	0.06307	0.02510	0.03030	0.03185
Viking 1-B	4240.175	39.83447	0.19646	0.06315	0.02504	0.03021	0.03177
Viking 1-C	4039.236	78.84999	0.24463	0.06225	0.03281	0.04000	0.04203
Viking 2	4279.447	42.23796	0.19943	0.06312	0.02552	0.03083	0.03241
Mars Pat.A-2	4173.813	46.10381	0.20389	0.06281	0.02617	0.03163	0.03325
Mars Pat A-4	4148.842	45.30595	0.20276	0.06281	0.02605	0.03149	0.03311
Mars Pat A-5	4277.645	48.54769	0.20717	0.06274	0.02649	0.03205	0.03368
Mars Pat A-8	4120.306	44.60879	0.20176	0.06289	0.02603	0.03146	0.03307
Mars Pat A-9	4252.377	47.62878	0.20583	0.06280	0.02642	0.03196	0.03359
Mars Pat A-10	4313.219	48.96156	0.20775	0.06276	0.02656	0.03213	0.03377
Mars Pat A-15	4275.854	48.17412	0.20669	0.06277	0.02645	0.03199	0.03362
Mars Pat A-3	4055.612	42.78205	0.19938	0.06289	0.02574	0.03109	0.03269
Mars Pat A-7	4077.499	43.79681	0.20067	0.06287	0.02589	0.03129	0.03290
Mars Pat A-16	4144.801	45.45740	0.20292	0.06283	0.02611	0.03156	0.03318
Mars Pat A-17	4048.341	42.72681	0.19928	0.06295	0.02578	0.03115	0.03275
Mars Pat A-18	4100.688	44.28025	0.20135	0.06283	0.02593	0.03134	0.03295

The graph in Figure 2 shows the difference between the results of μ for the *Viking 1-C* sample in relation to the other samples. The lack of some chemical elements data for the sample, as shown in Table 1, could have caused such difference. When the data for chemical composition are inserted in the WinXCOM program, the sum of the element percentages should be equal to 1, however, it did not happen for sample 1-C, as can be seen at Table 1. In this case, the program normalizes the distribution of the percentages of the elements automatically. Due to this normalization, the program generates, for some elements, a larger concentration, resulting in different values of m. For example, in the case of the *Viking 1-C* sample, Al concentration instead of 2.9%, turned to be 7%, and Si concentration instead of 20.5%, turned to be 49.7%.

On the Table 1, some values of *Viking 1-B*, *Viking 1-C* and *Viking 2* samples chemical composition don't appear. These data were not determined by the system, probably due to operational conditions on martian surface.

Figure 5 and Table 3 show that the values of μ calculated for the samples were also similar for energies greater than 25 keV (excepting *Viking 1-C* sample). Although the differences among the values of μ at low energies are not large, they clearly reflect the differences in their chemical compositions.

More calculations and analysis made on a group of Earth and Moon surface samples, using the techniques cited above, can be find in Moreira & Appoloni (2006).

CONCLUSION

The mass attenuation coefficients calculated for the Mars samples using the *Mars Pathfinder* mission data were highly similar to each other, as well as the coefficients calculated for the *Viking 1* and *2* missions, with exception for the *Viking 1-C* sample. But differences between the μ results were found when the APXS technique data was compared with the alpha particles scattering technique data.

The values of μ for energies larger than 50 keV, are practically the same, but present significant differences for energies smaller than 25 keV, which is exactly the energy range of operation of the X-ray fluorescence systems used in space probes for the study of chemical compositions of materials. This information must be taken into account in the measurements performed by this methodology, especially for the samples auto attenuation corrections.

REFERENCES

Economou, T. *Radiat. Phys. Chem.* 2001, *61*, 191–197.

Economou, T. E.; Turkevich, A. L.; Sowinski, K. P.; Paterson, J. H.; Franzgrote, E. J. *J. Geophys. Res.*1970, *75*, 6514-6523.

Franzgrote, E. J.; Paterson, J. H.; Turkevich, A. L.; Economou, T. E.; Sowinski, K. P. *Science* 1970, *167*, 376–379.

Gerward, N.; Guilbert, K. B.; Jensen, B.; Levring, H. *Radiat. Phys. Chem.* 2001, *60*,23–24.

Hubbell, J. H. *Radiat. Res.*1977, *70*, 58–81.

Hubbell, J.H. *Appl. Rad Isot.* 1982, *33*, 1269–1290.

Hubbell, J. H.; Seltzer, S. M. *NISTIR 5632 National Institute of Standards and Technology, Gaithersburg, MD 20899, USA.* 1995.

Hubbell, J. H.; Veigele, W. J.; Briggs, E. A.; Brown, R. T.; Cromer, D. T.; Howerton, R. J. *J. Phys. Chem.*1977, *4*, 471–538 erratum in *6*, 615–616.

Hubbell, J. H.; Gimm, H. A.; Overbo, I. *J. Phys. Chem.* 1980, *9*, 1023–1147.

Massambani, O.; Mantovani, M. S. M. *Marte: Novas Descobertas*, Diagrama & Texto: Instituto Astronômico e Geofísico, IAG/USP, São Paulo SP, 1997; pp 51-135.

Moreira, A. C.; Appoloni, C. R., *Appl. Radiat. Isot.* 2006, *64*, 1065-1073.

Paterson, J. H.; Turkevich, A. L.; Franzgrote, E. J. *J. Geophys. Res.*1965, *70*, 1311.

Paterson, J. H.; Franzgrote, E. J.; Turkevich, A. L.; Anderson, W. A.; Economou, T. E.; Griffin, H. E.; Crotch, S. L.; Sowinski, K. P. *J. Geophys. Res.*1969, *74*, 6120-6148.

Scofield, J.H. *Lawrence Livermore National Laboratory Report* 1973, UCRL 51326.

Turkevich, A. L.; Franzgrote, E. J.; Paterson, J. H. *Science* 1969, *165*, 277–279.

In: Planet Mars Research Focus
Editor: L. A. Costas, pp. 17-48

ISBN: 978-1-60021-826-2

Chapter 1

TIER-SCALABLE RECONNAISSANCE: A PARADIGM SHIFT IN AUTONOMOUS REMOTE PLANETARY EXPLORATION OF MARS AND BEYOND

***Wolfgang Fink*[(1),*], *Mark A. Tarbell*[(1)] *and Farrah M. Jobling*[(2)]**
[(1)]California Institute of Technology, Visual and Autonomous Exploration Systems Research Laboratory, Division of Physics, Mathematics and Astronomy, Mail Code 103-33, Pasadena, CA 91125, USA,
[(2)]University of Colorado School of Medicine, Dept. of Microbiology, Mail Stop 8333, Aurora, CO 80045

ABSTRACT

Robotic reconnaissance operations are called for in extreme environments such as space (including planetary atmospheres, surfaces, and subsurfaces), and hazardous or inaccessible areas on Earth. A fundamentally new planetary exploration mission concept, *Tier-Scalable Reconnaissance*, originated by Wolfgang Fink at Caltech and his collaborators, replaces engineering and safety constrained mission designs that perform local ground-level reconnaissance with rovers and landers, or global mapping with an orbiter. The tier-scalable paradigm integrates multi-tier (orbit ⇔ atmosphere ⇔ surface/subsurface) and multi-agent (orbiter(s) ⇔ blimps ⇔ surface/subsurface agents or sensors) hierarchical mission architectures, and enables reconnaissance in real time (e.g., monitoring of transient events) on global, regional, and local scales for high science return and fully autonomous robotic missions, providing independence from human intervention, yet permitting manual override at any level. A human operator may communicate to the orbiter(s), as well as command the airborne agents via the orbiters (and thereby command the ground-tier reconnaissance agents via the airborne agents).

* Corresponding Author: Wolfgang Fink, Ph.D., Visual and Autonomous Exploration Systems Research Laboratory; Division of Physics, Mathematics and Astronomy; California Institute of Technology; 1200 E California Blvd; Mail Code 103-33; Pasadena, CA 91125; Tel: 626-395-4587; Fax: 626-395-4587; Email: wfink@autonomy.caltech.edu, Website: http://autonomy.caltech.edu

Or, a highly automated inter- and intra-tier operation mode enables autonomous reconnaissance missions, e.g., when the communication time lag prohibits meaningful teleoperation, or during times of communication blackouts. To support such tier-scalable reconnaissance missions, a high degree of operation autonomy is required, such as: (1) automatic mapping of an operational area from different vantage points; (2) automatic feature extraction and target/region-of-interest identification within the mapped operational area; (3) automatic target prioritization for follow-up examination; and (4) subsequent autonomous navigation and relocation of entire tiers or agents within tiers. The airborne and ground-tier agents can be inexpensive enough for the deployment of numerous, expendable agents that can address specific science-driven questions within the operational areas with a complementary suite of instruments. Tier-scalable reconnaissance would overcome the inherent challenge of geologic planetary surface exploration: Space/airborne agents possess overhead perspectives at different length scales and resolutions. They would map out areas of interest and acquire terrain data, processed through automated feature-extraction, feature data analysis, and science prioritization algorithms, to choose potential targets for in-situ examination by sensors or ground-agents and for determining safe passages to these targets. The sensors and ground-agents would gather in-situ data that complement the remote sensing data obtained by the space/airborne agents. Tier-scalable reconnaissance not only introduces mission redundancy and safety, but enables distributed, science-driven, and less constrained reconnaissance (both spatially and temporally) of prime locations on Mars, Titan, Venus, and elsewhere.

Keywords: Tier-scalable reconnaissance, space exploration, hazardous environment, mission autonomy, orbiter, airship, blimp, balloon, rover, lander, atmosphere, surface, subsurface, sensor platform, agents, agent control, navigation, round-robin concurrent commanding scheme, distributed computing, planetary internet, sensors, micro electro mechanical systems (MEMS), sensor interoperability, sensor data fusion, feature extraction, feature classification, science goal prioritization, geographic information system (GIS), multi-layer information system (MLIS), automated global feature analyzer (AGFA), planetary geology, exobiology

INTRODUCTION

Robotic reconnaissance operations are called for in extreme environments such as space, including planetary atmospheres, surfaces, and subsurfaces. The concepts driving conventional robotic planetary missions of exploration are entirely inadequate for the next generation of missions that will be necessary to pave the way for human exploration to Mars.

Conventional robotic planetary exploration scenarios have favored single lander/rover missions, being driven primarily by safety and engineering constraints, at the expense of mission redundancy and science return. Although rovers are generally capable mobile platforms, they are incapable of exploring multiple sites on a planetary surface that are separated from each other by significant distances (a crucial feature for a geologist; Figs. 1-3).

Rovers are also unlikely to explore potentially hazardous but scientifically interesting surface and subsurface terrains.

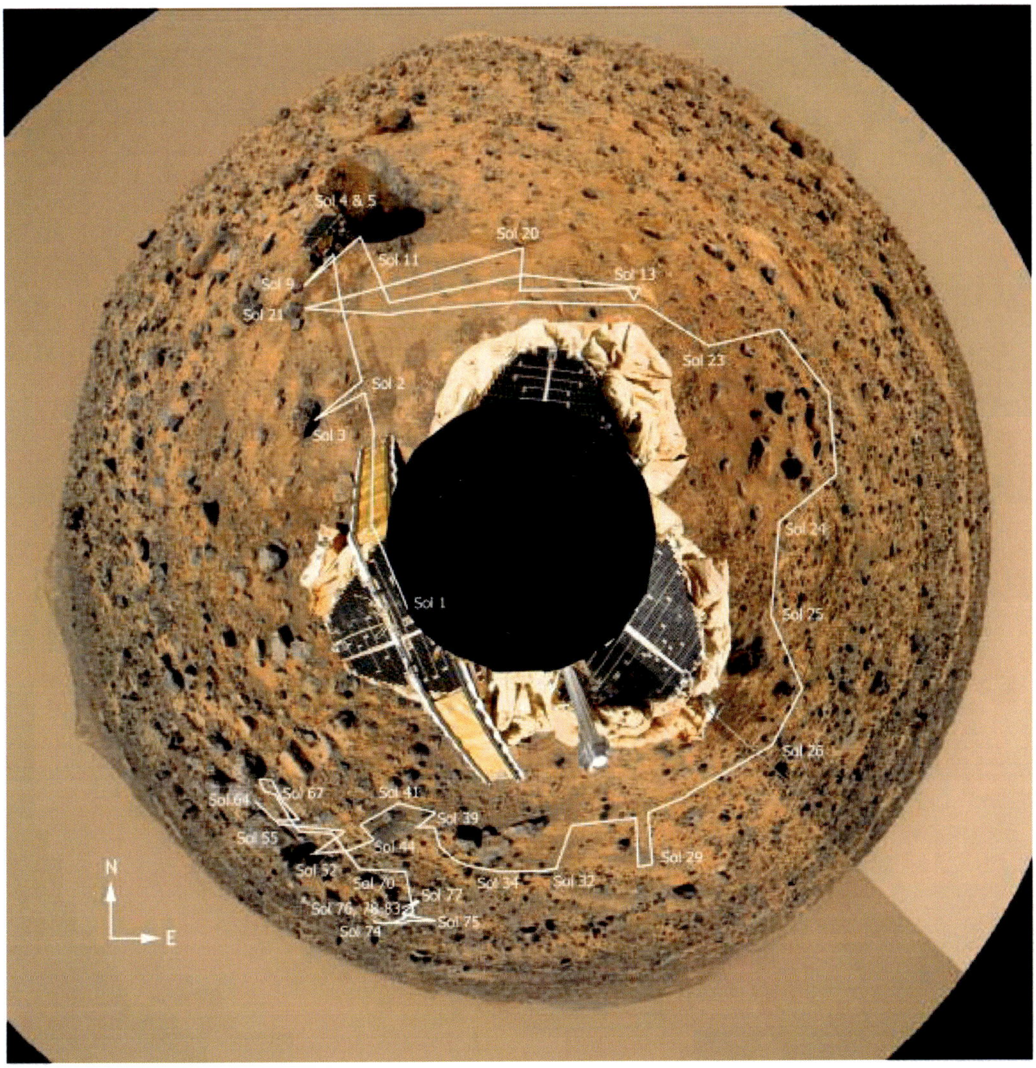

Figure 1. Delineation of the Sojourner rover traverse path during the 83 Sols (Martian days) of surface operations of the NASA Mars Pathfinder Mission. [Image courtesy of NASA/JPL-Caltech]

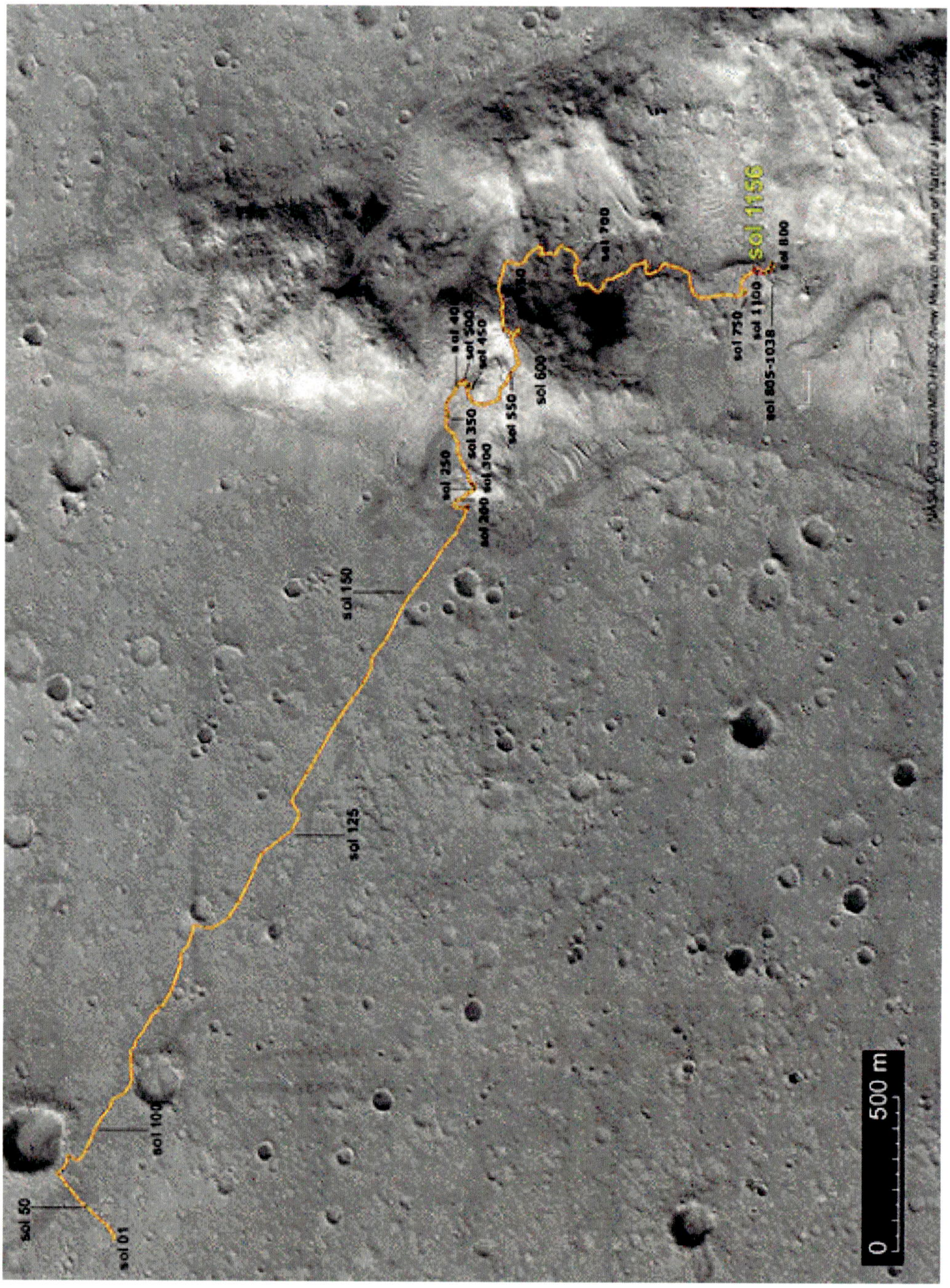

Figure 2. NASA Mars Exploration Rover Spirit's traverse map through Sol 1156. As of Sol 1162 (April 10, 2007), Spirit's total odometry was 7,092.29 meters (4.41 miles) since its landing at Gusev Crater on Mars on January 4, 2004. [Image courtesy of NASA/JPL/Cornell/MRO-HiRISE/New Mexico Museum of Natural History & Science]

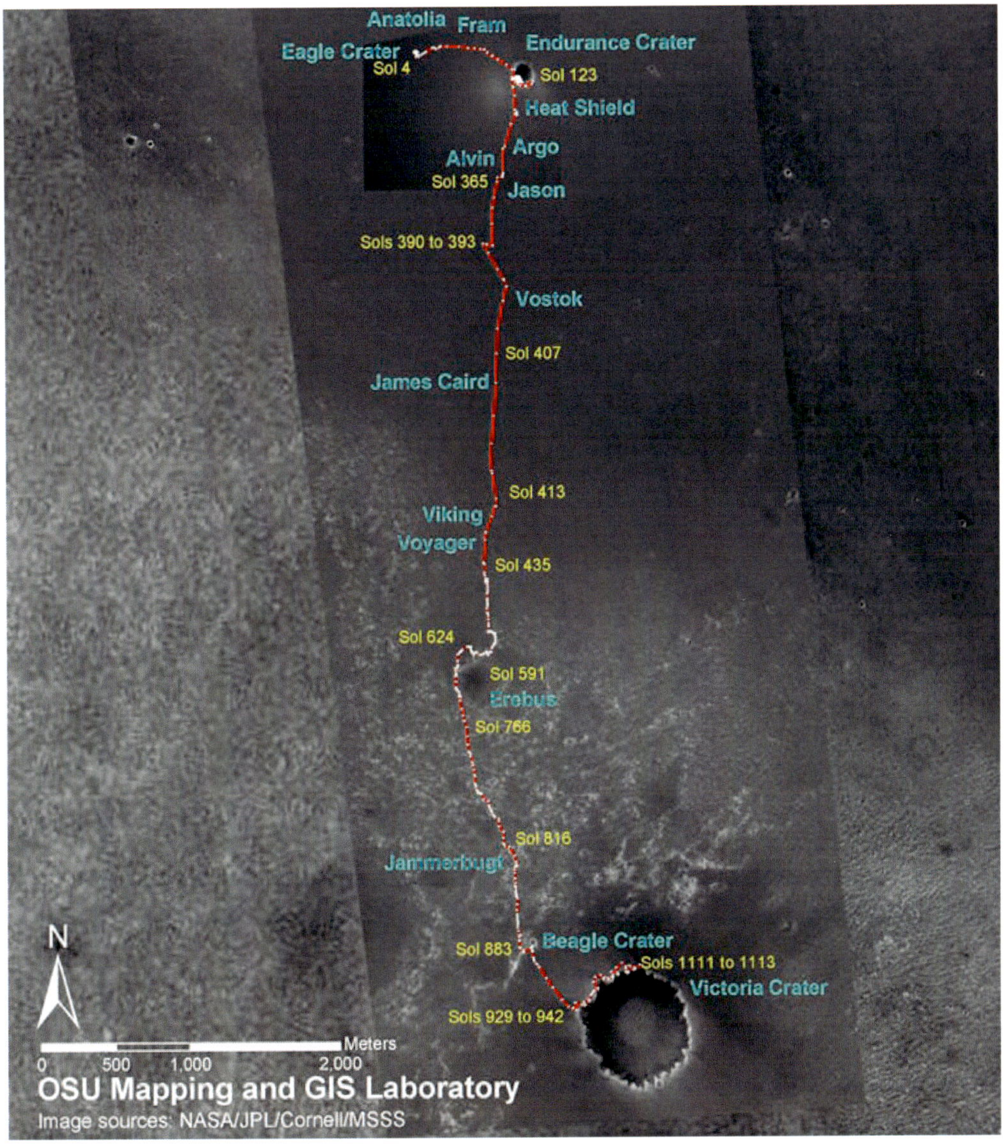

Figure 3. NASA Mars Exploration Rover Opportunity's traverse map through Sol 1113. As of Sol 1143 (April 12, 2007), Opportunity's total odometry was 10,443.41 meters (6.5 miles) since its landing at Meridiani Planum on Mars on January 25, 2004. [Image courtesy of NASA/JPL/Cornell/MSSS]

At the other extreme, orbiting satellites traditionally return immense global surveillance datasets, while often overlooking detailed information of local and regional significance. With the advent of the latest generation of orbiter missions, such as the ESA Mars Express and NASA Mars Reconnaissance Orbiter (MRO), high resolution optical instruments are now employed, such as the High Resolution Stereo Camera (HRSC) on Mars Express, and the High Resolution Imaging Science Experiment (HiRISE) on MRO. However, complete planetary surface coverage with these instruments is neither intended nor feasible (in a timely manner), and follow-up investigations of particular regions of interest (e.g., Figure 4) at high resolution are subject to the constraints of the satellite orbits, making monitoring of transient events difficult (e.g., Figs. 5-7), if not altogether impossible.

Figure 4. Impact crater located on Vastitas Borealis, a broad plain that covers much of Mars' far northern latitudes. The crater is 35 km wide with a maximum depth of approximately 2 km beneath the crater rim. The circular patch of bright material inside the crater is residual water ice. This image was taken with the High Resolution Stereo Camera (HRSC) aboard ESA's Mars Express. [Image courtesy ESA/DLR/FU Berlin (G. Neukum)]

To enable access to geologic sites not currently feasible on Mars or elsewhere, a completely new technological approach must be implemented. This is due to the need for mission redundancy, mission safety, greatly increased science return, and comparative analysis of spatio-temporal data, for optimal acquisition of geologic, geomorphologic, pedologic, (paleo)hydrologic, (paleo)climatic, and possible exobiologic information of Mars and other extraterrestrial targets. To effectively explore locations such as: (1) canyons, (2) mountain ranges, (3) chaotic and rugged terrain, (4) impact basins and craters, (5) volcanoes, and (6) liquid pools or lakes, in addition to other locales of scientific interest on the planetary bodies of the Solar System that are particularly crucial for exobiologic-oriented exploration in general, and sample return missions in particular, a radical shift from traditional mission concepts and designs is not only overdue, but indeed required.

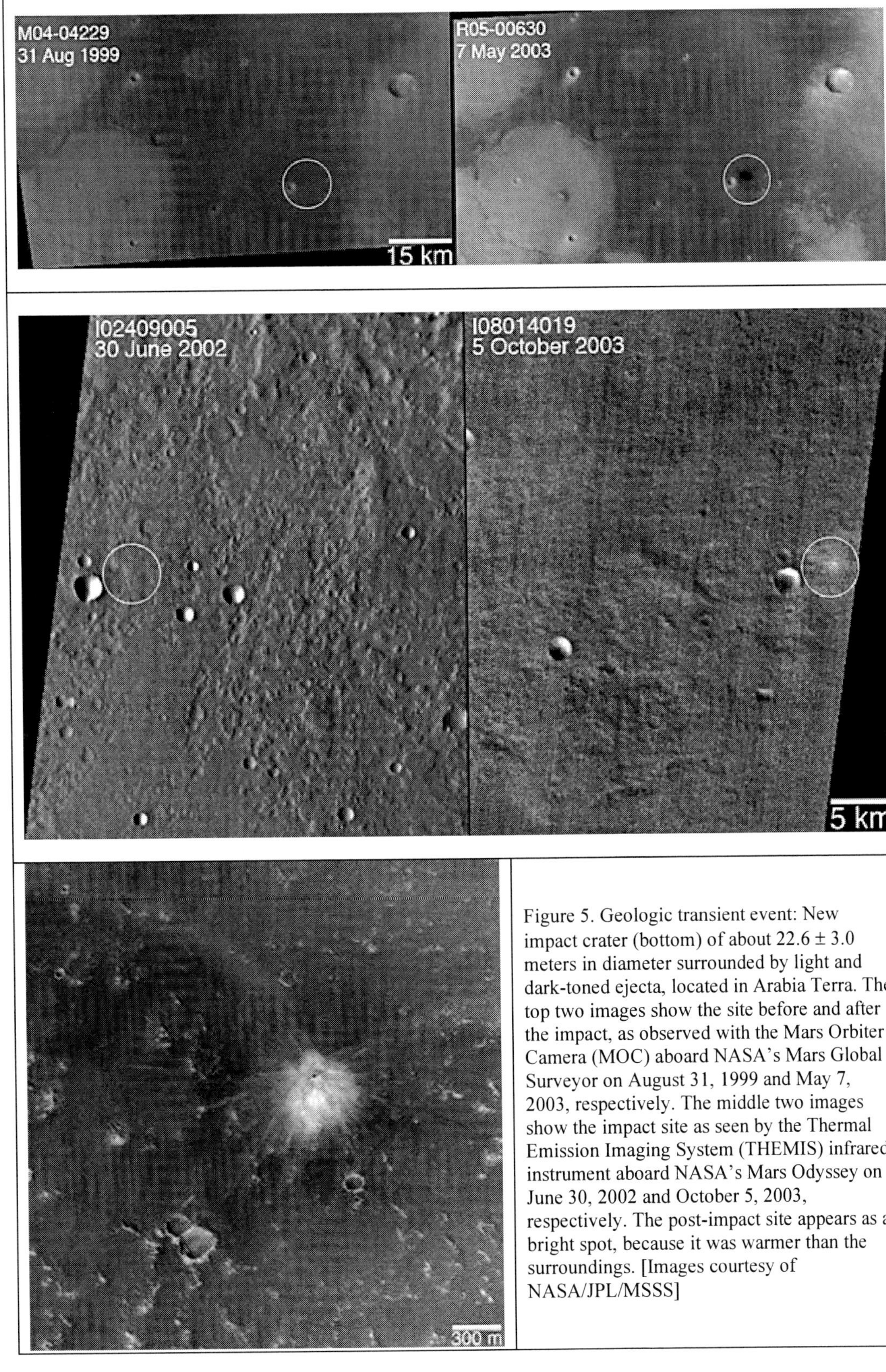

Figure 5. Geologic transient event: New impact crater (bottom) of about 22.6 ± 3.0 meters in diameter surrounded by light and dark-toned ejecta, located in Arabia Terra. The top two images show the site before and after the impact, as observed with the Mars Orbiter Camera (MOC) aboard NASA's Mars Global Surveyor on August 31, 1999 and May 7, 2003, respectively. The middle two images show the impact site as seen by the Thermal Emission Imaging System (THEMIS) infrared instrument aboard NASA's Mars Odyssey on June 30, 2002 and October 5, 2003, respectively. The post-impact site appears as a bright spot, because it was warmer than the surroundings. [Images courtesy of NASA/JPL/MSSS]

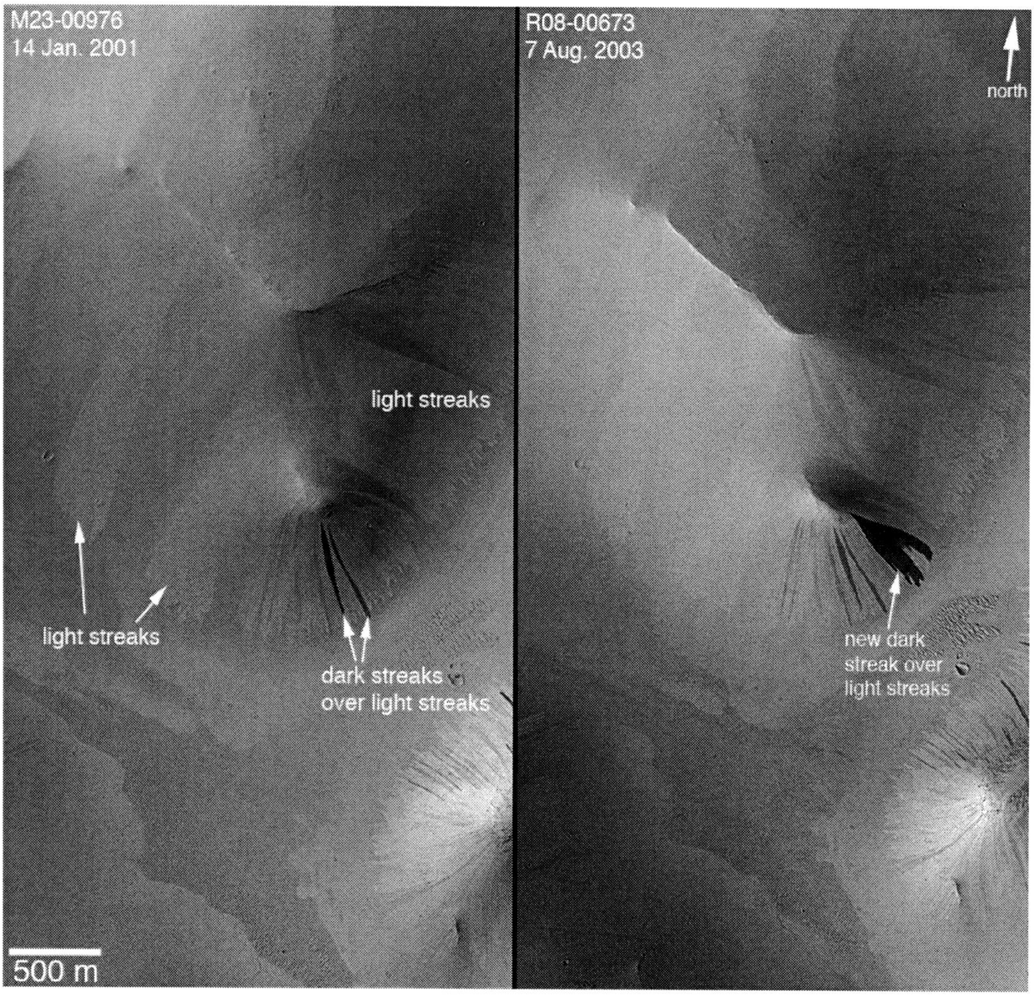

Figure 6. Geologic transient event: Light and dark slope streaks in Arabia Terra in an example where a new, large dark streak formed sometime between January 2001 and August 2003, observed by NASA's Mars Global Surveyor. The new dark streak formed on top of older light slope streaks. [Images courtesy of NASA/JPL/MSSS]

In this chapter, we report on a fundamentally new (scientific) reconnaissance mission concept, termed *Tier-Scalable Reconnaissance*, originated by Wolfgang Fink at the Visual and Autonomous Exploration Systems Research Laboratory (http://autonomy.caltech.edu) in the Division of Physics, Mathematics & Astronomy at the California Institute of Technology and his collaborators [1-9], of integrated multi-tier (orbit ⇔ atmosphere ⇔ surface/subsurface) and multi-agent (orbiter(s) ⇔ blimps ⇔ rovers, landers, drills, sensor grids) hierarchical mission architectures (Figure 8). This paradigm not only introduces mission redundancy and safety (as opposed to single-point failure paradigms), but enables intelligent, less constrained, and distributed science-driven planetary exploration on global, regional, and local scales, including comparative analysis of spatio-temporal data (obtained in transit), allowing for increased science return and paving the way towards fully autonomous robotic missions to Mars, Titan, Venus, and beyond (e.g., [1-10]).

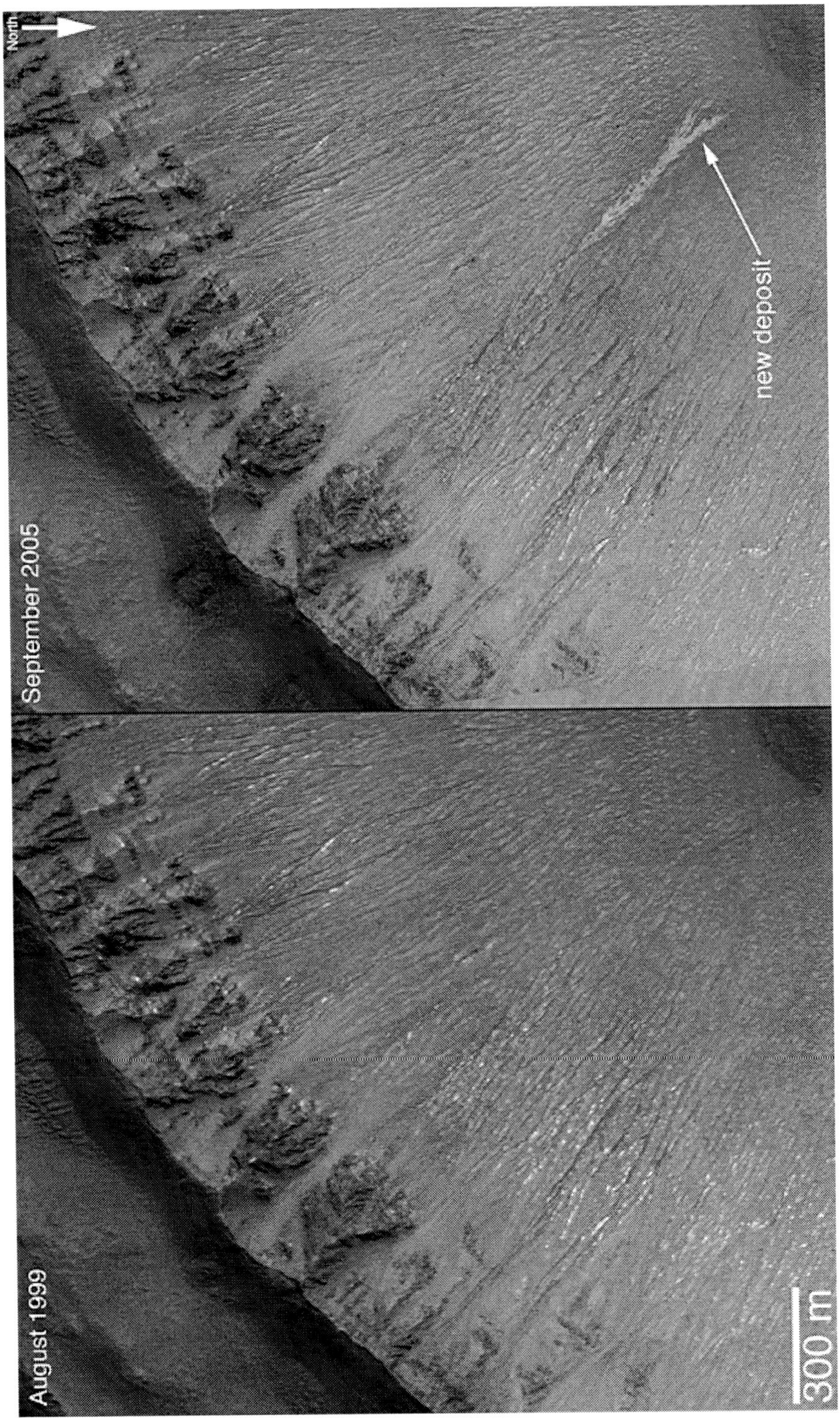

Figure 7. Geologic transient event: Light-toned gully deposit in a crater in the Centauri Montes Region, observed by NASA's Mars Global Surveyor. The displayed images show a comparison of the crater slope as it appeared in August, 1999 (left), with an image acquired in September, 2005 (right) after the change occurred. [Images courtesy of NASA/JPL/MSSS]

TIER-SCALABLE RECONNAISSANCE: ATMOSPHERIC, ARCHITECTURAL, AND OPERATIONAL CONSIDERATIONS

Atmospheric Considerations

To date, planetary missions have employed orbiters, landers, and rovers. Aerial platforms have not been employed beyond Earth with the exception of the Soviet Vega 1 and 2 missions to Venus. If a planetary body has a sufficiently dense atmosphere, the deployment of an airborne tier is warranted. In the past, airplanes/gliders (e.g., [11]), rotorcraft (e.g., [12, 13]), and, more recently, hybrid designs (e.g., a fixed-wing vertical take-off and landing (VTOL) aerobot [14]) have been proposed. Some of the major deficiencies with these types of aerial vehicles are: (1) the complexity of the vehicles themselves (especially a rotorcraft); (2) the complexity of their deployment; (3) no station-keeping in the case of the airplane/glider and only temporary station-keeping in the case of a rotorcraft; (4) need for propellant (either electric by means of solar power or radioisotope thermoelectric generators (RTG), or chemical/gas) for airplanes (except gliders) and rotorcraft; (5) (active) flight control system necessary; and (6) limited mission duration and termination of mission once vehicle lands/crashes on the surface (no repeated surface/subsurface probing possible, with the exception of the rotorcraft and the VOTL aerobot [14]).

In contrast, balloons, blimps, and airships offer many advantages over the above vehicle types. They are (1) simple in structure; (2) lightweight; (3) "easily" deployable (e.g., [15]), both in mid-air and from the surface; (4) buoyant without the need for propellant, (5) can operate without an active flight control system (e.g., wind-driven [16]), and (6) allow for extended mission durations, ranging from months to more than a year (e.g., [17-20]) and for repeated surface/subsurface sampling. Furthermore, equipped with either solar cells/batteries or RTGs, they can be electrically operated to support onboard instruments, data analysis capabilities, and active thrusting.

For airborne and potential airborne sample return missions using balloons/blimps, a basic estimation yields the following sample return mass and balloon/blimp size for Mars (for details on the derivation of the lift-equation, the parameters used for the calculations, and for the assumptions made, see [10]): For a Mars net sample return mass of 100 g at the surface, a blimp/balloon would need to be deployed with a radius >8 m (this includes the helium gas tank mass; without the tank this blimp/balloon would have a net liftable payload mass of about 15.8 kg).

The above results strongly suggest jettisoning the gas tank after deployment and inflation of a balloon/blimp in the Martian atmosphere and to possibly maintain a smaller tank for altitude changes and gas replenishment for the duration of the mission. Explorations of Venus and Titan with balloons/blimps/airships (e.g., [15-22]) are additional scenarios for tier-scalable missions in conjunction with orbiter(s) and ground-probing agents.

Architectural and Operational Considerations

A typical operations scenario for a tier-scalable reconnaissance system is illustrated schematically in Figure 8, which utilizes only 3 of a possible *n* tiers. Multi-tiered, multi-agent, hierarchical, integrated mission architectures allow for varying degrees of independence from human intervention, while also permitting manual override at any level. For example, a human operator may communicate with the orbiter(s) directly. Or, he may choose to command the airborne units via the orbiters. Or, in so doing, he may choose to also command the ground-level reconnaissance agents via the airborne units. Conversely, a highly automated operation mode may be used, enabling autonomous reconnaissance missions, as they are necessary, such as when the communication time lag prohibits meaningful teleoperation, or during times of communication blackout.

The often overstated and misunderstood term *autonomy* is defined in the context of this chapter as the high-level automation of planetary reconnaissance missions, including automated data acquisition, data feature extraction, data analysis, identification of science targets, science goal prioritization, execution of science goals, navigation, and guidance. As such, most currently deployed agents are not truly autonomous (with the exception of very basic and local obstacle avoidance), as they are for the most part Earth-commanded and teleoperated. Tier-scalable reconnaissance prepares the grounds for future, true mission autonomy. For an interesting novel approach to computational autonomy the reader may be referred to [23].

In the highly automated scenario, the orbiters command and control the airborne agents autonomously, and the airborne agents in turn autonomously command and control the ground-tier reconnaissance agents (Figure 8). This system integrates orbiters with balloons/blimps (airships) and ground-tier agents (rovers, fixed landers, sensor platforms, etc.). The airborne and ground-tier agents can be sufficiently inexpensive (in terms of mass, size, power, and operational resources) to allow for the deployment of numerous expendable agents that, collectively, can address specific science-driven questions within the operational areas of particular interest. Examples of such "inexpensive" agents are (MEMS-based) sensor platforms and mini-rovers akin to Minerva, the lander/mini-rover on the Japanese sample-return mission Hayabusa to asteroid Itokawa. Multiple ground-tier and airborne agents can collectively explore the same science target(s) with complementary suites of instruments.

The use of miniaturized ground agents is justified [9] because of the reduced driving requirements afforded by the tier-scalable reconnaissance concept [24]. Ground agents already are deployed in the vicinity of where they need to be. Therefore, driving is limited to "fine-tuning" the location of the ground-tier agents to perform in-situ sampling, as opposed to "long-distance" driving towards the location where they need to be. Hence, miniaturized agents with limited driving capabilities can be employed. They also can have limited onboard computing capabilities, by off-loading data analysis to higher tiers (e.g., orbiters, aerial platforms, or primary lander units) that can host higher-grade computing equipment [24]. Conceptually, a *primary* lander unit is a higher-grade computing engine and also acts as a communication relay (e.g., planetary internet router [25, 26]) between the airborne/spaceborne tiers and the miniaturized ground-tier agents. Thus, the tier-scalable reconnaissance paradigm, by its inherent architecture, is an extended distributed computing topology unto itself [24].

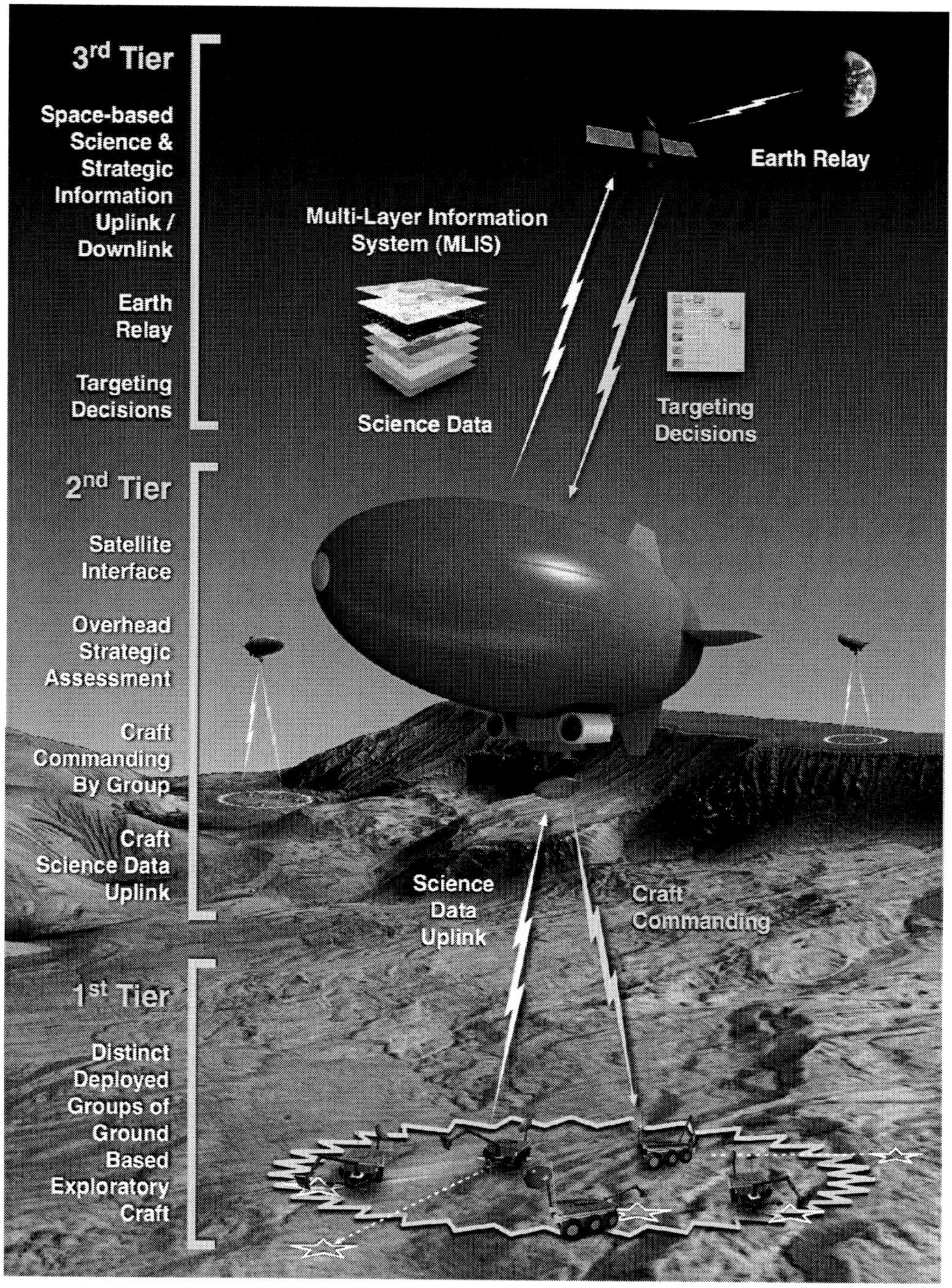

Figure 8. Tri-level hierarchical multi-agent architecture for autonomous remote planetary exploration on Mars (after [1-9]). 3-D oblique view, exemplifying orbiter-guided airborne agents (not drawn to scale) performing reconnaissance over Ophir Chasma. Part of the reconnaissance would include surveying the canyon walls, homing in on stratigraphic sequences, hovering above landslide and valley floor deposits, and identifying targets for subsequent deployment of ground-tier agents such as miniature rovers and immobile sensors (not drawn to scale). Target features of special scientific interest may include: (1) geomorphic/pedologic features and mineralogical/elemental signatures; (2) diversity of rock types; (3) elevated heat flow; (4) surface/near-surface water or moisture; and (5) volatile plumes (e.g., methane). [Ophir Chasma terrain background image courtesy ESA/DLR/FU Berlin (G. Neukum)]

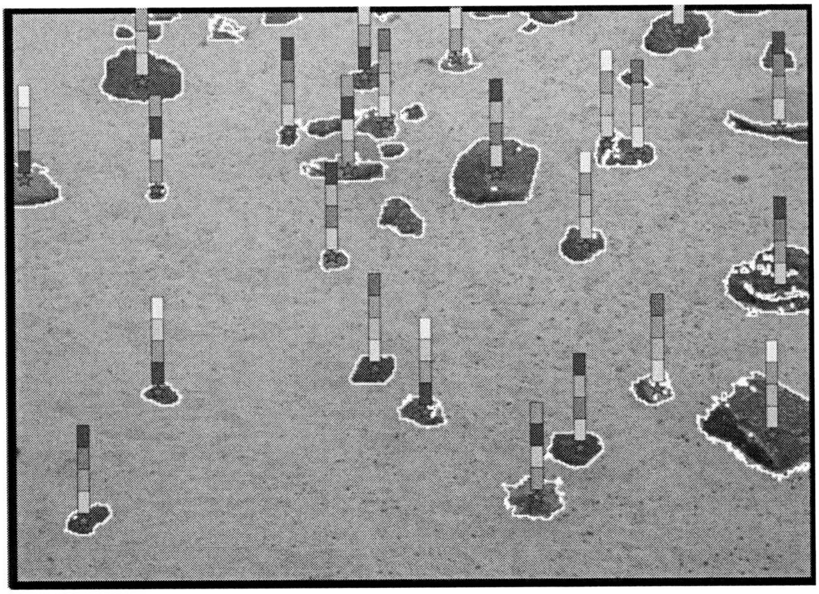

Figure 9. Automated Global Feature Analyzer (AGFA) [27] (formerly Automated Geologic Field Analyzer [28]) to perform automated mapping/imaging, feature extraction, feature analysis, and target prioritization within operational areas at different length scales and resolutions, depending on the vantage point (i.e., spaceborne, airborne, or ground). Depicted are identified targets (i.e., rocks) and their respective extracted feature vectors, derived by AGFA at the ground level. At this level, the targets can scale from microscopic soil particles to large rocks. For airborne and spaceborne vantage points, the targets scale from large rocks to mountain ranges.

Operational Autonomy

To support and control such tier-scalable reconnaissance mission architectures, a high degree of operational autonomy is utilized. Essential requirements of such operational autonomy are:

- *Automatic mapping of an operational area* from different vantage points (i.e., space, airborne, surface/subsurface).
- *Automatic feature extraction and target/region-of-interest identification* within the mapped operational area (e.g., with software packages such as the *Automated Global Feature Analyzer (AGFA)* [27] (formerly *Automated Geologic Field Analyzer* [28]), Figure 9).
- *Automatic target prioritization* for close-up examination (e.g., [27-34]), primarily by ground-tier agents. Target prioritization scenarios can range from simple feature-based or feature-clustering-based prioritization (e.g, [35-37]) to prioritization via context-based clustering (e.g., [38]). Recently, more advanced prioritization frameworks [29] have been developed using previously acquired, coarse feature/reconnaissance data that was pre-clustered using the above clustering methods. This advanced class of algorithms facilitates the selection of single or multiple targets, and the selection of instruments used for the close-up examination of these targets in an operational area for potential information gain about the

operational area. Alternatively, fuzzy logic-based inference systems (e.g., [30-34]) can synthesize and analyze diverse data sets (e.g., *Multi-Layer Information System (MLIS)* [1-3]) to identify prime targets for continued exploration (e.g., [39, 40]).

- *Automatic concurrent navigation* of agents within multiple tiers (see *Round-Robin Concurrent Commanding Scheme* below).

These requirements imply the optimal deployment of sensor platforms and individual sensors, sensor data fusion, and sensor interoperability.

Applications to Mars and Beyond

Integrated multi-tier, multi-agent hierarchical mission architectures are able to overcome the inherent challenge of traditional geologic planetary surface exploration [1-7]: airborne agents (orbiters in conjunction with balloons/blimps) possess overhead perspectives at different length scales and resolutions, which could provide guidance to ground-based agents (e.g., mobile rover units).

Tier-scalable reconnaissance missions introduce not only redundancy but also unprecedented mission fault-tolerance, and thus reliability and safety. They also enable spanning larger surface areas than previously possible – mimicking the way geologists explore regions on Earth – and therefore allow for increased science return. Several example scenarios are outlined below [1]:

- *Planetary bodies with atmospheres & non-extreme surface temperatures (e.g., Earth and Mars):* Orbiter-guided deployment and control of balloon/blimp units, which in turn deploy and control both mobile and immobile ground-based agents (3 tiers).
- *Planetary bodies with atmospheres & extreme surface temperatures (e.g., Venus and Titan):* Orbiter-guided deployment of balloon/blimp units. If surface temperatures permit, deployment and communication can be achieved with temperature-resistant immobile sensor webs or mobile agents (2-3 tiers).
- *Planetary bodies w/o atmospheres & extreme surface conditions (e.g., Mercury, Moon, Europa):* Orbiter-guided deployment of and communication with ground-based mobile agents and immobile sensor webs (2 tiers).

Multi-Agent Interaction: Round-Robin Concurrent Commanding Scheme for Tier-Scalable Reconnaissance

As the new multi-tier, multi-agent paradigm shift in scientific exploration missions unfolds, new demands in system command and control will need to be met. Using traditional means, commanding simultaneously even a few deployed reconnaissance agents becomes exceedingly complex. The logistical and computational demands of this new paradigm would scale quickly beyond the real-time processing capabilities available on current and near-future spaceborne, airborne, or in situ agents, sensors, or sensor platforms. This condition can be overcome, however, through the implementation of *round-robin concurrent commanding* [1].

The use of this novel commanding scheme within the tier-scalable reconnaissance paradigm allows the consideration of numerous new operational scenarios, opening many heretofore prohibited venues to scientific exploration and reconnaissance.

In the new paradigm, deployed ground-tier agents are commanded typically through the use of local airborne agents, such as blimps or balloons. The airborne agents contain onboard tracking (for most part possible via direct line-of-sight) and command systems for use with the ground-based agents. In this way, the airborne agents can identify possible targets for exploration by the ground-tier agents, as well as obstacles within the operational area.

With a wide field of view, the tracking and commanding systems onboard the airborne agents can optimally determine paths for the ground-tier agents, identifying both obstacles and interesting science targets. The ground-tier agents can then be commanded to navigate through the operational area to the designated targets, while avoiding the identified obstacles, rather than relying on their limited (horizontal) view of their environment. Each ground-tier agent includes its own instrument suite to collect surface and/or subsurface information; this is transmitted back to the airborne tier.

Algorithms for determining paths for a robot in an accurately characterized environment are well known in the design of robotics (e.g., A* [41, 42]). Using these conventional means, the issue of simultaneously commanding even a few deployed reconnaissance agents to their designated targets becomes exceedingly complex. The difficulty encountered herein is that with each additional simultaneously moving agent, the chance of collision increases geometrically. Likewise, the processing power required to compute and manage the agent trajectories increases geometrically as well.

Round-Robin Concurrent Commanding

The overhead perspective afforded by the airborne commanding tier greatly simplifies the path planning effort. The order of complexity of the task of commanding many simultaneously deployed mobile agents can be reduced from geometric to linear terms by adopting a *round-robin concurrent commanding* scheme [1]. In this scheme, all agents are not commanded simultaneously; instead, each is commanded in turn, one by one. Moreover, each agent is given a small initial command set to execute. In this way, one agent is being commanded actively while the others are executing their issued commands. Depending on the proximity of agents to each other and the precision of executing the overhead navigation commands (e.g., due to adverse terrain conditions such as sand slippage), an overhead (airborne) deployment reassessment is instituted so that refined trajectories are computed for each agent after every *n* command executions. Thus, each agent can be commanded in turn without threat of collision.

In a scenario of ground-deployed agents, in which each agent is mobile with a minimal suite of scientific instruments, such agents may conduct measurements en route to their designated targets. Should such measurements indicate that a condition of interest exists along the way, an agent may request that the commanding tier interrupt the agent's currently assigned trajectory to explore the anomaly. Or, an agent may find itself in an area of geography through which it cannot continue or traverse; in this case, it would request of the commanding tier an alternate trajectory to its target, or even a target reassignment. These resolutions are possible because the higher-level airborne commanding tier is able to reassess the deployment of the agents at the time of each command execution, refining the trajectories of each of the deployed agents at each step.

The functional decision tree of the new mission paradigm that is required to advance from Earth-based interaction to "tier-scalable" autonomy in scientific missions is illustrated schematically in Figure 10. The decision tree is illustrated from the point of view of the commanding (airborne) tier, which in turn has control jurisdiction over a deployed (ground-level) agent tier.

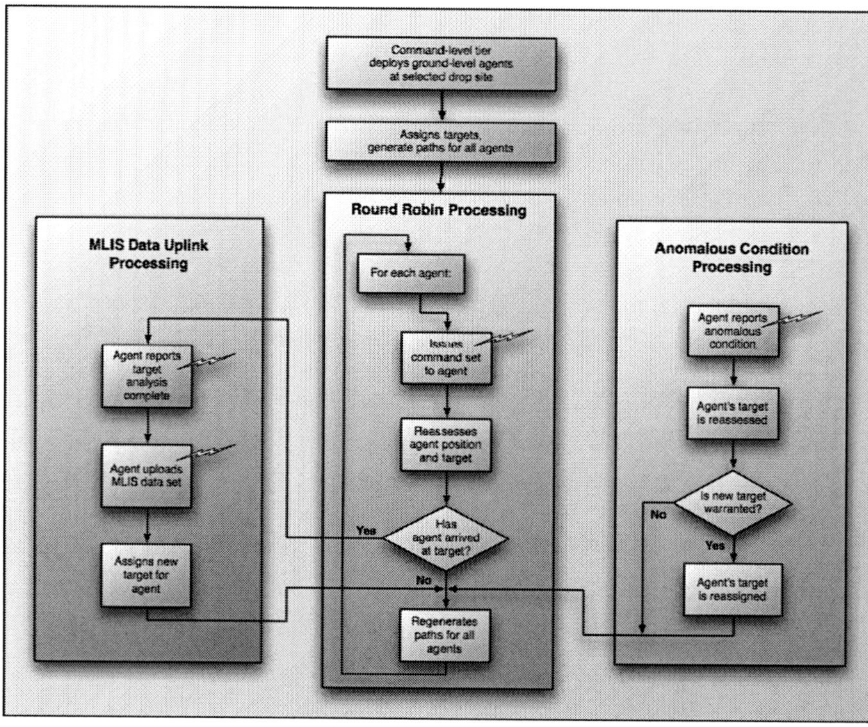

Figure 10. Functional decision logic of the new mission paradigm using *round-robin concurrent commanding* [1].

Operational Scenario applied to a Three-Tiered Reconnaissance System

The following image sequence (Figs. 11.0 - 11.17) details the operational scenario of a tri-level hierarchy using round-robin concurrent commanding [1].

- Figure 11.0 illustrates the three tiers used in the following example. Tier 3 is the highest level, and is here represented by an orbiter (satellite). This tier has direct jurisdiction over tier 2. Tier 2 is the middle tier, the airborne unit, represented by a blimp. It controls and commands the lower tier(s). Tier 1 is the lowest tier, here represented by reconnaissance agents deployed on the ground.
- Figure 11.1 shows the orbiter in tier 3 issuing prospective site information to the airborne unit in tier 2. The sites have been selected based on specific mission-critical criteria, including terrain, science targets, and other factors. The airborne unit travels

to, locates, and scans the requested destination sites for the suitability of deploying reconnaissance agents. Site scouting ends when a suitable site is located.

- Figure 11.2 depicts the scenario in which a suitable site has been located. The airborne unit initiates deployment or helps guide deployment of ground agents, in this case rovers. The parachute, which, depending on the respective planetary body, may also be a balloon, airbag, rocket, etc., descends to the surface at the designated site with its payload. Alternatively, the airborne unit approaches the surface to deploy its payload directly.
- Figure 11.3 depicts agent deployment on the surface. The airborne unit proceeds to relay the deployment status back to the orbiter.
- Fig 11.4 illustrates the process of target scouting for in-situ reconnaissance. The orbiter issues targeting information to the airborne unit, which then scans the drop zone area for specific targets that the deployed agents will travel to and analyze.
- Figure 11.5 shows the airborne unit in the process of assigning specific targets to each reconnaissance agent. The targets assigned to the agents are *first* targets; once this objective is achieved, a second target will be assigned, etc.
- Fig 11.6 begins the process of *Round-Robin Concurrent Commanding.* Here, the airborne unit issues the first set of instructions to the 1st agent. The instructions are generally twofold: (1) A rotation command is given, and (2) a movement command is given. The airborne unit need not wait for the agent to complete these instructions before commanding the other agents.
- In Fig 11.7, the airborne unit has finished issuing the first command to the 1st agent. It now moves on to the 2nd agent, and issues it its first command set, as in Figure 11.6. During the time the airborne unit is issuing the command set to the 2nd agent, the first agent is in the process of executing the command set it received.
- In Figure 11.8, the airborne unit is issuing the 3rd agent's first command set. During this time, the first and second agents are executing their command sets.
- In Figure 11.9, the final agent in this deployment is issued its first command set by the airborne unit. The other agents have received their first command sets, and are executing them, moving a short distance closer to their assigned targets.
- In Figure 11.10, one of the agents has arrived at its designated target. It performs its reconnaissance function (spectral analysis, chemical analysis, photography, etc.). It then initiates a MLIS/GIS [1-3] science data uplink to the airborne unit. The other agents are executing their command sets, en route to their designated targets.
- In Figure 11.11, the airborne unit relays the MLIS/GIS science data uploaded from one of the agents to the orbiter for analysis and/or further relay to Earth.
- In Figure 11.12, the airborne unit reassesses the position of each agent, since by this time, all agents have completed execution of their command sets. This serves as a "feedback" element for the airborne unit to appropriately calculate each agent's path, based on actual progress.
- In Figure 11.13, the airborne unit assigns a new target to the agent that arrived at its first target. The other agents are continuing to progress to their original target assignments.
- Figure 11.14 shows the airborne unit performing another iteration of *Round-Robin Concurrent Commanding* of agents to their respective targets. Agents are commanded, via command sets, to move toward their targets, etc. The above sequence (Figs 11.4 – 11.14) repeats until there are no more targets to investigate in this drop zone.

- Figure 11.15 shows the airborne unit relaying this fact to the orbiter.
- In Figure 11.16, the orbiter instructs the airborne unit to move from this drop zone to another operational area.
- Figure 11.17 shows the airborne unit has moved on to its second drop zone, as directed by the orbiter. The entire sequence then repeats at the new drop zone.

The same round-robin principle, detailed above, is equally applicable to the airborne tier and agents within, commanded by a higher tier, such as another blimp or orbiter.

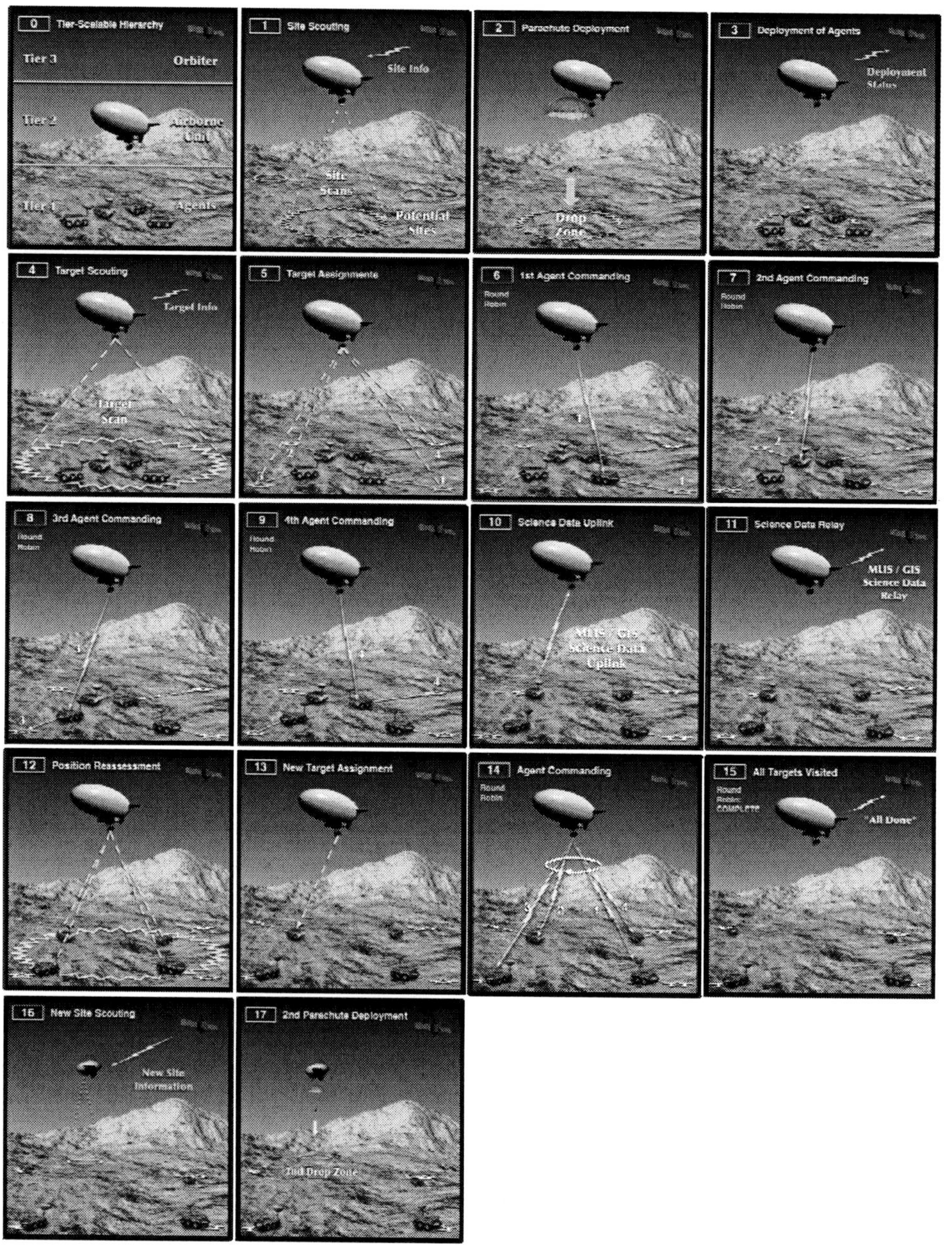

Figure 11. Operational depiction of three-tiered *round-robin concurrent commanding* [1].

MEMS: MINIATURIZATION TECHNOLOGY FOR TIER-SCALABLE RECONNAISSANCE

Tier-scalable reconnaissance mission architectures utilize not only a high level of operation autonomy (e.g., [27-34]), but also a new class of multi-purpose and reconfigurable instruments/sensors (e.g., intelligently reconfigurable snapshot hyperspectral imager [43, 44], and digital cameras with adapter-based microscopic and wide-angle imaging capability [45]) to render future missions more cost-effective (i.e., in terms of payload mass, size, and power requirements).

As with all space missions, payload mass, size, and power consumption remain the overarching constraints. Thus, to achieve the significantly higher benefits afforded by the tier-scalable reconnaissance architecture, the individual systems that make up the architecture ideally must themselves be miniaturized [9], if the overall mission payload is to remain within the range of current space missions. In fact, the miniaturization requirement may flow down to the component level, making extensive use of *Micro Electro Mechanical Systems (MEMS)* [9]. Since its advent in the late 1980s, MEMS development has progressed to the point that it now encompasses a diverse range of sensors and actuators [46]. MEMS devices, due to their inherently low mass, size, and power, are ideal for both space applications as well as sensor network applications. A number of MEMS devices have been specifically developed for the purpose of robotic planetary exploration [47, 48].

MEMS technologies have been developed for a diverse set of applications. In this chapter we briefly survey a few sensor technologies and their applications for space exploration (for more detail see [9]). For space exploration purposes, MEMS sensors can be categorized under two classes: *physical sensors* and *chemical sensors*. The early successes in MEMS technology were in the area of physical sensors, and more specifically in sensing the motion of a suspended proof mass for acceleration [49], rotation (e.g., MEMS micro-gyroscope) [50, 48], and vibration measurements [51, 52]. Another early application was the measurement of ambient pressure, which was achieved by monitoring the deflection of microfabricated membranes [53]. For in-situ planetary exploration applications, other physical measurements such as seismometry, temperature, and humidity measurements are also important. MEMS-based seismometers are essentially accelerometers with very low frequencies for their resonant elements [54, 55]. Temperature and humidity sensors could be used for instance in exobiological investigations for studying potential moisture rich areas such as on canyon floors on Mars as well as dark/bright slope streaks within Martian craters (to determine brines versus dust avalanches as the cause, e.g.: [56, 57]). MEMS devices have been demonstrated that can simultaneously measure both the temperature and the humidity [58].

A number of MEMS-based technologies exist in the arena of chemical sensors. For planetary exploration purposes, chemical sensors are required for detecting and characterizing compounds in solid, liquid, and gaseous forms. For instance, in the case of gas detection miniaturized methane sensors have been developed [59, 60] that could be utilized for the detection of methane outgassing on Mars and Titan. Miniature gas chromatographs [61, 62] have been developed that could potentially be coupled to miniature mass spectrometers [47, 63]. "Electronic Nose" instruments (i.e., chemical gas sensors) based on miniaturized chemi-resistors or chem-FETs (Field Effect Transistors) [64] have also been developed.

For the sensing of liquid species, the pre-eminent technique is Nuclear Magnetic Resonance (NMR) spectroscopy. Recently, a miniaturized Force Detected Nuclear Magnetic Resonance (FDNMR) spectrometer was demonstrated for the detection of water and organic materials [65]. In the case of solid-state chemical analysis, a novel, MEMS-enabled Atmospheric Electron X-ray Spectrometer (AEXS) has been developed that is capable of rapid elemental analysis in the atmospheric ambient [65].

Exobiological Implications of Tier-Scalable Reconnaissance

A current topic in Mars exploration is the possibility that microbes exist or existed on the surface or in the subsurface of Mars. The recent detection of methane (which is highly volatile) in the Martian atmosphere [66, 67] may point toward the existence of microbes on Mars among other alternatives such as geochemical reactions or volcanic activity. The typical determination of whether a gas is biogenic is made by measuring the ratio of two stable isotopes in the gas (e.g., [68]). A mass spectrometer able to measure the composition of the stable carbon isotopes in Martian methane with high sensitivity may be used aboard the airborne and the ground-tier agents to determine whether the Martian methane is biogenic.

It is possible that microbes may have existed on Mars, likely during the early evolution of the planet, when the temperatures were warmer. There is also evidence that water at mid- and low-latitudes may have been present on the Martian surface. These microbes may have survived to the present day, perhaps still living with extremely low metabolic rates at current subfreezing temperatures (e.g., [69, 70, 10, 68]).

Studies of microbial life in extreme environments on Earth are a step towards understanding how life could have arisen on planets such as Mars. These studies, among others, are a key in developing tools for the identification of such microbes on Mars for future explorations. The Atacama Desert is one such environment that serves as a model for Mars. It is an extreme, arid, temperate desert that extends from 20°S to 30°S along the Pacific coast of South America. This extreme environment has been used to test robotic microbe detection devices that may be employed on future missions to Mars [71, 72]. Another extreme environment used as a model for looking for life on Mars is the Greenland Ice Sheet Project 2 (GISP2) [73], which extracted a glacial ice core 3,053 meters long, rich with pockets of trapped methane.

Bacteria and Archaea have been found in all subfreezing terrestrial environments (e.g., [74-78]). Studies of terrestrial extremophiles (specifically the methanogens and Fe-reducing anaerobes are of interest) have involved extraction and examination of cells from cold water, ice, or permafrost. For example, in their study of microorganisms attached to iron-rich clay grains in the basal ice at GISP2, Tung et al. (2006) used F420 autofluorescence to identify methanogens [79].

Other instruments, such as the Mars Immunoassay Life Detection Instrument (MILDI), to look for chemical evidence of extant or extinct life in the Martian subsurface have been proposed and would utilize immunological reactions to specific antibodies to activate fluorescent stains to identify specific microbes [80]. Such instruments could be deployed through tier-scalable reconnaissance mission architectures, and Mars samples could be examined using the equipment without the need for them to be returned to Earth for analysis.

The European Space Agency is currently developing the first in a series of Aurora Flagship missions, called ExoMars [81, 82], to further characterize the biological environment on Mars as a steppingstone for robotic missions and human exploration. ExoMars will comprise a Mars orbiter, a descent module, and a Mars rover. The ExoMars rover will be equipped with the "Pasteur" payload, a collection of instruments dedicated to exobiology and geology. The proposed instrumentation includes a camera, an infrared spectrometer, and a radar system for the study of surface and subsurface targets of interest. The ExoMars rover will also be equipped with instruments that will test the surface environmental characteristics. A key component of the ExoMars rover is a drill capable of obtaining samples up to a depth of two meters along with a sample preparation and distribution system, which will allow for the processing of preserved samples without contamination. Once the samples are obtained, they can be further analyzed using a microscope, an X-ray diffractometer, a gas chromatograph mass spectrometer, a "Mars Organics and Oxidants Detector," and an antibody-based life-marker chip [82].

PRIME CANDIDATE SITES FOR HIGH-RISK SCIENTIFIC EXPLORATION

Non-traditional autonomous missions to remote planetary bodies will be necessary [1-7, 10, 83-86] primarily to allow less constrained access to scientifically interesting targets on planetary bodies of the Solar System not currently feasible with conventional mission designs, including:

- Geologic transient events, such as:
    - New impact cratering events (e.g., Figure 5)
    - Dark/light slope streaks (e.g., Figure 6)
    - New gully deposits (e.g., Figure 7)
- Canyons (e.g., Valles Marineris on Mars, or Devana Chasma, a big rift valley on Venus)
- Mountain ranges (e.g., Thaumasia highlands on Mars, Isthar Terra on Venus)
- Chaotic terrain (e.g., source areas of the circum-Chryse outflow channel system on Mars, Conamara Chaos on Europa)
- Impact basins and craters (e.g., Holden Crater on Mars [87])
- Volcanoes of diverse sizes and shapes (e.g., Mars and Venus)
- Polar ice caps (e.g., Mars)
- Suspected (water) ice deposits within impact craters (e.g., Mars, Figure 4) or impact basins (e.g., Mercury and Moon [10]). Water ice deposits are of particular interest for potential human settlement.
- Putative hydrothermal sites (e.g., Mars [85, 86])
- Sites of suspected magmatic-driven uplift and associated tectonism and possible hydrothermal activity (e.g., plume-related activity such as hypothesized for the central part of Valles Marineris and the Warrego Valles rise on Mars [40, 88, 89])
- Regions indicating potential recent hydrologic (Mars) or hydrocarbon (Titan) activities

- Liquid pools of ammonia-water mixtures associated with cryovolcanism or a recent impact cratering event (e.g., Titan, Triton, Enceladus)
- Liquid hydrocarbon accumulation (lakes) on the surface (e.g., Titan).

These and many other targets of scientific interest on the planetary bodies of the Solar System are particularly crucial for exobiologic-oriented exploration in general [69, 4, 85, 86], and sample return missions in particular [69, 10, 83].

Tier-Scalable Reconnaissance: Martian Deployment Scenario

For the case of Ophir Chasma (within Valles Marineris) (Figure 8), the following deployment of a tier-scalable reconnaissance mission is envisioned [1, 2, 10, 4-7]: Orbiter(s) with an embedded knowledge base (e.g., *Multi-Layer Information System (MLIS)* [1-3]) scout areas of scientific interest at a global (Valles Marineris) to regional (Ophir Chasma) scale. They subsequently deploy airborne agents (such as blimps) in mid-air above Ophir Chasma for further tier-scalable reconnaissance. This includes identifying and homing in on prime targets of scientific interest, using both pre-mission and orbiter-based information. The airborne agent(s) attempt to detect scientifically interesting features, such as volatile releases (methane plumes or water vapor) and elevated heat flow, or transient geologic events (e.g., landslides), hydrologic events (e.g., water seeps), atmospheric events (e.g., reoccurring fog in a specific part of the canyon system), and unique rock assemblages (e.g., [90, 91]). The airborne agent(s) would then home in on these features for close-up reconnaissance (e.g., [19]). They would map out areas of interest and acquire terrain data, processed through feature-extraction, feature data analysis, and science prioritization algorithms (e.g., [27-34]), to choose potential targets for in-situ examination by subsequently deployed sensors or surface/subsurface-agents and for determining safe passages to these targets. The sensors and ground-agents would gather in-situ data at the respective targets that complement the remote sensing data obtained by the space/airborne agents. For example, the ground-agents would help identify, characterize, and map out sources of volatile plumes (e.g., methane sourcing from a vent in a hydrothermally active area) for future sample return missions [10, 83].

Conclusion

Multi-tier, multi-agent autonomous robotic planetary atmosphere, surface and subsurface reconnaissance will lead to an improved understanding of the various histories (e.g., geologic, geomorphic, pedologic, aqueous, climatic, and possible biologic) of Mars and other extraterrestrial targets, through the tier-scalable geologic approach. This new paradigm in planetary reconnaissance allows for confirmation of working hypotheses and theories [92] such as in the case of Mars, whether (a) the mountain ranges contain a greater diversity of rock types than just volcanic; (b) sites of suspected hydrothermal activity are indeed hydrothermal environments [85, 86]; (c) prime candidate sites of potential life-containing habitability actually contain extant or fossil life or life forms [39, 93, 94, 85, 10]; or (d) close examination of surface and subsurface rock materials with sensors suitable for microscopic observation and chemical analysis of coatings (e.g., varnish) on weathered rock materials might reveal important data on possible soil microenvironments, live microbes, or fossil forms (e.g., [94-96, 45]). Moreover, tier-scalable autonomous reconnaissance missions afford a first-of-a-kind opportunity to scout, discover, and characterize potential habitats for possible life [69, 10, 4, 83] or future human settlement.

Following the published works by Fink et al. [1-3], NASA recently has been soliciting proposals calling for technology development for multi-tier sensor webs: *"Sensor webs of the future may include space-based, airborne, and in-situ sensors, all working together in a semi-closed loop system in which "smart" sensors sense what is happening per their designed sensing capabilities and feed that information into a control system. Based on the sensor inputs, the control system then modifies the environment (instrument pointing, data collection on or off, etc.) and causes the sensors to take in and provide new information to the control system."* (excerpt from Science Mission Directorate NASA Research Announcement "Advanced Information Systems Technology" Solicitation: NNH05ZDA001N-AIST).

Moreover, in testimony to Congress in May 2005, NASA Administrator Michael Griffin included the following statement: *"In the future, NASA plans to develop a "sensor web" to provide timely, on-demand data and analysis to users who can enable practical benefits for scientific research, national policymaking, economic growth, natural hazard mitigation, and the exploration of other planets in this solar system and beyond."*

This followed the release of the February 2005 publication *The New Age of Exploration: NASA's Direction for 2005 and Beyond* that stated: *"NASA will develop new space-based technology to monitor the major interactions of the land, oceans, atmosphere, ice, and life that comprise the Earth system. In the years ahead, NASA's fleet will evolve into human made constellations of smart satellites that can be reconfigured based on the changing needs of science and technology. From there, researchers envision an intelligent and integrated observation network comprised of sensors deployed to vantage points from the Earth's subsurface to deep space."*

Furthermore, NASA's Mars Exploration Program in Washington has now fully embraced the tier-scalable reconnaissance concept, originated by Fink et al. [1-6, 8], as stated in a JPL News Release from October 06, 2006 [97]: *"This is a tremendous example of how our Mars missions in orbit and on the surface are designed to reinforce each other and expand our ability to explore and discover," said Doug McCuistion, director of NASA's Mars Exploration Program in Washington. "You can only achieve this compelling level of exploration capability with the sustained exploration approach we are conducting at Mars through integrated orbiters and landers."* And Steve Squyres, the Principal Investigator for the two Mars Exploration Rovers Opportunity and Spirit, adds in the same JPL News Release [97]: *"The combination of the ground-level and aerial view is much more powerful than either alone," said Steve Squyres of Cornell University, Ithaca, N.Y. [...] "If you were a geologist driving up to the edge of a crater in your jeep, the first thing you would do would be to pick up the aerial photo you brought with you and use it to understand what you're seeing from ground level. That's exactly what we're doing here."*

Individual components of the tier-scalable mission architecture, originated by Fink et al. [1-9], are either under development or have already been tested and proven "in the field." These include orbiters, balloons/blimps/airships, and ground-based agents such as rovers and landers as well as immobile sensor webs. The biggest challenge, however, appears to be not so much the hardware but the "intelligent" software that would enable all the components of a multi-tier multi-agent mission to be integrated and function autonomously. This is not to be confused with Artificial Intelligence (AI)-based (i.e., rule-based) automation efforts, such as the Autonomous Sciencecraft Experiment (ASE) out of JPL for Earth-orbiting satellites, that have to be told what to notice and how to then act accordingly, as explained in the NASA News Release from October 26, 2006 [98]: *"We programmed it to notice things that change [...] and take appropriate action."* In the same News Release [98], JPL now realizes: *"We have four satellites orbiting Mars and two rovers on the ground. They could work together."*

To study and validate the tier-scalable reconnaissance concept under more realistic (outdoor) conditions we have begun development and implementation of an outdoor test bed at the Visual and Autonomous Exploration Systems Research Laboratory at Caltech, consisting of remote controllable robotic platforms (Figure 12) and blimps [99, 8, 9]. Furthermore, we are developing, implementing, and field-testing software (e.g., [27-29]) that would allow the orbiters, blimps, and rovers both to communicate with one another and to navigate and explore the planetary terrain with greatly reduced (and ultimately without) help from mission control on Earth, thus affording more mission autonomy/flexibility and increased science return.

Figure 12. 4WD remote controllable robotic platform as a representative mobile ground-tier agent of the tier-scalable reconnaissance mission test bed [99, 8, 9].

We envision that multi-tier, multi-agent hierarchical mission architectures for remote planetary reconnaissance, as described here, are feasible within a 10-15 year timeframe. A foretaste of what is to come is the anticipated interaction of NASA's MRO (and possibly ESA's Mars Express) with the Mars Exploration Rovers Spirit and Opportunity (Figure 13) as well as with lander and rover missions such as Phoenix and Mars Science Laboratory (MSL). Moreover, ESA's first Aurora Flagship mission ExoMars [81, 82], by design, will comprise a Mars orbiter, descent module, and a Mars rover. Integrated orbiter-airship missions, especially suitable for the exploration of Titan, Venus, and Mars, are envisioned to be feasible within a decade from now. Tier-scalable reconnaissance represents a paradigm shift in autonomous remote planetary exploration of Mars and beyond.

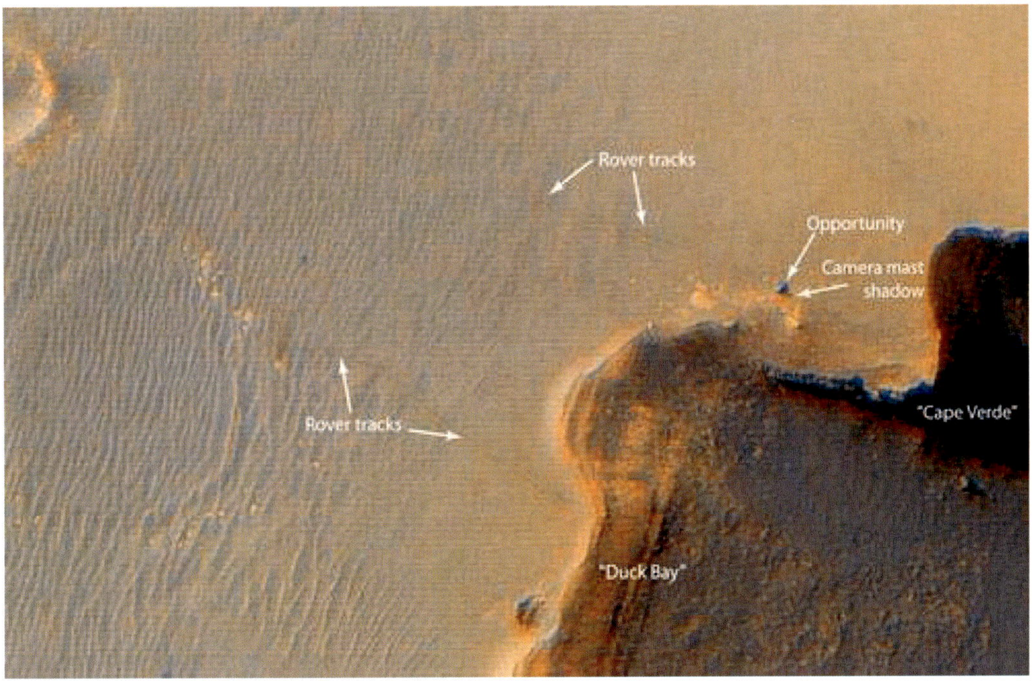

Figure 13. Overhead image of operational area of NASA's Mars Exploration Rover Opportunity near the rim of "Victoria Crater," taken by NASA's Mars Reconnaissance Orbiter in October 2006. [Image courtesy NASA/JPL/UA]

REFERENCES

[1] Fink, W.; Dohm, J. M.; Tarbell, M. A.; Hare, T. M.; Baker, V. R. Next-Generation Robotic Planetary Reconnaissance Missions: A Paradigm Shift. *Planetary and Space Science.* 2005, *53*, 1419-1426.

[2] Fink, W.; Dohm, J. M.; Tarbell, M. A.; Hare, T. M.; Baker, V. R. Next-Generation Robotic Planetary Surface/ Subsurface Reconnaissance Missions: A Paradigm Shift. *36th Lunar and Planetary Science Conference Abstracts [CD-ROM]*, Lunar and Planetary Institute, Houston, TX, 2005, abstract 1977.

[3] Fink, W.; Dohm, J. M.; Tarbell, M. A.; Hare, T. M.; Baker, V. R. Next-Generation Robotic Planetary Reconnaissance Missions: A Paradigm Shift. *Geochimica et Cosmochimica Acta.* 2005, *69(10S),* A533.

[4] Fink, W.; Dohm, J. M.; Tarbell, M. A.; Hare, T. M.; Baker, V. R.; Schulze-Makuch, D.; Furfaro, R.; Fairén, A. G.; Ferré, T. P. A.; Miyamoto, H.; G. Komatsu, G.; Mahaney, W. C. Multi-tier Multi-agent Autonomous Robotic Planetary Surface/Subsurface Reconnaissance For Life. *37th Lunar and Planetary Science Conference Abstracts [CD-ROM]*, Lunar and Planetary Institute, Houston, TX, 2006, [abstract 1433].

[5] Fink, W.; Dohm, J. M.; Tarbell, M. A.; Hare, T. M.; Schulze-Makuch, D.; Furfaro, R.; Baker, V. R. Tier-Scalable Reconnaissance for Remote Planetary Exploration. *25th International Space Development Conference*, Los Angeles, CA, 2006.

[6] Fink, W.; Dohm, J. M.; Tarbell, M. A.; Hare, T. M.; Baker, V. R.; Schulze-Makuch, D.; Furfaro, R.; Fairén, A. G.; Ferré, T. P. A.; Miyamoto, H.; G. Komatsu, G.; Mahaney, W. C. Autonomous Tier-Scalable Reconnaissance Missions For Remote Planetary Exploration. *Proc. of 4th International Planetary Probe Workshop,* Pasadena, CA, 2006.

[7] Fink, W.; Dohm, J. M.; Tarbell, M. A.; Hare, T. M.; Baker, V. R.; Schulze-Makuch, D.; Furfaro, R.; Fairén, A. G.; Ferré, T. P. A.; Miyamoto, H.; G. Komatsu, G.; Mahaney, W. C. Tier-Scalable Reconnaissance Missions For The Autonomous Exploration Of Planetary Bodies. *IEEE Aerospace Conference Proceedings*, #1199, Big Sky, MT, 2007.

[8] Fink, W. Caltech's Visual and Autonomous Exploration Systems Research Laboratory. News Media Releases on Tier-Scalable Reconnaissance. [http://autonomy.caltech.edu/autonomy/tierscalable.html].

[9] Fink, W.; George, T.; Tarbell, M. A. Tier-Scalable Reconnaissance: The Challenge of Sensor Optimization, Sensor Deployment, Sensor Fusion, and Sensor Interoperability. *Proceedings of the SPIE Defense & Security Symposium,* 2007, *6556(1)*, 655611.

[10] Schulze-Makuch, D.; Dohm J. M.; Fairén, A. G.; Baker, A. G.; Fink, W.; Strom, R. G. Venus, Mars, and the Ices on Mercury and the Moon: Astrobiological Implications and Proposed Mission Designs. *Astrobiology*. 2005, *5*, 778-795.

[11] Colozza, A. J. Preliminary Design of a Long-Endurance Mars Aircraft. *AIAA 90-2000, Proc. of AIAA/SAE/ASME/ASEE 26th Joint Propulsion Conference*. 1990, Orlando, FL.

[12] Savu, G; Trifu, O. Photovoltaic rotorcraft for Mars missions. *Joint Propulsion Conf. and Exhibit, 31st*. 1995. San Diego, CA. AIAA-1995-2644, ASME, SAE, and ASEE.

[13] Young, L. A.; Aiken, E. W.; Gulick, V.; Mancinelli, R.; Briggs, G. A. Rotorcraft as Mars Scouts. *IEEE Aerospace Conference Proceedings*, Big Sky, MT, 2002.

[14] Song, H.; Underwood, C. A Mars VTOL Aerobot – Preliminary Design, Dynamics and Control. *IEEE Aerospace Conference Proceedings*, Big Sky, MT, 2007.

[15] Nott, J. Ballutes: Launching Aerobots without Compromises. *Proceedings of the 4th International Planetary Probe Workshop.* 2006, Pasadena, CA.

[16] Nott, J. Titan – The Ultimate Destination for Aerobots. *Proceedings of the 4th International Planetary Probe Workshop.* 2006, Pasadena, CA.

[17] Hall, J. L.; Kerzhanovich, V. V.; Yavrouian, A. H.; Jones, J. A.; White, C. V.; Dudik, B. A.; Plett, G. A.; Mennella, J.; Elfes, A. An aerobot for global in situ exploration of Titan. *Advances in Space Research*, 2006, *37(11)*, 2108-2119.

[18] Jones, J. A.; Wu, J. J. Montgolfiere Aerobots for Titan. *Proceedings of the 4th International Planetary Probe Workshop.* 2006, Pasadena, CA.

[19] Shrotri, K.; Khalid, A.; Gunduz, M. E.; Manyapu, K.; Sumer, Y. F.; Schrage, D. P. MARVIN – Near Surface Methane Detection on Mars. *IEEE Aerospace Conference Proceedings*, Big Sky, MT, 2007.

[20] Elliott, J. O.; Reh, K.; Spilker, T. Concept for Titan Exploration Using a Radioisotopically Heated Montgolfiere. *IEEE Aerospace Conference Proceedings*, Big Sky, MT, 2007.

[21] Nishimura, J.; Hinada, M.; Yajima, N.; Fujii, M. Venus balloons at low altitudes. *Advances in Space Research*, 1994, *14(2)*, 261-271.

[22] Noor, A. K.; Cutts, J. A.; Balint, T. S. Platforms for discovery: Exploring Titan and Venus. *Aerospace America*/June 2007, 32-37.

[23] De Gyurky, S.; Tarbell, M. A. The Cognitive Dynamics of Computer Science: Cost-Effective Large Scale Software Development. Wiley-IEEE Computer Society Pr., 2006.

[24] Fink, W.; Tarbell, M. A. Tier-Scalable Reconnaissance – A Paradigm Shift in Autonomous Robotic Planetary Exploration. *7th IAA International Conference on Low-Cost Planetary Missions*, Pasadena, CA, 2007.

[25] CNN.com SCI-TECH with IDG.net News Release from February 28, 2001. Interview: Building the interplanetary Internet. [http://archives.cnn.com/2001/TECH/internet/02/27/mars.network.office.idg/index.html].

[26] SPACE NEWS Business Report News Release from April 26, 2004. Cisco Takes Its internet Router to Space. [http://www.space.com/spacenews/archive04/ciscoarch_042104.html].

[27] Fink, W.; Datta, A.; Dohm, J. M.; Tarbell, M. A.; Jobling, F. M.; Furfaro, R.; Kargel, J. S., Schulze-Makuch, D.; Lunine, J. I.; Baker, V. R. Automated Global Feature Analyzer (AGFA) – A Driver for Tier-Scalable Reconnaissance. *IEEE Aerospace Conference Proceedings*, #1273, Big Sky, MT, 2008.

[28] Fink, W.; Datta, A.; Baker, V. R. AGFA: (Airborne) Automated Geologic Field Analyzer. *Geochimica et Cosmochimica Acta*, 2005, *69(10S),* A535.

[29] Fink, W. Generic Prioritization Framework for Target Selection and Instrument Usage for Reconnaissance Mission Autonomy. *Proceedings of IEEE World Congress on Computational Intelligence (WCCI)*, Vancouver, Canada. 2006, 11116-11119.

[30] Furfaro, R.; Dohm, J. M.; Fink, W.; Schulze-Makuch, D.; Fairén, A. G.; Tarbell, M. A.; Hare, T. M.; Baker, V. R. Multi-Layer Fuzzy Logic-based Expert System for Conducting Tier-scalable Planetary Reconnaissance. *37th Lunar and Planetary Science Conference Abstracts [CD-ROM]*, Lunar and Planetary Institute, Houston. TX, 2006. [abstract 1257].

[31] Furfaro, R.; Dohm, J. M.; Fink, W. Fuzzy Logic Expert System for Tier-scalable Planetary Reconnaissance. *9th International Conference on Space Operations, AIAA*, Rome, Italy. 2006.

[32] Furfaro, R.; Dohm, J. M.; Fink, W. Autonomy in Planetary Exploration: Fuzzy Expert System for Tier-Scalable Reconnaissance. *25th International Space Development Conference*, Los Angeles, CA, 2006.

[33] Furfaro, R.; Dohm, J. M.; Fink, W.; Kargel, J. S.; Schulze-Makuch, D.; Fairén, A. G.; Ferré, T. P. A.; Tarbell, M. A.; Hare, T. M.; Komatsu, G.; Palmero-Rodriguez, A. J.; Baker, V. R.; Miyamoto, H. *38th Lunar and Planetary Science Conference Abstracts [CD-ROM]*, Lunar and Planetary Institute, Houston. TX, 2007, [abstract 1372]. Searching for Life on Extraterrestrial Bodies: Fuzzy Autonomous Systems for Planetary Reconnaissance.

[34] Furfaro, R.; Dohm, J. M.; Fink, W.; Kargel, J. S.; Schulze-Makuch, D.; Fairén, A. G.; Palmero-Rodriguez, A. J.; Baker, V. R.; Ferré, P. T.; Hare, T. M.; Tarbell, M. A.; Miyamoto, H.; Komatsu, G. The search for life beyond Earth through fuzzy expert systems. *Planetary and Space Science.* 2007 (in press).

[35] Duda, R. O.; Hart, P. E. Pattern Classification and Scene Analysis, John Wiley & Sons, 2nd edition, 2000.

[36] Bishop, C. M. Neural Networks for Pattern Recognition, Clarendon Press: Oxford, 1995.
[37] Williams, C. K. I. An MCMC Approach to Hierarchical Mixture Modelling. In *Advances in Neural Information Processing Systems*. Solla, S. A.; Leen, T. K.; Mueller, K. R. eds. NIPS 1999 Volume 12; MIT Press: 2000; 680-686.
[38] Fink, W.; Castano, R.; Davies, A.; Mjolsness, E. Clustering Algorithm for Mutually Constraining Heterogeneous Features. *Technical Report JPL-ICTR-01-5*, 2001.
[39] Dohm, J. M.; Ferris, J. C.; Barlow, N. G., Baker, V. R.; Mahaney, W. C.; Anderson, R. C.; Hare, T. M. The Northwestern Slope Valleys (NSVs) region, Mars: A prime candidate site for the future exploration of Mars. *Planetary and Space Science*. 2004, *52*, 189-198.
[40] Dohm, J. M.; Tanaka, K. L.; Hare, T. M. Geologic map of the Thaumasia region of Mars. *US Geol. Survey Map I-2650*, 2001.
[41] Hart, P. E.; Nilsson, N. J.; Raphael, B. A Formal Basis for the Heuristic Determination of Minimum Cost Paths. *IEEE Transactions on Systems Science and Cybernetics SSC4.* 1968 *2*, 100–107.
[42] Hart, P. E.; Nilsson, N. J.; Raphael, B. Correction to "A Formal Basis for the Heuristic Determination of Minimum Cost Paths". *SIGART Newsletter.* 1972 *37*, 28–29.
[43] Johnson, W. R.; Wilson, D. W.; Fink, W.; Humayun, M.; Bearman, G. Snapshot hyperspectral imaging in ophthalmology. *J. Biomed. Opt.* 2007, *12(1),* 014036.
[44] Bearman, G. H.; Johnson, W. R.; Fink, W.; Wilson, D. W. An Intelligently Reconfigurable Snapshot Imaging Spectrometer For Planetary Exploration. *38th Lunar and Planetary Science Conference Abstracts [CD-ROM]*, Lunar and Planetary Institute, Houston, TX, 2007, [abstract 1103].
[45] Fink, W.; Mahaney, W. C.; Kuhlman, K. R. Adapter-based Microscopic and Wide-angle Imaging Capability For Digital Cameras For Planetary Exploration and Astrobiology. *38th Lunar and Planetary Science Conference Abstracts [CD-ROM]*, Lunar and Planetary Institute, Houston, TX, 2007, [abstract 2397].
[46] The MEMS Handbook: MEMS Applications. M. G. Hak, ed. CRC Press: Florida, 2006.
[47] George, T. MEMS/NEMS development for Space Applications at NASA/JPL. *Proc. SPIE* 2002, *4755*, 556.
[48] George, T. Overview of MEMS/NEMS Technology Development for Space Applications at NASA/JPL. *Proc. SPIE.* 2003, *5116,* 136.
[49] Yazdi, N.; Ayazi, F.; Najafi, K. Micromachined inertial sensors. *Proc. IEEE.* 1998, *86*, 1640.
[50] Weinberg, M.; Connelly, J.; Kourepenis, A.; Sargent, D. Microelectromechanical instrument and systems development at the Charles Stark Draper Laboratory, Inc. *Proc. 16th DASC. AIAA/IEEE.* 1997, *2*, 33.
[51] Gupta, A.; Singh, R.; Jain, V. K.; Kumar, V. Development of micromachined vibration sensors for MEMS applications. *Proc. SPIE.* 1999, *3903*, 153.
[52] Ollier, E.; Philippe, P.; Chabrol, C.; Mottier, P. Micro-Opto-Mechanical Vibration Sensor Integrated on Silicon. *J. Lightwave Tech.* 1999, *17*, 26.
[53] Petersen, K. E. Silicon as a mechanical material. *Proc. IEEE.* 1982, *70*, 420.
[54] Banerdt W. B.; Pike, W. T. A Miniaturized Seismometer For Subsurface Probing On Mars. *Conf. Proc. Geophysical Detection of Subsurface Water on Mars, Lunar and Planetary Institute*. 2001.

[55] Suzuki Y.; Tai, Y-C. Micromachined high-aspect-ratio parylene beam and its application to low-frequency seismometer. *Proc. 16th Annual International IEEE Conference on Micro Electro Mechanical Systems*. 2003, *486*.
[56] Ferris, J. C.; Dohm, J. M.; Baker, V. R.; Maddock III, T. Dark slope streaks on Mars: Are aqueous processes involved? *Geophys. Res. Lett. 29*(10), 1490, doi:10.1029/2002GL014936.
[57] Head, III, J. W.; Marchant, D. R.; Dickson, J. L.; Levy, J. S.; Morgan, G. A. Slope Streaks in the Antarctic Dry Valleys: Characteristics, Candidate Formation Mechanisms, and Implications for Slope Streak Formation in the Martian Environment. *38th Lunar and Planetary Science Conference Abstracts [CD-ROM]*, Lunar and Planetary Institute, Houston, TX, 2007, [abstract 1935].
[58] Fang, Z.; Zhao, Z.; Wu, Y.; Zhang B.; Wang, Y. Z. Integrated temperature and humidity sensor based MEMS. *Proc. IEEE Int. Conf. Information Acquisition*. 2004, *84*.
[59] Sosin, S.; Moldovan, C.; Iosub, R. Design and simulations for a calorimetric micro-sensor for methane detection. *Proc. Int. Semicond. Conf. CAS*. 2004, *2*, 381.
[60] Mitzner, K. D.; Sternhagen, J.; Galipeau, D. W. Development of a micromachined hazardous gas sensor array. *Sens. & Actuat. B: Chem*. 2003, *93*, 92.
[61] Lu, C-J.; Steinecker, W. H.; Tian, W-C.; Oborny, M. C.; Nichols, J. M.; Agah, M.; Potkay, J. A.; Chan, H. K. L.; Driscoll, J.; Sacks, R. D.; Wise, K. D.; Pang S. W.; Zellers, E. T. J. First-generation hybrid MEMS gas chromatograph. *Roy. Soc. Chem. Lab. Chip*, 2005, *5*, 1123.
[62] Kolsear E. S.; Reston, R. R. Review and Summary of a Silicon Micromachined Gas Chromatography System. *IEEE Trans. Comp., Packag., Manuf. Tech*. 1998, *66*, 481.
[63] Geear, M.; Syms, R. R. A.; Wright S.; Holmes, A. S. Monolithic MEMS quadrupole mass spectrometers by deep silicon etching. *J. Microelectromech. Syst*. 2005, *14*, 1156.
[64] Gong, J.; Chen, Q.; Fei, W.; Seal, S. Micromachined nanocrystalline SnO2 chemical gas sensors for electronic nose. *Sens. Actuat. B. Chem.*, 2004, *102*, 117.
[65] George, T.; Urgiles, E.; Toda, R.; Wilcox, J. Z.; Douglas, S.; Lee, C-S.; Son, K.; Miller, D.; Myung, N.; Madsen, L.; Leskowitz, G.; El-Gammal, R.; Weitekamp, D. MEMS-based Micro Instruments for In-Situ Planetary Exploration. *Proc. SPIE*. 2005, *5836*, 20.
[66] Formisano, V.; Atreya, S.; Encrenaz, T.; Ignatiev, N.; Giuranna, M. Detection of methane in the atmosphere of Mars. *Science*. 2004, *306*, 1758–1761.
[67] Krasnopolsky, V. A.; Maillard, J. P.; Owen, T. C. Detection of methane in the martian atmosphere: evidence for life? *Icarus*. 2004, *172*, 537–547.
[68] Price, B. P. Microbial life in glacial ice and implications for a cold origin of life. *FEMS Microbiology Ecology*. 2007, *59 (2)*, 217–231.
[69] Schulze-Makuch, D.; Irwin, L. N. Life in the Universe: Expectations and Constraints. Springer, Berlin 2004.
[70] Schulze-Makuch, D.; Irwin, L.N.; Lipps, J. H.; LeMone, D.; Dohm, J. M.; Fairén, A. G. Scenarios for the evolution of life on Mars. *J. Geophys. Res. Planets*. 2005, *110*, E12S23, doi:10.1029/2005JE002430.
[71] Cabrol, N. A.; Chong-Diaz, G. C.; Wettergreen, D.; Stoker, C. R.; Dohm, J. M.; Keaten, R.; Schwher, K.; Gulick, V. C.; Landheim, R.; Lee, P.; Roush, T. L.; Zent, A. P.; Herrera Lameli, C.; Jensen Iglesia, A.; Pereira Arredondo, M.; Dunfield, G. C.; Pedersen, L.; Bualat, M. G.; Christian, D.; Mina, C.; Sims, M. H.; Thomas, H. J.;

Tucker, D. J.; Zbinden E. Science results of the Atacama Nomad rover field experiment, Chile: implications for planetary exploration. *J. Geophys. Res.* 2001, *106 (E4)*, 7664-7675.

[72] Cabrol, N. A.; Bettis III, E. A.; Glenister, B.; Chong, G.; Herrera, C.; Jensen, A.; Pereira, M.; Stoker, C. R.; Grin, E. A.; Landheim, R.; Thomas, G.; Golden, J.; Saville, K.; Ludvigson, G.; Witzke B. Nomad Rover Field Experiment, Atacama Desert, (Chile) II. Identification of Paleolife Evidence Using A Robotic Vehicle: Lessons and Recommendations for a Mars Sample Return Mission. *J. Geophys. Res.* 2001, *106 (E4)*, 7639-7663.

[73] Greenland Ice Sheet Project 2 (GISP2) [http://www.gisp2.sr.unh.edu/].

[74] Priscu, J. C.; Christner, B. C. Earth's icy biosphere. *Astrobiology.* 2004, *4*, 327–344. Microbial Diversity and Bioprospecting (Bull AT, ed). ASM Press, Washington, DC.

[75] Gaidos, E.; Lanoil, B.; Thorsteinsson, T.; Graham, A.; Skidmore, M.; Han, S.-K.,; Rust, T.; Popp, B. A viable microbial community in a subglacial volcanic crater lake, Iceland. *Astrobiol.* 2004, *4*, 327-344.

[76] Bulat, S. A.; Alekhina, I. A.; Blot, M. Petit, J.-R.; de Angelis, M.; Wagenbach, D.; Lipenkov, V. Y.; Vasilyeva, L. P.; Wloch, D. M.; Raynaud, D.; Lukin, V. V. DNA signature of thermophilic bacteria from the aged accretion ice of Lake Vostok, Antarctica: implications for searching for life in extreme icy environments. *Int J Astrobiol.* 2004, *3*, 1–12.

[77] Sheridan, P. P.; Miteva, V. I.; Brenchley, J. E. Phylogenetic analysis of anaerobic psychrophilic enrichment cultures obtained from a Greenland glacier ice core. *Appl Environ Microbiol.* 2003, *69*, 2153–2160.

[78] Miteva, V. I; Sheridan, P. P.; Brenchley, J. E. Phylogenetic and physiological diversity of microorganisms isolated from a deep Greenland glacier ice core. *Appl Environ Microbiol.* 2004, *70*, 202–213.

[79] Tung, H. C.; Price, P. B.; Bramall, N. E.; Vrdoljak, G. Microorganisms metabolizing on clay grains in 3-km-deep Greenland basal ice. *Astrobiology.* 2006, *6*, 69–86.

[80] Steele, A.; McKay, D.; Allen, C.; Thomas-Keprta, K.; Warmflash, D.; Pincus, S.; Schweitzer, M.; Priscu, J.; Sears, J.; Hedgecock, J.; Avci, R.; Fogel, M. Mars Immunoassay Life Detection Instrument for Astrobiology (MILDI). *Lunar and Planetary Science.* 2001, *32*: 1684.

[81] ESA Aurora Exploration Programme ExoMars [http://www.esa.int/SPECIALS/Aurora/SEM1NVZKQAD_0.html].

[82] Vago, J.; Gardini, B.; Kminek, G.; Baglioni, P.; Gianfiglio, G.; Santovincenzo, A.; Bayón, S.; van Winnendael, M. ExoMars - Searching for Life on the Red Planet. *ESA Bulletin.* 2006, *126*, 16-23.

[83] Schulze-Makuch, D.; Dohm, J. M.; Fairén, A. G.; Baker, V. R.; Fink, W.; Strom, R. G. Sample Return Missions to Mars, Venus, and the Ices on Mercury and the Moon. *37th Lunar and Planetary Science Conference Abstracts [CD-ROM]*, Lunar and Planetary Institute, Houston, TX, 2006, [abstract 1324].

[84] Dohm, J. M.; Fink, W.; Schulze-Makuch, D.; Fairén, A. G.; Baker, V. R.; Furfaro, R.; Tarbell, M.; Hare, T. M. Tier-scalable Reconnaissance To Test Overarching Geological Theories and Locate Prime Targets on Mars. *25th International Space Development Conference Abstracts [CD-ROM]*, Los Angeles, CA, 2006.

[85] Schulze-Makuch, D.; Dohm, J. M.; Fan, C.; Fairén, A. G.; Rodriguez, J. A. P.; Baker, V.R.; Fink, W. Exploration of Hydrothermal Targets on Mars. *Icarus*. 2007, *189*, 308-324.

[86] Schulze-Makuch, D.; Dohm, J. M.; Fairén, A. G.; Fink, W.; Fan, C.; Rodriguez, J. A. P.; Baker, V. R. Prioritizing Putative Hydrothermal Sites on Mars. *38th Lunar and Planetary Science Conference Abstracts [CD-ROM]*, Lunar and Planetary Institute, Houston, TX 2007, [abstract 1735].

[87] Grant, J. A.; Glotch, T. D.; Grotzinger, J. P.; Irwin, III, R. P.; McEwen, A. S.; Milliken, R. E.; Tornabene, L. L.; Weitz, C. M.; HiRISE Team. Stratigraphy in Holden Crater as Revealed by HiRISE. *38th Lunar and Planetary Science Conference Abstracts [CD-ROM]*, Lunar and Planetary Institute, Houston, TX, 2007, [abstract 1435].

[88] Dohm, J. M.; Ferris, J. C.; Baker, V. R.; Anderson, R. C.; Hare, T. M.; Strom, R. G.; Barlow, N. G.; Tanaka, K. L.; Klemaszewski, J. E.; Scott, D. H. Ancient drainage basin of the Tharsis region, Mars: Potential source for outflow channel systems and putative oceans or paleolakes. *J. Geophys. Res.* 2001, *106(32)* 943-32, 958.

[89] Dohm, J. M.; Casavant, R. R.; Strom, R. G.; Zimbelman, J. R.; Scott, D. H. Latent outflow activity for western Tharsis, Mars: Significant flood record exposed. *J. Geophys. Res.* 2001, *106(12)*, 12301-12314.

[90] Christensen, P. R.; Bandfield, J. L.; Bell III, J. F.; Gorelick, N.; Hamilton, V. E.; Ivanova, A.; Jakosky, B. M.; Kieffer, H. H.; Lane, M. D.; Malin, M. C.; McConnochie, T.; McEwen, A. S.; McSween, Jr., H. Y.; Mehall, G. L.; Moersch, J. E.; Nealson, K. H.; Rice, Jr., J. W.; Richardson, M. I.; Ruff, S. W.; Smith, M. D.; Titus, T. N.; Wyatt, M. B. Morphology and composition of the surface of Mars: Mars Odyssey THEMIS results. *Science*. 2003, *300*, 2056-2061.

[91] Gendrin, A.; Mangold, N.; Bibring, J. P.; Langevin, Y.; Gondet, B.; Poulet, F.; Bonello, G.; Quantin, C.; Mustard, J.; Arvidson, R.; LeMouelic, S. Sulfates in Martian Layered Terrains: The OMEGA/Mars Express View. *Science*. 2005, *307*.

[92] Baker, V. R.; Maruyama, S.; Dohm, J. M. A theory of early plate tectonics and subsequent long-term superplume activity on Mars. *Superplume International Workshop*, 2002, 312-316.

[93] Fairén, A. G.; Dohm, J. M.; Uceda, E. R.; Rodríguez, A. P.; Baker, V. R.; Fernández-Remolar, D.; Schulze-Makuch, D.; Amils, R. Prime candidate sites for astrobiological exploration through the hydrogeological history of Mars. *Planetary and Space Science*. 2005, *53*, 1355-1375.

[94] Mahaney, W. C.; Milner, M. W.; Netoff, D. I.; Malloch, D.; Dohm, J. M.; Baker, V. R.; Miyamoto, H.; Hare, T. M.; Komatsu, G. Ancient wet aeolian environments on Earth: clues to presence of fossil/live microorganisms on Mars. *Icarus*. 2004, *171*, 39-53.

[95] Mahaney, W. C.; Dohm, J. M.; Baker, V. R.; Newsom, H. E.; Malloch, D.; Hancock, R. G. V.; Campbell, I.; Sheppard, D.; Milner, M. W. Morphogenesis of Antarctic paleosols: martian analogue. *Icarus*. 2001, *154*, 113-130.

[96] Kuhlman, K. R.; McKay, C. P. Occurrence of rock varnish at Yungay, Atacama Desert, Chile. *38th Lunar and Planetary Science Conference Abstracts [CD-ROM]*, Lunar and Planetary Institute, Houston, TX, 2007, [abstract 2251].

[97] Jet Propulsion Laboratory News Release from October 06, 2006. NASA's Mars Rover and Orbiter Team Examines Victoria Crater. [http://www.jpl.nasa.gov/news/news.cfm?release=2006-121].

[98] Science@NASA News Release from October 26, 2006. A Growing Intelligence Around Earth. [http://science.nasa.gov/headlines/y2006/26oct_sensorweb.htm].

[99] Fink W.; Tarbell, M. A. Tier-scalable Reconnaissance Mission Test Bed: Implementation of Ground-Tier. *38th Lunar and Planetary Science Conference Abstracts [CD-ROM]*, Lunar and Planetary Institute, Houston, TX 2007, [abstract 2410].

In: Planet Mars Research Focus
Editor: L. A. Costas, pp. 49-91

ISBN: 978-1-60021-826-2

Chapter 2

THE SEARCH FOR ORGANICS IN EXTRATERRESTRIAL ENVIRONMENTS: LESSONS FOR MARS EXPLORATION

C. Rodier[1,*], A. Buch[2] and C. Szopa[3]

[1] Laboratoire de Synthèse et Réactivité des Substances Naturelles, Poitiers, France
[2] Laboratoire de chimie et Génie des Procédés, Ecole Centrale Paris, France
[3] Service d'Aéronomie-IPSL, Université Pierre et Marie Curie-Paris, Verrieres le Buisson Cedex, France

ABSTRACT

The determination of the nature and the spreading of the organic compounds in the universe are of particular importance for revealing the chemistry that led to the emergence of life on Earth. In spite of the great knowledge we acquired from spectroscopic observations, the extraterrestrial organic molecules detected remain rather simple and in a limited amount. However analyses of meteorites composition, as well as laboratory and theoretical works, let assume that more complex molecules, including species relevant to life, could be present in extraterrestrial environments and more particularly in the solar system. It thus appears that in situ analyses of solar system bodies of interest for astrobiology are necessary to evidence extraterrestrial prebiotic chemistry or life forms. In this case, any strategy for investigating extraterrestrial environments should: (i) focus on compounds of astrobiological interest; (ii) avoid contamination of the samples; (iii) cross various information to improve the interpretation of the analytical results.

Obviously, the main molecules of interest are those which are readily synthesised under plausible prebiotic conditions, resistant to diagenesis, and which play a major role in biochemistry. Among them, amino acids, the building blocks of proteins in all the

* All correspondence should be addressed to: Claude Rodier, Laboratoire de Synthèse et Réactivité des Substances Naturelles, UMR 6514, 86022 Poitiers, France; Tel: +33 (0)5 49 45 35 90; Fax: +33 (0)5 49 45 35 01; E-mail: claude.rodier@univ-poitiers.fr

living organisms, are currently the most studied even if they have not been detected so far and remain to be found. In the future in situ organic analyses, discriminating molecules produced abiotically from biotic ones will be added to the previous goals of identification and quantification of organic compounds. Two main ways seem currently suited for this objective: (i) the search for homochirality, and more particularly in amino acids; (ii) the detection of chemical biomarkers, by chemical and isotopic analysis.

The search for so many different organics in extraterrestrial environments is a challenging task regarding the difficulties faced to analyze low concentrated and small size samples without any contamination from the sample to the analysts and conversely. After focusing on the choice of the better targeted organic compounds for the in situ space exploration, the authors propose to review the past and future in situ analyses of organics, with a more particular interest for the future Mars Science Laboratory and Exomars missions to Mars. Indeed, these missions will carry Gas Chromatograph-Mass Spectrometer instrumentation, partly developed by the authors, which will be coupled with a chemical reactor. This association should allow the in situ analysis of complex organics which cannot have been detected by former missions, and also the chiral analysis of theses species. Hence, it could reveal for the first time traces of extraterrestrial prebiotic chemistry or even trace of life.

Introduction

The existence of life in the Universe elsewhere than on Earth, the origin of organic matter leading to the emergence of life on Earth, the homochirality of the molecules of the living, and the possible existence of life on Mars in its early stage, or even today, are among the most fundamental current questions in the field of astrobiology.

The emergence of life is a question of a tremendous complexity (Davies, 1998; Horneck, 1995). Every organism on Earth is constituted of organic molecules. From now, we have no evidence of the origin of the prebiotic/biotic organic material on Earth; it could have been endogenous, exogenous or both (Melosh, 1988; Chyba et al., 1990; Chyba and Sagan, 1992; Orgel, 1998; Pierazzo and Chyba, 1999; Lindsay et al, 2005). Too many events occurred on Earth since its formation and erased the first organics traces and their evolution. Thus, more information is needed to answer the great question of the origin of life. One way to get through would be to look outside our planet as potential traces of organics are probably less altered from their original state than on Earth (Brack, 1996; Brack, 1997; McKay and Stoker, 1989).

When looking for life on other planets, some assumptions are useful. One is to suppose that the extra-terrestrial life-forms are based on carbon chemistries, while it is possible that non carbon-based life exists such as silicium-one for example. The second is the presence of liquid water as it provides an excellent environment for the formation of complex organic molecules that could eventually lead to the emergence of life.

On the basis of these assumptions, it seems that Mars is today the best candidate currently known, waiting for the discovery of planets like exoplanets. The premise for the appearance of life on Mars is based on similarities between the early histories of Mars and the Earth (McKay and Stoker, 1990; McKay and Davis, 1991; Pollack et al., 1987) and, especially, on geomorphological evidence for the presence of liquid water on its surface at various times in the past (Baker et al., 1991, Parker et al., 1989; Parker et al., 1993). If so, Mars is a good candidate for having once developed life or a prebiotic state, since life on Earth could have started during this period of time (McKay, 1996). It is even thought that life could have emerged on Mars and been transferred to Earth by an asteroid. Recent evidence for sedimentary processes and standing liquid water on Mars from the Mars Exploration Rovers support the idea that condition for life may have once occurred on the planet. Even if life became extinct or never arose, the subsequent cooling of Mars and the lack of extensive tectonic recycling may have enabled molecular evidence of biotic or prebiotic activity to be preserved in ancient sediments.

Man's curiosity about Mars spans more than a century. In 1854, William Whewell, theorized that Mars had seas, land and possibly life forms. However, serious investigation of life on Mars started with the Viking lander experiments in 1977, some results of which are still raising debate (Levin, 1997; Levin and Levin, 1998). Dozens of spacecraft, including orbiters, landers, and rovers, have already been sent to Mars by the United States, Europe, the Soviet Union and Japan to study the planet's surface, climate, and geology. About two-thirds of all spacecraft have failed before completing or even beginning their missions. Landing on Mars is thus the first challenge of the highly difficult but fascinating study of organics on Mars.

Thanks to previous missions and expertise in the field of space research, the exobiology communities will prepare future ambitious missions dealing with organics but also homochirality; with all the constraints those goals implied. Instrumentation available to detect trace levels of key biological compounds have improved dramatically since the Viking missions and many of these methods can be miniaturized so they can be accommodated into a spacecraft. The critical issue is to provide the most definitive results in answering the life on Mars question. For future missions interpretation of results will be a key point with retrieval of data as a major step in space research.

ORGANICS IN SPACE

Organic Molecules in the Universe

The organic molecules, originally named that way because it was thought that they could only be synthesized by biological processes, were thought to be exclusively present on the Earth for a long time. But at the end of the 19th century and beginning of the 20th, the discovery of the capability to produce such molecules abiotically, associated to the emergence of spectroscopic techniques to probe the composition of materials showed that organic molecules were also present in space. It is today observed that organic materials are widespread in the solar system and in the Universe. The major source of information on this spreading obviously consists in the spectroscopic observations of the universe in the domains of infra-red and radio wavelengths. Then, in situ measurements or analysis of samples returned from solar systems environment bring important local information. Finally, laboratory simulations of the chemistry of various environments of the Universe provide important clues on the existence of molecules which were not detected yet, but which could be present in the studied environments. Generally, these simulations generate organic materials more complex than those currently detected. Such species are difficult to detect with observations either because they are not in the physical state of gas or ice, or because their concentration is too low compared with the current limits of detections of the analytical techniques used today. It can also be underlined that such simulations provide an idea on chemistry occurring in places of the universe which are difficult to access today (e.g. planetary sub-surfaces).

Spectroscopic observations provide evidences for the presence of numerous organics in the interstellar space medium (ISM) which contains the original materials of the solar systems. More than 120 molecules containing up to 11 atoms have been detected in the interstellar space (Winnewisser, 1997). The organic fraction contains many precursors to biological molecules such as formaldehyde, formic acid and acetaldehyde. The number of molecules identified in these clouds increases each year and various species are now listed (Ehrenfreund and charnley, 2000). This material includes hydrocarbons, nitriles, alcohols, and it is assumed to be created from ices of simple composition (e.g. H_20, NH_3, CO) condensed on silicate grains, under the action of radiations present in the ISM (Allamandola et al;, 1999; Strazzulla and Palumbo, 2001; Cottin et al., 2001) and thermal processes (Schutte, 1993) associated to the variation of temperatures in the interstellar dust clouds (see for example Cottin et al., 1999). But from laboratory experiments, it is expected that the material produced should be much more complex than the part which is currently detected with observations, because a refractory solid organic material is created under these experiments. This material is made of organic molecules difficult to characterize, regarding their polymer like structure, but some precursor of amino acids have been already detected (Munoz-Caro et al., 2002; Bernstein, 2006). Glycine is also suspected in the ISM thanks to spectroscopy measurements but this identification is still debated (Kuan, 2003).

In planetary systems observed to be under formation, the organic molecules detected are currently identical to those observed in the ISM. Progresses done in the observation of exoplanets will soon allow obtaining the composition of planetary atmospheres outside the solar system. Gaseous sodium was already observed in the atmosphere of an exoplanet located very close to its star (Charbonneau et al., 2002) but this remains an exception due to the current limitations of this type of observations.

As a consequence, our knowledge about organic material outside our planet mainly deals with the solar system bodies. The known organics are rather located in the outer solar system. Indeed, they are detected in the atmospheres of giant planets (Encrenaz, 2005). These planetary atmospheres contain significant amounts of methane from which occur a photochemistry producing more complex organics. This mechanisms induced by this chemistry leads to a gaseous to solid phase transition, thus producing solid particles in these atmospheres which are the core of atmospheric aerosols producing the colourful features observed in the visible range (West et al., 1986). Besides the outer planets, some of their satellites raise a great interest for organics. The most famous one is Titan which looks like the Earth, meaning a solid surface with liquid areas and a dense atmosphere, but which gets an atmosphere mainly composed of nitrogen and methane at lower temperatures than the Earth. As for the giant planets, a photochemistry is induced in the atmosphere which creates organic aerosols at the scale of the satellite (Lebonnois, 2002). As methane comes from the surface or the interior of the satellite, and that the aerosols settle down to the soil, this creates a carbon cycle on Titan which is of particular interest since part of this cycle could have occurred in the primitive Earth atmosphere, thus contributing to the providing of organics for the prebiotic chemistry. For the other satellites of the outer planets, an eventual presence of organics is still difficult to demonstrate because their remote probing is difficult as they are only solid bodies from which we have only information from the observation of their surface. However, some reasonable assumptions let us hope the presence of such molecules in several satellites. Among them, Europa is important because under its thick ice surface layer, a liquid underground ocean of water could exist and chemistry could occur in liquid water under the action of underwater volcanoes. Some hopes to detect Europa organics exist by the observations of plumes of material released by the satellite because of gravitation tides. Phoebe could also include organics since with the last pictures taken by Cassini, it can be assumed that Phoebe could be an extinct comet trapped by the gravitation of Saturn. Iapetus is also intriguing because of the colour contrast between two of its sides. The dark side colour is interpreted by some authors as the colour of organic solid particles deposited at the surface of the satellite. Concerning the icy satellites, some experiments tend to show that these ices can be processed by the interplanetary radiation, or episodic impact, and this could generate organic compounds of interest but this assumption was not proved until today.

The last category of bodies known to carry organic molecules is small bodies of the solar system, meaning comets and asteroids. From observations and laboratory simulations, comets are expected to carry numerous organic species (Cottin et al., 1999) they could have brought to Earth during the heavy bombardment of the Earth which occurred about 4 billion years ago. This organic material is supposed mainly to come from the interstellar material because comets should be remnant of the planetesimals formed in the earliest stage of the solar system. Therefore, they could include both organic ices but also refractory compounds formed in the ISM. Concerning the asteroids, it is today known from the analysis of a class of meteorites found on Earth, named carbonaceous chondrites, that asteroids can contain organic material. This material is generally found under the form of kerogens in big meteorites (Bernstein, 2006), But recent measurements done on interstellar dust particles collected in space (Sandford et al., 2006) could show that the organic material could be richer than what we think today.

As a consequence, researches led on the distribution of organics compounds in the universe show that they are widespread. The structure of these compounds can be very complex and molecules important for the living organisms are retrieved in space, out of the Earth. But in spite of this significant progress done in this field, no environment compatible with life, as we know it, was discovered so far.

Life and Organic Molecules on Mars?

The particular interest raised by the search for organic on Mars comes from the fact that the conditions for the emergence of life, as we imagine them today, could have been met in the primitive Mars. Thus, Mars is currently the only example of such a planetary object we have except the Earth. Search for Life can be performed at different levels. The visible aspects of life relate to the morphology of an organism or the characteristic structures built by organisms can be taken into account. Indeed, ancient organisms from any of extraterrestrial environments may have left a detectable fossil record (Farmer and Desmarais, 1994; Westall, 1999). However, most life detection methods use biogeochemical techniques for determining the biomolecules that make up an organism.

The presence of carbon on Mars can be inferred from the existence of carbon in the martian meteorites EETA79001 (Wright et al., 1989; Beker et al. 1997) and ALH84001 (McKay et al., 1996; Grady et al., 1994; Clemett et al., 1998; Meierhenrich et al., 2001; McCollom, 2003). Important sources of carbon on Mars (and Earth), including prebiotic molecules from comets, meteorites and interstellar dust particles, have been proposed (Kissel and Krueger, 1987; Kvenvolden et al., 1970; Maurette, 1998; Oró, 1961). Indeed, space appears to be a potential chemical reactor that can provide a broad scale of organic molecules, including biological building blocks and their chemical precursors. Intensive studies of the chemical composition of the meteorite samples also have been carried out. The study of the martian Murchison meteorite revealed more than 500 organic compounds, including several nucleic acid bases and more than 80 amino acids (Cronin and Pizzarello, 1983; Cronin and Pizzarello, 1988; Sephton, 2002; Meierhenrich *et al.*, 2004). In addition, amino acids are abundant (the total amino acids in Murchison reaches 700 nmole g^{-1}) and some of them present enantiomeric excesses (Cronin and Pizzarello, 1983, 1997, 1999). Poly aromatic hydrocarbons (PAHs), are also interesting targets. Indeed, PAHs are probably the most abundant carbonaceous gas phase molecules in space (Tielens et al., 1999; Yan et al., 2005). PAHs have been identified in meteorites, interplanetary dust particles (IDPs) and in comets (Clemett et al., 1998; Ehrenfreund and Charnley, 2000; Sephton, 2002).

Among all these molecules, potential organics on Mars should then contain chiral components with a biological configuration such as amino acids and compounds with structures characteristic of a biosynthetic origin (i.e biomarkers). However, oxidising agents at the martian surface would influence preservation of organic matter. The survival of complex organic molecules will depend upon their chemical stability under martian conditions (Kanavarioti and Mancinelli, 1990). Not all organic molecules are equally susceptible to oxidation (Levin and Levin, 1998). Some important biomolecules, including amino acids (Bada and McDonald, 1995) and purines, are stable in an oxidising environment, whereas some others, such as nucleotides, ribose, deoxyribose and pyrimidines, are not (Larralde et al., 1995). Calculations based on the stability of amino acids lead to the conclusion that remnants of these compounds might have been preserved buried beneath the oxidizing surface layer. The carbon skeleton of some lipid molecules is also resistant to degradation and can remain almost intact for billions of years. Some lipids are found only in specific groups of organisms and serve as biological markers for these groups. For example, long chain carboxylic acids with even over odd carbon predominance are characteristic of higher plants (Eglinton and Hamilton, 1967) whereas branched iso and anteiso carboxylic acids are characteristic of bacterial activity (Ambles et al., 1994). Those fatty acids are known to form vesicles when the hydrocarbon chains contain more than ten carbon atoms. Such vesicle have been identified in the Murchison meteorite (Deamer, 1985). Complex carboxylic acids are also supposed to be present on Mars (Benner et al., 2000). Thus, carboxylic acids could be stable and give clue to the origin of martian organic matter. In addition, on Earth, the organic matter undergoes a complex process of degradation and condensation (humification) to give chemically stable macromolecules known as solvent-insoluble kerogens (de Leeuw and Largeau, 1993). Thus macromolecules have also to be taken into account in seek for life (Helfer, 1990). Such organic biomarkers are very stable as they occur in ancient terrestrial sediments from the Precambrian (Imbus and McKirdy, 1993; Brocks et al., 1999; Summons et al., 1999).

The organics to be sought in martian material are, in order of importance: (i) volatile low molecular weight compounds: hydrocarbons, alkanoic acids, and peroxy acids present in the oxidising atmosphere; (ii) medium molecular weight compounds: hydrocarbons, alkanoic acids, amino acids, purines; and (iii) macromolecular components: kerogens, oligo- and polypeptides (Figure 1).

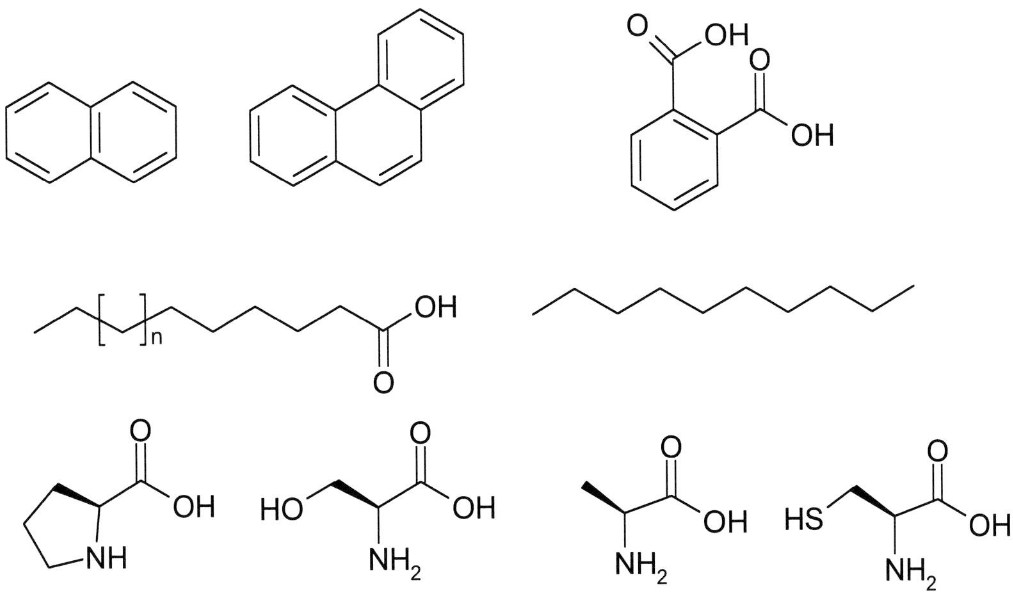

Figure 1. Some potential organic molecules on Mars.

In addition, the finding of organics in the martian surface and near surface environment would indicate either a biotic source or an imported meteoritic source, since the oxidising atmosphere of Mars precludes the present formation of prebiotic organics. More information is thus needed to better understand the origin of martian organics. Chiral analysis would permit to distinguish between prebiotic meteoritic organics and indigenous biogenic organics. Indeed, abiotic synthesis leads to racemic mixture (L and D isomers in the same concentration) whereas on Earth, biomolecules in general are largely chiral (L or D) and, moreover, almost exclusively homochiral (Figure 2). As Pasteur realized, understanding the origin of homochirality may be the key to understand the origin of life. On Earth, both biotic and abiotic theories have been proposed (Bonner, 1991). According to the former, life initially was based on achiral molecules and/or racemates and the use of specific enantiomers came about through evolution, whereas the latter theories propose that a tendency toward homochirality was inherent in chemical evolution. The finding of L-enantiomeric excesses in the amino acids of meteorites supports the latter concept. Accordingly, the accumulation of amino acids and other chiral compounds on the early Earth through the accretion of dust, meteorites and, possibly, cometary material would have provided organic matter with small enantiomeric excesses, but ones sufficient for amplification by further chemical evolution. Thus, homochirality can be a crucial signature for life. Although homochiral molecules gradually revert to racemic mixtures with time, in a dry environment such as that on Mars, the homochirality of molecules 3–4 billion years old may still be detected (Bada and McDonald, 1995).

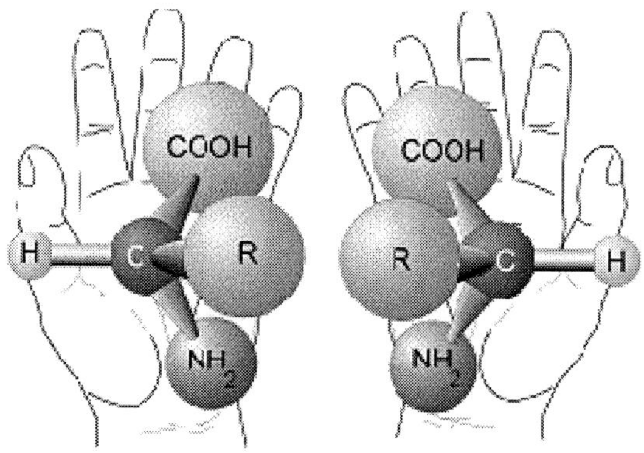

Figure 2. Example of L and D enantiomers (amino acids)

In future missions the ratio of enantiomeric molecules in a martian sample has then to be used as a test for biotic or abiotic origin (Bada and McDonald, 1996). To confirm the results, additional data must be added. Thus, determination of the carbon isotope ratio could also allow establishing the presence of biotic/abiotic organic matter. Indeed, biotic organic compounds are enriched with ^{12}C isotope whereas abiotic organic or inorganic matters are much more concentrated with ^{13}C isotope. Then the presence of organics and their origin will be better understood. However, nowadays no martian space mission has been able to detect organic molecules with more than two carbons.

Search for Biotic and Prebiotic Activity on Mars with Space Missions

Mars was considered, until recently, as a damned planet for space missions because most of space missions to Mars failed before reaching the planet, especially at the beginning of the space exploration era. Among the numerous launched missions, only just a few ones were devoted to organics. The first successful fly-by mission to Mars was the NASA Mariner 4 probe (1964) which provided pictures of the surface of Mars which definitely ruled out the existence of macroscopic life on the red planet (Leighton et al., 1965). The first successful probes to land on the surface were two Soviet probes, Mars 2 and Mars 3 (1971), but both lost contact within seconds of landing. One decade after Mariner 4, it was decided to search for microscopic life forms on Mars and it resulted in the first astrobiological mission to Mars, Viking.

Attempts to Find Life on Mars up today: Between Optimism and Pessimism

In 1976 the Viking mission was consisted of two orbiters, each having a lander. The Viking landers relayed the first colour pictures of Mars. They also mapped the surface of

Mars. The search for organics on Mars began with the Viking landers GC-MS (Gas Chromatograph Mass Spectrometer) which was looking for organics on the surface and subsurface (Margulis et al., 1979). They did not detect even a trace of organic matter at the ppb-level (Biemann et al., 1977). Since the meteorites and interplanetary dust should carry organics to Mars, it had been suggested that they were actively destroyed. The experiments did indicate the presence of OH radicals (or peroxides derived from them) in the martian soil, diffusing into the dry soil and acting as a strong oxidizing agent which could lead to the organics destruction. The Viking mission had also permitted an important characterization of the environmental constraints on possible martian organisms (Zasova et al., 2002). The relative failure of the Viking mission led to the idea that life could not be present on Mars, resulting in a loss of interest to investigate Mars, especially to search for life. More than 20 years will pass before the red planet will see a new space probe successfully reach it, after the failure of 3 probes in the 80's. It was Mars Global Surveyor in 1996.

Indeed, the Soviet probes Phobos 1 and 2 were sent in 1988 to study Mars, unfortunately Phobos 1 lost contact on the way to Mars, and Phobos 2, while successfully photographing Mars and its moon Phobos, failed just before it was set to release two landers on Phobos' surface.

Following the 1992 failure of Mars Observer orbiter, Mars Global Surveyor was launched in 1996. This mission was a complete success regarding the mapping mission. The Thermal Emission Spectrometer (TES) experiments provided some major information concerning the volcanic mineralogy (Christensen et al., 2003 and 2004). It has been shown that volcanic mineralogy varied from basalt to andesitic (Christensen et al., 2001). Basalt mainly located in ancient southern hemisphere highlands was composed of plagioclase feldspar, clinopyroxene, olivine ($(Mg,Fe)_2SiO_4$) and sheet silicates. Andesite mainly located in younger northern plains was composed by plagioclase feldspar and high silicate volcanic glass. TES has also shown that aqueous mineralization is limited to regions with ambient or hydrothermal conditions. In the equatorial region it has been detected the presence of gray crystalline hematite (α-Fe_2O_3) in three locations which appear to be a probable sedimentary environment. Even if no carbonate has been detected by TES it could be assumed that under these three locations, liquid water was stable near the surface of mars for a long-term, probably million ycars (Christensen et al., 2001b). Indeed one of the more plausible processes which imply deposit of hematite is a chemical precipitation from Fe-rich water. TES also highlighted that in the dark region of Mars volcanic mineral such as pyroxene, feldspar and minor olivine are unweathered. Moreover no presence of weathering product has been found. Then geologic history seems dominated by cold and dry climate (Christensen et al., 2001) which helps to the preservation of organics.

Only a month after the launch of the Surveyor, there was Mars Pathfinder, carrying a robotic exploration vehicle, which landed in the Ares Vallis on Mars.

Then Mars 96 mission, an international mission led by Russia, was lost just after the launch. Among the scientific instruments an American instrument MOx was developed to characterize the reactive nature of the Martian soil (McKay et al., 1998).

In 2001, Mars Odyssey orbiter was successfully launched. Odyssey's Gamma Ray Spectrometer (GSR) has detected significant amounts of elemental hydrogen (Feldman et al., 2005) in the upper metre or so of Mars' regolith and in two different locations. One is the northern lowlands (Mitrofanov et al., 2002) and the second in the southern highlands (Mitrofanov et al., 2002; Boynton et al., 2002) (figure 3 and 4). Detected hydrogen seems to be identified as ice-water. In the south pole subsurface ice-water could be located at high elevation whereas in the north, ice-water seems to be at low elevation with much greater content (Mitrofanov et al., 2003). From these data it has been concluded that in the polar caps, subsurface regolith could be a substantial reservoir for martian water. The Thermal Emission Imaging System (THEMIS), which was able to differentiate solid CO_2 and water frost to sub-kilometer scale, has confirmed the water presence in the polar cap (Christensen et al., 2003). Especially in the north polar location when the seasonal CO_2 cap recedes.

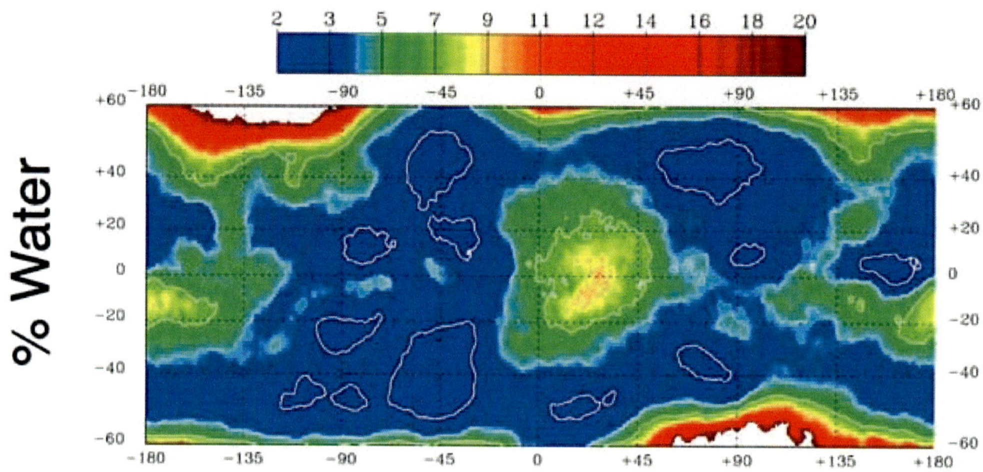

Figure 3. Maps of Water-Equivalent Hydrogen (WEH) found in the regions around the equator of Mars in cylindrical projection between ±60. The map is based upon gamma ray data collected for the element hydrogen. (from Feldman et al., 2005).

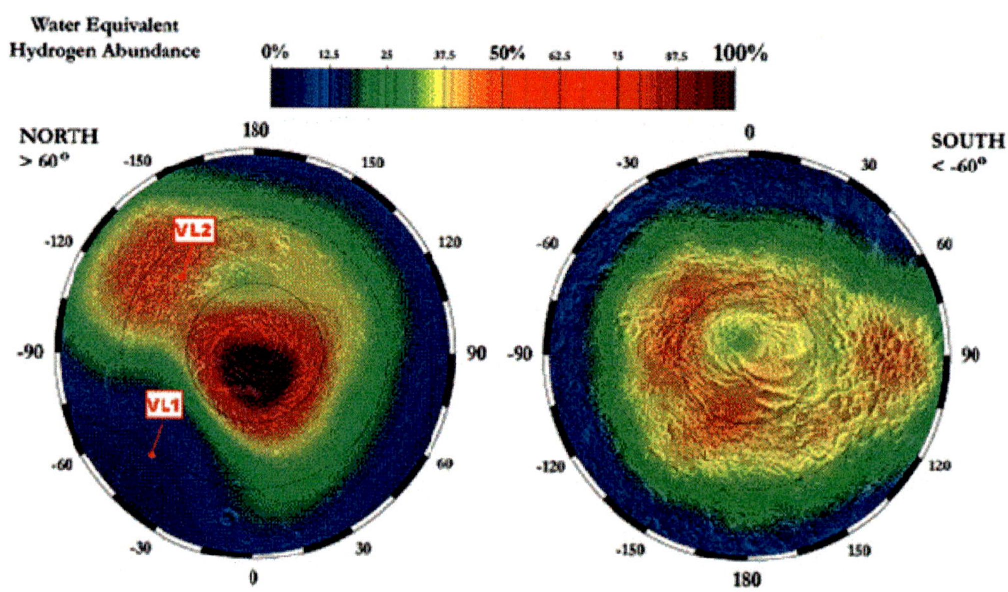

Figure 4. Maps of Water-Equivalent Hydrogen (WEH) found in the north and south hemisphere of Mars. Viking 1 (VL1) is shown to have landed in terrain with about 2% of water, while the water content of the Viking 2 (VL2) lander site is about 10%. (from Levin, 2004).

In 2003, the ESA launched the Mars Express craft consisting of the Mars Express Orbiter and a lander. Goal of the lander BEAGLE II was to search for organic material and other volatiles on and below the surface of Mars in addition to study of inorganic chemistry and mineralogy (Sims et al., 1999). BEAGLE II failed during descent. However, imaging spectrometer OMEGA in the orbiter, made it possible to obtain specific and direct observation of the Mars polar regions about water, carbon dioxide and dust. Results were supposed to help to determining the variability of these ices and studying the process of condensation, metamorphosis, and sublimation of the volatils compounds. The OMEGA experiment found also the presence of clays which are markers of liquid water standing for a long time at the surface of Mars. Indeed, formation of clay minerals would mean a long-lasting wet episode, with large surface aqueous reservoirs and alkaline water resulting from chemical alteration. It could also have been formed primarily in the subsurface by hydrothermal activity, cratering, supplying subsurface water to the impacted minerals or during the cooling of the mantle. Formation of sulfate needs important amount of water and acidic environment. For example calcium sulfate requires a sulfur rich environment with presence of water and pyroxenes or feldspars. Moreover, phillosilicates seem to have been formed during Naochian era. If these phillosilicates have been formed in the subsurface, then Mars environment might have been tenuous, cold and dry, but if they have been formed at or close the surface then early martian atmosphere might have been dense. Therefore, if living organisms exist or have existed, location where phillosolicates have been detected could be the area where living and biochemical development took place. OMEGA has confirmed the presence of sulfates and phyllosilicates indicative of aqueous alteration (Bibring et al. 2006; Poulet et al., 2005). Sulfate rich areas have been detected within Valles Marineris, Terra Eridiani or Juventae Chasma (Langevin et al., 2005; Gendrin et al., 2005). Phyllosilicate are detected in only a very restricted numbers of areas such as Nili Fossae, Nili Syrtis Mensae, crater floor in northern Syrtis Major or MawrthVallis (Feldman et al., 2005). In addition places where phillosilicates have been observed are areas with low level of the further surface alteration. Then, record of biological molecule could be preserved in these areas.

HRSC, OMEGA and MARSIS instruments also showed that there are large reservoirs of frozen water at the surface (pole) and in the underground of Mars (Hansen et al., 2005). Fitted model of the OMEGA spectra confirmed the presence of water ice in the south polar cap. Three types of terrains were identified (Douté et al., 2007, Bibring et al., 2004). This opens possibilities to the existence of liquid water reservoirs which could contain micro-ecosystems. Another intriguing discovery was made with the PFS instrument. Methane was found in the atmosphere. This detection of the first martian organic molecule, also done from Earth with observations, raises question about its origin (Kerr, 2004; Formisano et al., 2004). As methane cannot persist in the martian atmosphere for more than a few hundred years, its presence suggests either that it is being replenished by some unidentified volcanic or geologic process (Oze and Sharma, 2005), or that some kind of extremophile life form similar to some existing on Earth is metabolising carbon dioxide and hydrogen and producing methane. These small quantities of methane, and formaldehyde are both claimed to be hints for life, as these particles would quickly break down in the martian atmosphere (Krasnopolsky et al.; 2004).

However, the first great improvement in the probability to find life on Mars was made by the famous Mars Exploration Rovers (MER), which are still investigating the surface of Mars. Even if these probes were not devoted to search for life, the scientific payload being composed of experiments to characterize the mineralogy of the red planet, these rovers and the measurements they performed allowed to prove that large amount of liquid water really ran off at the surface of Mars for long periods (mineral deposits..). Moreover, from a technological point of view, they demonstrated the superiority of a mobile platform on a static one because the geologic indices that enabled to demonstrate the presence of water were found at a certain distance of the landing sites. Moreover, they demonstrated that rovers can effectively live for a long time at the surface of Mars. Landing sites were Gusev Crater and Meridiani Planum. These places were respectively a possible place of a former lake and a place where presence of hematite suggests a possible former water presence. Opportunity was the fourth probe to successfully land on Mars in fifthteen tries and achieved the first in situ characterization of sedimentary rocks in Meridiani Planum (Squyres et al., 2004; Knoll et al., 2005). It did not resolve the question of whether Mars ever supported life, but it provided fresh perspectives on both habitability and biogenesis on Mars. Among the most significant science returns has been the conclusive evidence that liquid water existed at some time in the past at both landing sites. Mini-TES, a Michelson interferometer, showed evidence in the Gusev Crater for minor (<5%) carbonate and bound water. It had been assume that the soils and rock have a similar composition due to the decomposition of the rock perhaps mixed with materials from others sources such as any aqueous materials (Rieder et al., 2003; Christensen et al., 2004). From Spirit instruments it has been shown that the rock analysed at the Gusev place was originated from the same lava.

These rocks interior are soft and concentrated in sulfur, chlorine and bromine. Presence of sulphur, chlorine and zinc in the bright dust and dark soil are attributed to precipitate of volcano outgassing (Clark and Baird, 1979). Presence of bromine at Gusev and Meridiani place is a possible indicator of liquid water. Indeed bromide mineral are very soluble in water. And after evaporation of water bromine is deposited. Presence of nickel in the soil, in the rock and in the bright surface can be analyzed as be originated from meteorites. Indeed, concentration of nickel find at the Gusev place can be correlated with the concentration of nickel in CI chondrite Meteorite. Soil of the Gusev place are formed by a rock component and dust component (Goetz et al., 2005). Conclusion of the Mössbauer Spectrometer analysis is airborne dust is formed of primary mineral and possibly hydrated mineral. Primary mineral could be olivine, pyroxene ($(Mg,Fe)SiO_3$) and magnetite crystallites (Fe_3O_4). They are formed from basaltic rock by a physical process which occurs since the planet formation. The possibly hydrated mineral could be constituted of oxidized cristallites of ferric and nanocrystaline oxide. It could be residu from the early water rich period but for now there are no certainty. Jarosite (iron sulfate compound) has been detected by opportunity without ambiguity (Klingelhoefer et al., 2004; Squyres et al., 2004). Presence of this mineral is very informative concerning the Mars chemistry. Indeed, jarosite is only formed in the presence of liquid water and precipitate from acidic solution. Then, presence of sulfate would mean that sulphuric acid plays an important role at Meridiani place. Indeed, sulphuric acid could be implied in the sediment deposition and the basalt wearhering (Burns, 1987; Clark and Baird, 1979).

Thus Mars Exploration Rover mission has proved that Meridiani place was arid, acidic and strongly oxidizing. All these chemical parameter are strongly disadvantageous for life (Fernandez-Remolar et al., 2005). Apart from extremophiles able to live with strong oxidizing environment and microorganism able to live in acidic environment or at low water activity, life, as we know on Earth, would have been a significant challenge to life outbreak. Moreover, even if it has existed a life early originated on Mars, it has been subject to late heavy bombardment. Then, if no safe place has existed on Mars, originated life has certainly been destroyed by this bombardment. Meridiani could not be representative of the Mars's surface, but this place is for now the better place for life than Gusev crater.

As an intermediate conclusion, it can be stated that after the Viking missions, hope was lost to find life on Mars, but the missions of the two last decades raised a new hope for astrobiology. The conclusions drawn by all the missions are that the present day Martian surface is cold, dry, and chemically harsh. However the improvements made on the knowledge of the red planet and on the Earth living organisms (extremophiles) raised the interest to search for life on Mars again. Indeed, Mariner and Viking images of channels and possible ancient lake basins indicated that Mars may have been more hospitable to life early in its planetary history (Car, 1996). Past missions helped also to better understand the oxidizing surface and the subsurface mineralogy and to precisely define landing sites for future space missions. Concerning the martian meteorites analyzed, no correlation has been found with the spectral properties of any martian regional volcanic unit; their site of origin must then be elsewhere on the planet.

The huge amounts of collected data from the past missions are however not including information about organic compounds. In the future, instruments designed to search for a broader range of organic compounds on Mars, especially those that are important in life on Earth, will require an improved ability to extract and detect non-volatile organic species in samples that are better protected from oxidants and ultraviolet radiation (Soffen., 1976).

A Status of our Current Knowledge on the Possibilities to Find Life or Organics on Mars Today

On the basis of the results obtained with the recent space missions, it seems today probable that a prebiotic chemistry, or even life, can have taken place on Mars. Indeed, all the recipients considered as necessary to produce life should have been present on Mars approximately at the same epoch at which life is expected to have appeared on Earth: water, a probable dense atmosphere of CO_2, organics brought by external sources (i.e. meteorites and comets) (Bland and Smith, 2000). Moreover, current exogeneous sources of organics, mainly consisted of interstellar dust particles, exist. Therefore, we can wonder why we did not detect any organic at Mars (except atmospheric CH_4) until today, especially with the Viking instruments which were specifically devoted to this task.

With the progresses made in various domains, the interpretation of the results obtained with the Viking experiments can be today regarded as not excluding the presence of life and organics. Regarding life, it is expected that the environment of the probe is probably too harsh for living organisms as we know them to survive. Original recent theories however think about organisms using H_2O_2 for their metabolism, making them not sensible to this oxidant. They suggest that if they were present in the samples collected by the Viking probes, they would have been destroyed by the processes used to evaluate the presence of life, which all include a heating step of the sample which could destroy the organisms. The non-detection of organics by the Viking pyrolysis gas chromatography mass spectrometry (pyr-GC-MS) experiment suggested chemical rather than biological interpretations for the reactivity of the martian soil. A comparison between pyr-GC-MS and total organics has been conducted for a variety of Mars analog soils. In iron-contain soils (jarosites from Rio Tinto and Panoche

Valley) and the Mars simulant (palogonite), oxidation of the organic material to carbon dioxide (CO_2) by iron oxides and/or their salts drastically attenuates the detection of organics. The release of 50-700 ppm of CO_2 and of 0.01-1.0% of water vapour (H_2O) by pyr-GC-MS in the Viking analysis may indicate that an oxidation of organic material took placed. Therefore, the martian surface could have several orders of magnitude more organics than previously thought (Navarro-Gonzales et al., 2006).

The non detection of organics at the surface of Mars could also be attributed to the environmental conditions. These conditions could be destructive for the organic matter. The proposed processes of destruction are the radiation decomposition and the oxidation processes.

Radioactive radiation coming from the radionucleides decomposition, cosmic rays or solar energetic particles can be sufficiently energetic to degrade organic material. Recent studies dealing with the resistance of amino acids seems to demonstrate that this type of radiation can affect the amino acids, particularly if they are present at the surface, but once they are buried, the shielding effect of the Martian soil can be sufficiently efficient to limit this degradation with time (Kmineck and Bada, 2006). A simple model shows that below 2 m depth, the action of this type of radiation could be negligible. Secondly, the UV radiation at the surface of Mars can also contribute to the degradation of the organic materials. Indeed, the atmosphere of Mars is too tenuous to filter UV radiations above 200 nm wavelength (Patel et al., 2003). Experimental studies carried out either in laboratory (Stoker and Bullock, 1997; Stalport, subm.) or in Earth orbit (Barbier et al., 2002), showed that at these wavelengths, organic materials, including amino acids, were degraded when exposed directly to these radiations. However, when buried under a thin layer of minerals, the shielding effect is so efficient that no degradation occurs. Therefore, radiations can contribute to the destruction of the organic material when directly exposed to the radiations at the surface of Mars, but this contribution is not well quantified. Moreover all the experiences led, show that organics present in the sub-surface or inside rocks can survive to radiations due to the shielding effect of rocks.

Concerning oxidation, several possible sources of oxidation were so far identified. Most of them were listed just after the arrival of the Viking missions to Mars (Hunten, 1979). They are hydrogen peroxides (H_2O_2) generated by the atmospheric chemistry, and recently detected (Clancy, 2004), and superoxides possibly present in the soils (Tsapin, 2000; Yen et al., 2000). However, the action of these real or hypothetical oxidants on the organic materials is unclear. Hence, the entire destruction of organics at the surface is not evident. For example, Benner (2000) shows from a theoretical model that organics could be oxidized by H_2O_2, but the material would evolve to stable organics. These organics could survive at the surface, and it was shown that they could not have been detected with the Viking instrumentation which was not suited to this type of chemical species. Indeed, to be detectable by Viking experiments, organic compounds or their derivatives compounds should be volatile after pyrolysis. For example salt of benzenecarboxylic acids, oxalic acid and acetic acid are not volatile. In addition, some product of pyrolysis could be not volatile. When they are volatile such as CO_2, CO, H_2O, they were detected as being martian atmosphere (Benner et al., 2000). Moreover, as for the radiation, the influence of these oxidants on the organic materials depends on the location of the organics. Indeed, superoxides must be activated by photochemistry. Also, the processes they induce is active only at the surface of the planet. Concerning H_2O_2, it is expected that it could diffuse in the subsurface, making it capable to alter buried organics. But the depth at which H_2O_2 can diffuse is not known. Models of such a process provide a wide range of possible depths from the meter to several kilometres (Zent, 1998). Therefore, the potential influence of these oxidants needs to be validated by in situ measurements at Mars.

As a consequence, it seems today that organics should be present on Mars, the challenge being to find where. A consensus in the community exists on the fact that we should have more chances to find organics by digging in depth. This will be one of the challenges of the future mission to go to Mars.

What Future for the Search of Life on Mars

The three fundamental requirements for a satisfactory search for life on Mars are a landing site with exobiological potential, sampling requiring mobility, and an integrated instrument package providing observational, spectroscopic and analytical techniques to reduce the potential for ambiguous interpretation of the results.

For all these exploration missions, many instruments and analysis techniques have been developed to explore atmosphere, surface and sub-surface. Here we focus on techniques with exobiological interest associated with martian missions. There is obviously a strong need for an unequivocal chemical identification of organic compounds within material recovered from extraterrestrial samples.

Analytical Tools Used in the Past and Current Space Missions to Search for Life and Organics

The most powerful tool for qualitative and quantitative identification of chemicals of an unknown nature is the chromatographic method. However chromatographic methods require larger investments in equipment and performance to be carried on space missions, but benefit from a number of great advantages over the physical spectroscopic methods alone. Chromatography able to separate less than nanograms of chemicals; it allows the reliable identification of unknown compounds, especially if it is combined with a mass spectroscopic detection unit. Because Mass Spectrometer (MS) and Gas chromatography (GC) instruments fulfil the spacecraft's severe constraints on mass, size, mechanical strength, automation, and energy consumption, these techniques remain the most frequently used for in situ chemical analysis. In 1962, Mars Lander mission (Oyama, 1963) started to consider the possible space utility of GC. But the first mission which carried a gas chromatography system was the Viking mission.

Gas chromatographic techniques have been successfully used for the exploration of Venus and Mars, and they are being used in the atmospheric Huygens probe when the Cassini-Huygens mission reaches Titan. It must be noted that for the future EXOMARS mission liquid chromatography (capillary electrophoresis) will be used as chromatographic techniques with the Mars Organic Detector (MOD).

Viking GC-MS experiment analysed the composition of the martian atmosphere and determined compounds released when soil was pyrolysed or heated (Rushneck et al., 1978). The biological experiment was formed of one oven where sample was filled in, a Gas Chromatograph and a Mass Spectrometer (Bieman et al., 1977). Transfer of organic compounds from the sample oven to GC was performed by a hydrogen flow. However, it is now known that the Viking GC-MS would not have detected refractory organic compounds (Benner et al., 2000) or amino acids associated with over a million bacterial cells in a gram of soil (Glavin, 2001). With this system future missions could however determine if there is carbon in martian sample as BEAGLE II was supposed to do. Biotic organic matter burns in the 200 to 500°C range whereas an abiotic organic compound and pure carbon compound burn at higher temperature. Then, stepped combustion could permit to heat a solid sample (soil or rock) in steps of increasing temperature with presence of oxygen. Any carbon compound present would burn at a special temperature and would give carbon dioxide analyzed by the mass spectrometer. However, no molecular data could be obtained by this device. Gas such as CO_2, resulting from combustion of organics and volatile compounds are directly analysed by GC (Szopa et al., 2001) whereas higher molecular weight or polar compounds need first to be extracted and then to be derivatized previous to Gas Chromatography (GC) analyses.

First, pyrolysis in mild conditions could allow the detection of some of our targets such as kerogens and polypeptides. This technique was applied to the Atacama soil (Chili), analogues of Mars (Navarro-González et al., 2006). However, the use of a chemical reagent during pyrolysis could give additional data as it will protect thermolabile compounds and allow the chromatographic separation of both polar and apolar targets in the same run. Pyrolysis in the presence of tetramethylammonium hydroxide (TMAH) was first developed for the analysis of phenolic polymers. It affords both an efficient cleavage of polar bonds and the subsequent methylation of OH and NH_2 groups, hence the release of less polar, GC-amenable compounds. It was successfully applied to a number of natural macromolecules including biopolymers, humic substances, and kerogens (for a review see Challinor, 2001) and recently provided evidence for the survival of proteinaceous moieties within a 140-million-years-old kerogen (Mongenot et al., 2001). Pyrolysis in the presence of TMAH was used for the first time on chondritic material (Remusat et al., 2005). When compared to conventional pyrolysis, it results in a better detection of the oxygenated products thanks to in situ methylation. Indeed, methyl and dimethyl esters of aromatic acids and diacids are identified, demonstrating the occurrence of ester linkages in the macromolecules. Thermochemolysis could be used in future missions for macromolecules but for other targets additional instrument must be added such as extraction and derivatization cell (Rodier et al., 2005).

Extraction techniques for a large number of organic molecules bound in soil or in a rock matrix are well developed in laboratories. The efficiency of extraction, after one run, is generally estimated at 80% for most compounds. To adapt extraction to a space experiment several parameters such as time, temperature and solvent are studied and optimised (Buch et al., 2003). There is no universal solvent. Regarding the polarity of the different targets, compromise between solvents used for amino and carboxylic acids has been proposed. Ultrasonication significantly decreases the time of extraction compared to static heating. This also provides the advantage of lowering the temperature of extraction and increasing the extraction yields. However, the propanol extraction method is not sensitive for concentrations lower than those found in the Murchison meteorite. An optimisation of the extraction procedure is undertaken to be efficient in many extraterrestrial environments (Buch et al., 2004). Sublimation can also be used on amino acid, amines, nucleobases and other organic compounds contained on a soil which have appreciable vapor pressure at temperature higher than 150°C. By this way, targeted compounds could be isolated from the matrix and no sample preparation and no wet chemistry is necessary. Volatile compounds are then recovered thanks a cold finger. Sublimation is part of the UREY experiment of EXOMARS.

After recovery of the polar organics from the samples, analysis of main targets by GC requires chemical derivatization procedures (Figure 5). Polar organic compounds that contain labile hydrogen are typically of low volatility due to their tendency to self-associate or to associate with polar liquids or solid media through hydrogen bond formation. Derivatization increases polar analyte volatility. To face space constraints the derivatization procedure also has to be low energy consuming, one step reaction and sensitive. Dimethylformamide-Dimethylacetal (DMF–DMA), already used in the COSAC experiment, allows the separation of several volatile compounds and amino acids (Meierhenrich et al., 2001; Rodier et al., 2002). Regarding the difficulties faced during in situ analyses and the lack of knowledge concerning organics present in future studied samples, choosing a too specific process could lead to miss information. Thus, to perform the derivatization of a wide range of target molecules the derivatization agent has to be non specific. To avoid drawbacks of non specific agents, such as interferences in chromatograms, separation on devoted GC columns is optimized. Chemometric analyses could also help to interpret results when coelutions occurred (Pietrogrande et al., 2005). Silylation was proposed to analyze a wider scope of the targeted molecules (Rodier et al., 2001 and 2002). Silytation procedure was optimised (Figure 6) and tested on Mars analogue (Figure 7). However, enantioseparation on chiral columns with this derivatization procedure is not yet achieved. If capillary electrophoresis is chosen as a separative technique, derivatization is also needed to obtain fluorescent derivatives (Skelley et al., 2005).

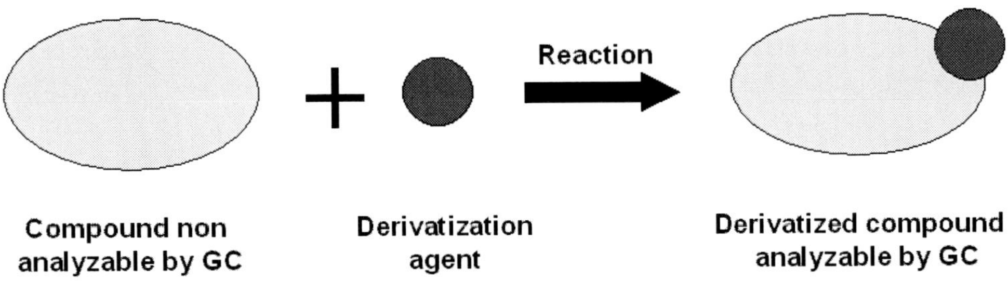

Figure 5. Principle of derivatization.

For chirality assessment of amino acids the derivatization step must in addition avoid racemization. After volatilization of the amino acids, chiral separations can be performed by diastereomer formation or by direct measure on chiral stationnary phases. The first possibility involves derivatization of amino acids with a chiral agent. The resulting diastereomers are subsequently separated on an achiral stationary phase. Among the important characteristics are the chiral center of the agent being closely juxtaposed to the stereogenic elements, absence of racemization during derivatization, absence of kinetic resolution, absence of diastereomeric fractionation upon sample injection, as well as the quantitative enantiomeric purity of the derivatizing agent. To fulfil space constraints, the derivatization reaction has also to be automated and the derivatization agent must be stable under space temperature and radiation without any racemization. The most widely utilized method of diastereomeric reaction is the esterification of carboxyl groups with chiral alcohols (Koenig and Benecke, 1980; Powers et al., 1994), with subsequent acylation of the remaining hydroxyl or phenolic hydroxyl groups. Another commonly employed approach is the acylation of the hydroxyl groups with chiral acid chlorides, or carbonation with chiral chloroformates after esterification of the carboxyl group (Powers et al., 1994). The amidation of carboxyl groups with chiral amines after ethoxymation of the hydroxyl group is also possible (Carlson and Gyllenhaal, 1990). Regarding this analytical challenge, only chiral chloroformate has been evaluated (Rodier et al., 2001). Figure 8 shows the enantiomeric separation on a 5% phenyl dimethyl polysiloxane column of a mixture of amino acids achieved by menthyl chloroformate derivatization.

A direct approach could also be used with a non-racemic chiral stationary phase. Chiral separations are mainly performed on chiral stationary phases capable of hydrogen bonding, coordination and inclusion, such as amino-acid derivatives, terpene-derived metal-coordination compounds and modified cyclodextrins, respectively (Schurig, 2001). Thus each derivatization process occurring without racemization can be studied. The three main types of derivatization process, such as alkylation, acylation-alkylation and silylation, have been tested. Acylation-alkylation has already been evaluated for the amino acid enantiomeric ratios measurement (Rodier et al., 2002; Zampolli et al., 2006).

In MOD analyser to perform a chiral separation, the buffer which is running through the capillary zone electrophoresis contains a chiral selector: γ-cyclodextrin. Indeed, amino acids are able to complex with cyclodextrin and their mobility is affected. Cyclodextrin must be chosen in function of their cavity size which must provide a better complexation with one of the enantiomers. This technique has been tested and validate on the analysis of the extract of the Murchison meteorite. Indeed, results obtained were equivalent or better than those obtained by HPLC or GC/MS methods (Hutt et al., 1999).

After extraction and/or derivatization, GC separation is conducted on capillary columns that comply with space requirements: vibration, radiation and thermal cycles at vacuum condition simulating usual space mission conditions met during the transfer flight. Chromatographic conditions must present a compromise between short retention times, low column temperature and good resolution for almost all organic compounds. Only packed columns were used in the past. Viking GC columns were filled with 60-80 mesh tenax-GC (2,6-diphenyl-P-phenylene oxide) coated with polymetaphenoxylene. This column made in stainless steel was suitable for alcohol, amines and hydrocarbons separation and able to separate trace amount of organic compounds with an excess of water or carbon dioxide (Novotny et al., 1975). Two stainless steel columns were also filled with 100-200 mesh Porapak. The advantage of this polymer was that it is very stable up to 250°C and was not destroyed by vacuum or radiation. Carrier gas was helium with a flow rate of 13.5 ml/min at 24°C. GC system did not detect life presence but it has been able to determine the Mars atmosphere composition. Nowadays PLOT (Porous Layer Open Tubular) columns have recently been developed, even chiral ones, using a better immobilisation process with covalent coupling and copolymerisation, which enable them to withstand space travel, especially vibrations during the launch. For the MOD analyser with electrophoresis, the sample is moved using microfabricated pump to capture chambers that contains affinity matrices. Separation of amino acids is carried out by using zwitterionic antibodies fixed on a polymeric matrix, for nucleobase there is a polymeric matrix contained specific nucleobases antibodies, for sugars a boronic acid matrix which reacts with a diol and finally for acids and bases anions and cations exchangers are used.

For the detection, even though thermal conductivity detectors have proved to be efficient in Viking mission, Mass spectrometer must be used for exobiology purpose. Indeed mistake in the interpretation of enantiomeric excess in Laboratory was attributed to several coelutions (Cronin and Pizzarelo, 1999). Then a fingerprint of the detected molecules is needed. For capillary electrophoresis confocal laser-induced fluorescence is used as detector (Skelley et al., 2005) as coupling with mass spectrometer is not yet achieved for space purpose.

Thus, GC-MS associated with extraction-chemical derivatization would enable the characterization of most of the extraterrestrial target compounds. To improve these results, before launching a mission, analyses should be performed on extraterrestrial analogues. Microfabricated chip and portable capillary electrophoresis has been tested on Mars like soil from Atacama Desert (Chile) and from Panoche Valley (CA, USA) and it ability to analyze amino acid at 70 part per trillion to 100 ppb has been proved. Extraction and derivatization prior to GC-MS have also been tested on such analogues (Figure 6 and 7).

All those instrumentations must be tested with martian analogues as the structure of the regolith can influence the analytical results. Indeed yield of extraction, quality of chromatographic separation and limit of detection depend on the mineralogy. The study of amino acids in carbonaceous chondrites has raised the problem of heterogeneity of the sample and effect of mineral matrix on the precision measurement for several amino acid concentrations (Cronin and Chang, 1993; Ehrenfreund and Charnley, 2000). Thus size of samples and mineral matrix will influence the whole analytical process.

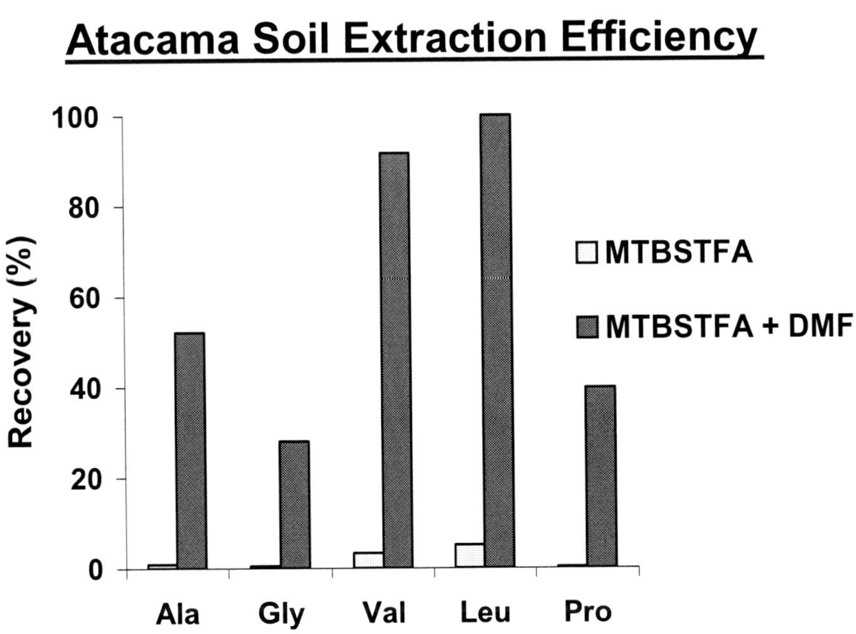

Figure 6. Comparison of the amino acid recoveries obtained from 0.5 cc of doped Atacama soil after chemical extraction and derivatization in 0.5 cc MTBSTFA (yellow) and in a 0.5 cc mixture of MTBSTFA and DMF (blue). Abbreviations: Ala, alanine; Gly, glycine; Val, valine; Leu, leucine; Pro, proline.

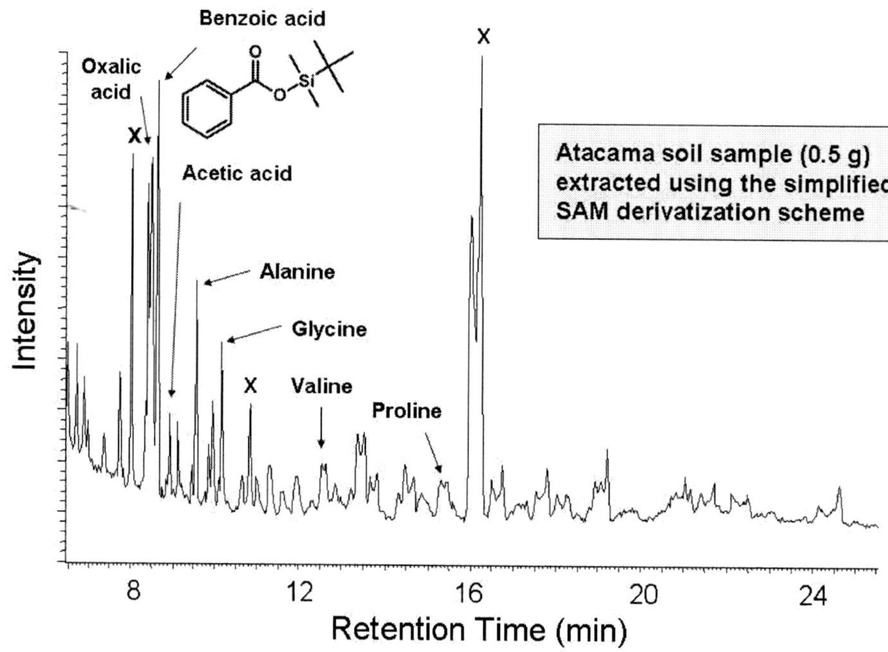

Figure 7. GCMS analysis of organic compounds extracted from an Atacama Desert soil sample after chemical derivatization in a mixture of MTBSTFA and DMF using the SAM prototype derivatization cell and GC column at GSFC. The peaks labeled X in the top chromatogram could not be identified by their mass fragmentation patterns.

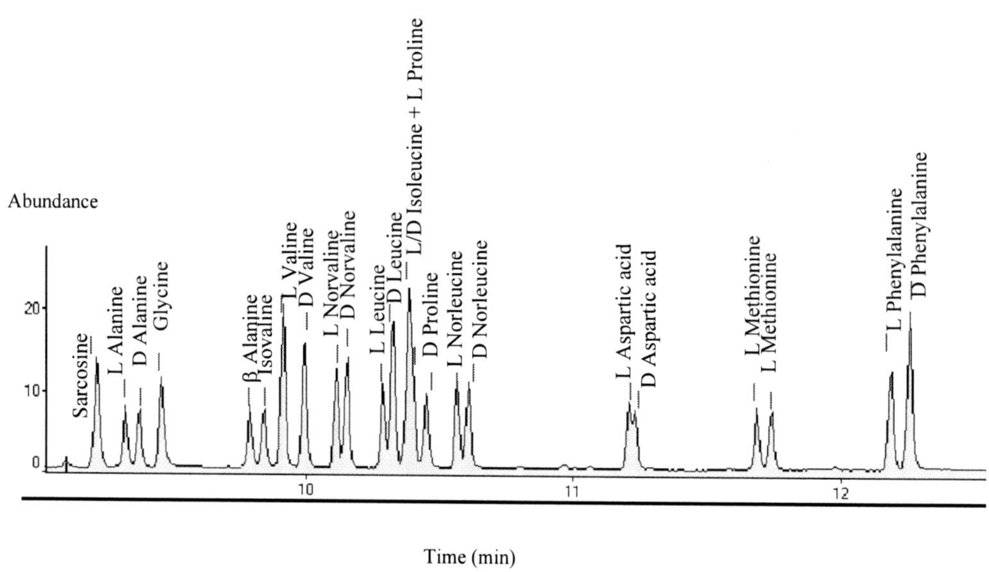

Figure 8. Separation of amino acids as diastereomers after treatment of the analytes with the optically active menthyl chloroformate in aqueous ethanol. A 15 m x 0.25 mm CPSIL5 fused-silica WCOT column was used, operated in the split mode (1:20) at 20°C/min from 120 (held 4 min) to 280°C with an inlet internal helium pressure of 19.5 kPa.

Future Mission of Exobiological Interest

The next generation of surface exploring probes to Mars is currently under development. These probes are specifically devoted to astrobiology researches, with the aim to find indices of life (Figure 9).

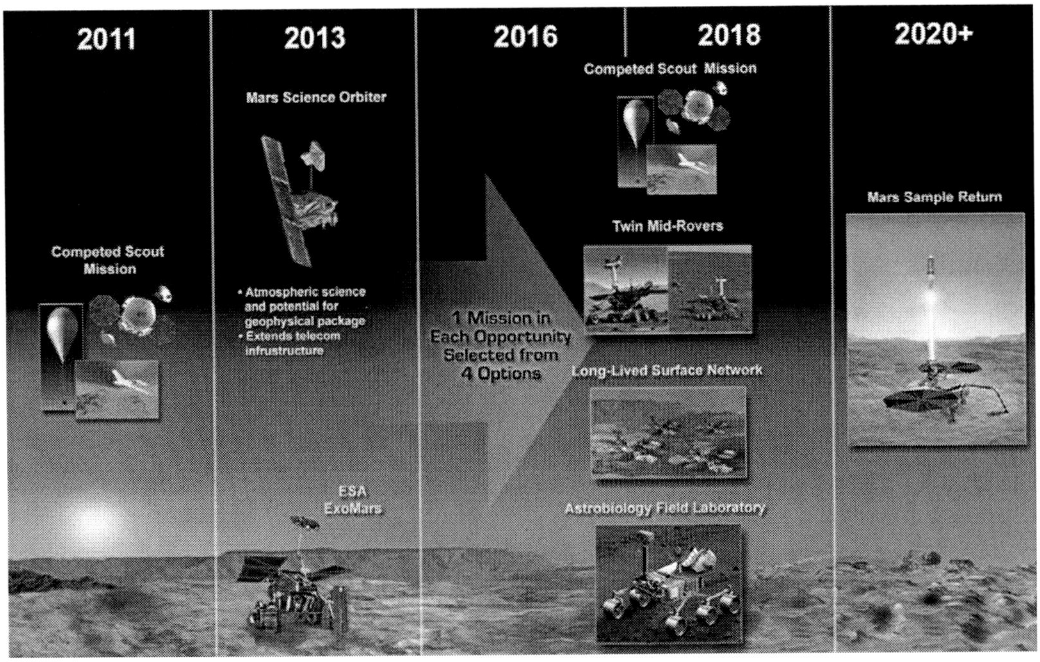

Figure 9. Mission queue. Proposed next decade Mars Program flight architecture. From MEPAG (2006).

The most advanced program is the NASA Mars Science Laboratory 2009 mission. This mission is composed of a rover looking like the MERs, but that will be approximately twice bigger and which should have more autonomy by the use of a nuclear source of energy.

The second main project is the ESA Exomars program. It is still under definition but its development is starting. This mission is composed of a rover that will have similar goals to those of the MSL mission. However, this will be the first attempt for Europeans to land a rover on a planetary surface, resulting in a more limited rover than the MSL one. It will be smaller, resulting in a reduced mass of the science payload (approx 10 kg).

However several other missions will bring additional data such as Mars Reconnaissance Orbiter and Phoenix.

Mars Reconnaissance Orbiter has been successful launched in 2005. Scientific objectives of MRO are the following: Search for evidence of past or present life, understand the climate and volatile history of Mars, understand the geology and geophysics of the martian surface and subsurface and assess the nature and inventory of resources on Mars in anticipation of human exploration (Graf et al., 2005). However, the main exobiological interest goal of MRO is the search the upper crust for evidence of water and following the past mission try to understand the role of water on Mars and its implications for possible past or current biological activity. Scientific instrument associated with these aims is the Compact

Reconnaissance Imaging Spectrometer for Mars (CRISM) and the SHAllow Subsurface Soundind RADar (SHARAD).

CRISM instrument is a visible and near-infradred high spatial and spectral resolution spectrometer to identify minerals, especially carbonate and those likely formed in the presence of water, in surface areas on Mars. SHARAD is a second generation of radar to see if water ice and if layers deposited by geological processes in the ancient history of Mars are present at depths greater than one meter up to 1km.

Phoenix mission is a rebirth of a previous project (MSP'01) cancelled in 2000 following the lost of Mars Climate Orbiter and MPS pair. One of the Goals of Phoenix mission is to determine whether life ever arose on Mars and to characterize the geology of Mars. To reach these goals, phoenix will analyse volatile compounds, such as water, and organic matter at high latitude where Mars Odyssey had detected presence of freezing water in subsoil and studied the recent climates variation of Mars (Shotwel, 2005). Phoenix will also assess the habitability of the martian polar environment by conducting chemical experiments that will test the soil for life-giving elements such as carbon, nitrogen, phosphorus, and hydrogen. In order to understand the relationship between presence of ice to oxidants, organics, the deuterium/hydrogen ratio and other chemical compositional gradients and horizons (concentrations of salts or hydrated minerals) will be studied. After that the Robotic Arm (RA) will dig through the dry soil to reach the ice layer, bring the soil and ice samples to the lander, samples will be treated by the Microscopy, electrochemistry and conductivity analyser (MECA). MECA is formed by four main parts: Wet Chemistry Lab (WCL), Optical microscope, Atomic force microscope and thermal, electrical conductivity probe (TECP). With its 26 sensors WCL vial contain pure water and reagent. It will allow to measure pH, temperature, dissolved O_2 and CO_2, redox potential, NO_3^-, NH_4^+, I-, Na^+, K^+, Mg_2^+, ClO_4, Ca^{2+} and to perform cyclic voltammetry and anodic stripping. Thermal evolved gas analyzer (TEGA) is formed by a differential scanner calorimeter (DSC) for the thermal analysis of the martian soil and a mass spectrometer (EGA) which will analyse the released volatile compounds after heating. The mass spectrometer will be also able to analysis atmospheric samples and then enabling detailed atmospheric composition analysis. With a sensitivity levels down to 10 parts per billion, TEGA will be able to determine ratios of various isotopes of hydrogen, oxygen, carbon, and nitrogen. Then, these results could provided clues to origin of the volatile molecules, and possibly, biological processes that occurred in the past.

Mars Science Laboratory 2009 (MSL)

The NASA Mars Science Laboratory mission will be the first mission specifically devoted to astrobiological investigation of Mars since the Viking missions in the 1970's. It is today under construction, and it is planned to be launched in late 2009. It was selected by NASA to investigate the surface of Mars, seeking for trace of biological or prebiotic activity, but also to investigate the surface environment in the frame of the future coming of man on Mars. The probe consists in a rover based on the model of the MERs, but it is twice bigger in order to onboard a bigger scientific payload. The capability for the rover to move is also important to investigate several places. As it was demonstrated by the MERs, which found evidences of water only outside their landing site, mobility should be a key to perform measurements in places having different properties and which could have kept traces of life.

The scientific payload includes instruments devoted to study the mineralogical context of the places where measurements are done. Among these instruments, ChemCam and CheMin should be rather devoted to characterize salts precipitated from liquid water. The Dynamic of Albedo Neutrons (DAN) experiment will search for water in the soil of Mars by detecting neutrons. The core of the measurements of astrobiological interest is represented by the Sample Analysis at Mars (SAM) experiment (Cabane et al. 2001 and 2004; Szopa et al., 2004). This instrumental suite will include 3 instruments devoted to characterize the chemical composition of the atmosphere and of solid samples collected at the surface. The Tunable Laser Spectrometer (TLS) is a laser absorption spectrometer especially devoted to measure the atmospheric CH_4, H_2O and H_2O_2 and their isotopes. Measurements of methane are important to determine its origin which could be biotic. H_2O_2 is also important to quantify in order to determine its possible influence on organic materials and future equipments to be brought by future astronauts.

The Quadrupole Mass Spectrometer (QMS) is a mass spectrometer that will analyze both the atmosphere, to determine its accurate global composition through time, and the gases evolved from solid samples collected by the probe, including potential organics present in the soil. The last instrument is the Gas Chromatograph (GC) that will analyze the same components as the MS.

The use of gas chromatography is important since the chemical species potentially present in the solid samples could be numerous. Hence, the chromatographic separation should be of importance for the identification of the species. Moreover, it also constitutes an instrument independent from the MS. With this aim, the GC includes 6 independent chromatographic channels with a chromatographic column and its thermal conductibility detector (Cabane et al., 2001 and 2004; Szopa et al., 2004).

For the first time in space GC, the GC device will include thermal desorption injectors. These injectors trap the gaseous species which are quickly released by fast heating. This device allows the enrichment of the sample, relatively to the gas used to transfer the sample from the sampling system to the analyser, and also allow limiting the injection pulse, resulting in a better chromatographic resolution. An example of chromatogram obtained with the SAM GC brassboard with this type of injection is given in figure 10. It shows the interest of this system comparatively to the use of a large gaseous sample loop with gives larger peaks width which increases the peaks overlapping and the problems for compounds identification. Obviously, the peaks identification will also be done by using mass spectrometric signal. Indeed, SAM GC will be systematically coupled with the SAM MS that will provide structural information on the analyzed compounds. A stringent identification of all the compounds will thus be performed by combining the GC and MS data. An example of signal resulting from the coupling of the SAM GC and MS brassboards is given in figure 11. It can be seen in this figure that the signals collected with the two detectors are quite similar in term of retention time. The sensitivity of the MS is also much greater than that of the TCD. This result was expected because in laboratory, the difference in the sensitivity of the two detectors is about 3 orders of magnitude. However, we can see that the chromatographic peaks observed are much wider in the case of the MS signal. This means that the split device used to couple the two instruments induces a peak broadening. This device should be improved for the flight model and we expect to improve the MS limit of detection.

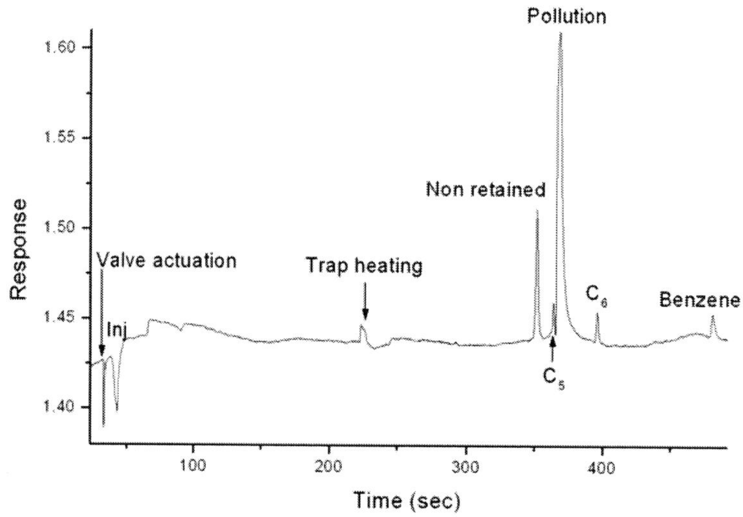

Figure 10. Chromatogram obtained with an analytical channel of the SAM GC brassboard using a thermal desorption injection. Analytical details: column=Chirasil dex CB (30 m x 0.25 mm); analyzed sample: gaseous mixture of pentane (C5), hexane (C6) and benzene diluted at 1000 ppm in helium.

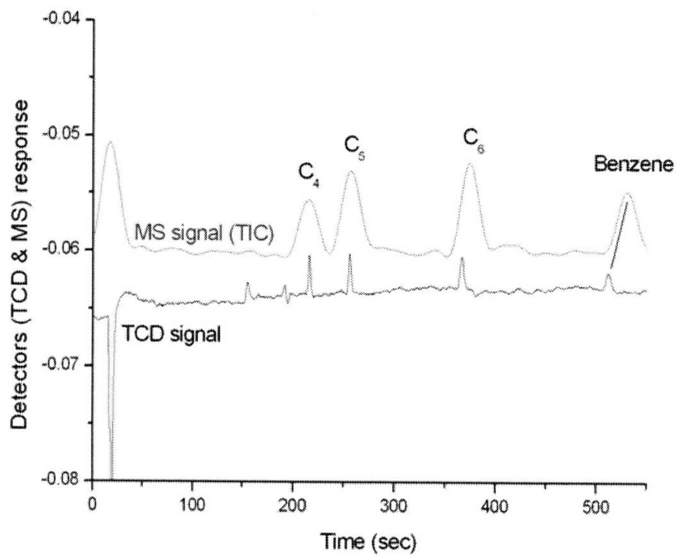

Figure 11. Chromatogram obtained with an analytical channel of the SAM GC brassboard using a direct gaseous injection. Signal was recorded both with the GC detectors (TCD) and the SAM MS. The MS signal is the total ion current. The MS signal was shifted in time to correlate the dead time of the two signals. The intensity of the MS signal was rescaled to compare the two signals.Analytical details: column=MXT 20 (30 m x 0.25 mm); analyzed sample: gaseous mixture of butane (C4), pentane (C5), hexane (C6) and benzene diluted at 1000 ppm in helium.

In addition, extraction and derivatization of organic compounds contained in the soil (if they exist) will be performed in the same step. Soil should be contacted with the derivatizing agent (MTBSTFA-DMF or DMF-DMA) at constant temperature (70-80°C) during a fixed time (15-30 min) (Rodier et al., 2001). After this step solvent will be evaporated at martian atmosphere in order to avoid pollution of the chromatogram by an excess of derivatizing agent. Then, sample will be heated and derivatized organic compounds injected in the GC.

Finally, a last experiment named Radiation Assessment Detector (RAD) will measure the high energy radiations at the surface to evaluate the potential influence of radiations on future inhabited mission.

EXOMARS

In the frame of the future European Mars mission EXOMARS, ESA has selected several instruments package able to search for extinct or extant life on Mars: MOMA and UREY. This experiment showed the possibility to synthesise organic compounds such as amino acid starting from the hypothetical condition of primitive Earth.

MOMA is an experiment devoted to the detection and identification of organic molecules, including molecules of interest for life or prebiotic chemistry. It is based on a gas chromatograph-mass spectrometer instrumentation, as the SAM experiment of the MSL mission. The gas chromatograph will include several analytical channels (5 maximum) devoted to separate different sets of organics and inorganic species, including enantiomers. The sample preparation will include a derivatization process for the analysis of refractory or very polar compounds such as amino acids. It is expected that some advances to be made in the study of derivatization procedures for space application should improve the process already used in COSAC and SAM experiments. The gas chromatograph will be able to work as a stand alone instrument thanks to specific chromatographic detectors, and it will be also coupled with the mass spectrometer.

The main originality of the experiment resides in the mass spectrometer instrument. Indeed, this MS will be capable to achieve MS-MS analyses, meaning that MS fragments of big molecules susceptible to be analyzed will be also analyzed by MS. This will enable to refine the identification of chemical functional groups and of the mother molecule. Moreover, the MS will be associated with a laser. This laser will be used for the ablation of molecules present in the sample, and the ablated molecules will be analyzed with the mass spectrometer. This instrument will be much lighter than its predecessors.

Instrument package is named UREY in recognition of the landmark Stanley L. Miller and Harold C. Urey experiment (Miller et al., 1959). This experiment showed the possibility to synthesise organic compounds such as amino acid starting from the hypothetical condition of primitive earth. UREY package is composed of four instruments: a subcritical water extraction (SCWE), a sublimation apparatus: Mars Organic Detector (MOD), a lab-on-a-chip microcapillary electrophoresis (CE) and a Mass Oxidant Instrument (MOI). UREY experiment is developed on the following scheme. Three type of sample can be accepted by the analyzer (MOD): aqueous sample from the water extractor device, melt water and sublimated sample that come from a cold finger. Sublimation can be used on amino acid, amines, nucleobases and other organic compounds contained on a soil which have appreciable vapor pressure at temperature higher than 150°C. By this way, targeted compounds could be isolated from the matrix and no sample preparation and no wet chemistry is necessary. Volatile compounds are then recovered thanks a cold finger. Then, the sample is moved using microfabricated pump to capture chambers that contains affinity matrices specific to five classes of biomolecules. These five classes of molecules are: amino acids, nucleobases, sugars organic acids and bases. After preconcentration, biomolecules are analyzed and identified by capillary electrophoresis.

MOD analyzer contains five captures chambers that allow separating the five classes of organic compounds. This separation is sequentially done as presented on the figure 12. Confocal laser-induced fluorescence is used as detector (Skelley et al., 2003). Contrary to the sublimed PAH which are naturally fluorescent and in order to be analyzed by a fluorescence spectrograph, amino acid, amines and nucleobases must be derivatized to achieve fluorescein labeling. Once derivatized, sample is drive in the capillary electrophoresis where separation occurs based on the charge/size ratio. Tested on amino acids analysis, Skelley et al., (2005) have shown that they could be detected with ppb to part per trillion sensitivities from 1 g soil sample. Skelley team announces a limiting sensitivity 1,000-fold improvement over Viking experiment (Skelley et al., 2006).

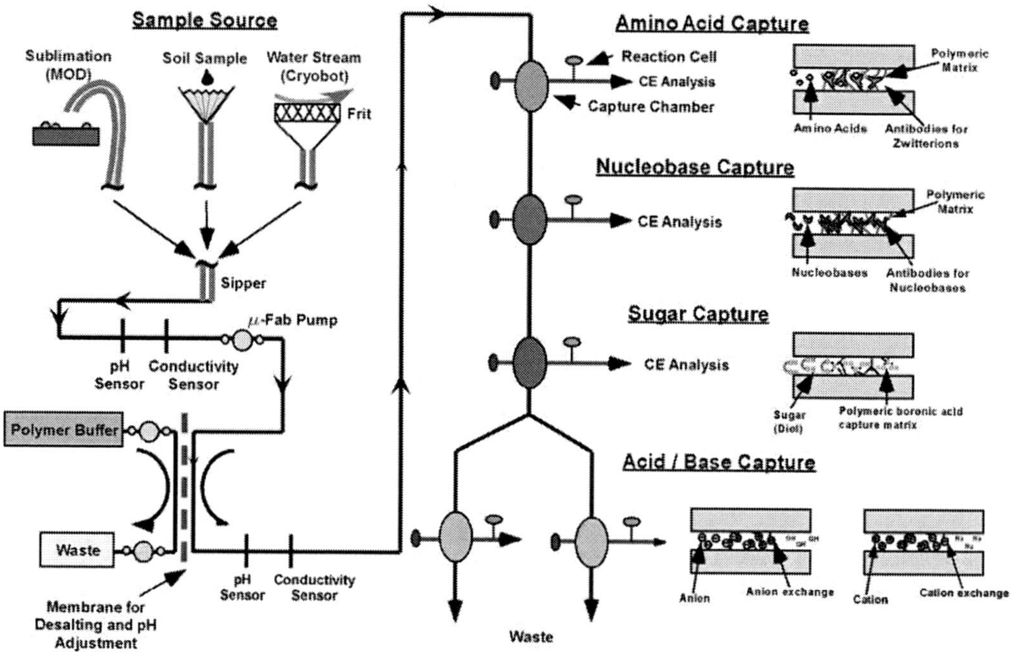

Figure 12. Schematic of the MOD chemical analyser (Skelley et al., 2003).

Specific Molecular Identification of Life (SMILE) experiment was proposed as a part of Pasteur payload onboard the EXOMARS mission (Vago et al., 2006). SMILE was developed by an Union Kingdom team (Sims et al., 2005) to detect a wide range of biomarker molecules such as hopanes, isoprenoids derived from prokaryotic and eukaryotic membranes, amino acids and their enantiomeric excess. SMILE is based on a molecular receptor and imprinted polymers (protein such as antibodies) that have the ability to recognize and bind targeted organic compounds. Various analytical methods allow to detect binding of biomarker to the molecular receptor:

- optical interferometer
- fluorescence marker detection with the use of imaging system.
- Electrochemical readout using impedancemetry

Thanks this electrochemical module, SMILE will provide information about pH, conductivity and redox potential. Associated with a capillary electrophoresis system (MUSE) (Eckhard et al., 2003), and a polymerase chain reaction (PCR) thermo-cycling system (MARS), SMILE becomes SMILE-Pro plus which allows also the determination of the chiral partitioning and allows the detection of hereditary macromolecules such as DNA/RNA.

Mars Atmospheric Sensor (MAOS) is based on the heritage of MOx experiment aboard the lost Mars'96 mission. MAOS experiment is a chemometric array sensor designed to identify the nature and the origin of oxidants in the martian surface materials. Indeed, some of martian oxidant may originate in the atmosphere and hence be able to diffuse through the regolith (Huten, 1974). With this aim MAOS should measure the differential DC resistivity of thin film composed of various materials as a function of dust abundance, UV flux and temperature. Finally MAOS should be able to identify responsible molecules of oxidation of martian materials.

Sample Return

Concerning the search for trace of life, a complete list of organic possibly present on Mars have to be done and require a returned sample. Indeed, sample return will provide information that in situ mission will not be able to get. Indeed, there are several limitations with the instruments dedicated for the in situ measurements. For example mass, volume, power and space are limited on board of spacecraft. Sensitivity, repeatability, precision and unexpected results are directly limited by the previous restriction. However, since Apollo no sample return mission has been launched.

In 1998 NASA initiated a new science plan for Mars Sample Return (MSR). Launch of MSR was first planned for 2003 and 2005. But project was replanning and replaced by other missions and development projects. In 2003 a Rapid Impactor Sample Return (RISR) on Mars has been presented. Scenario of RISR was to strike the martian surface with an impactor which throws up debris. 1.5 kg of these surface materials will be captured by the return sample vehicle and returned to Earth. Since 2003, Russian aviation and Space Agency project a sample return mission on Phobos (PSR), a satellite of Mars, and an extended capability of some Main Belt (MB) asteroids (Figure 13). Main scientific objectives of this mission will be to collect soil samples from Phobos and to bring the samples back to Earth for comprehensive scientific research into Phobos, Mars and martian space (Marov et al., 2004). In addition of the sample back other scientific objectives are included. These researches are in situ and remote studies of Phobos and asteroid class bodies with the aims to better understand the problem of the origin of Phobos and the genetic relation with Mars. Other studies of the environment of Mars such as plasma, radiation environment and dust component are also planned.

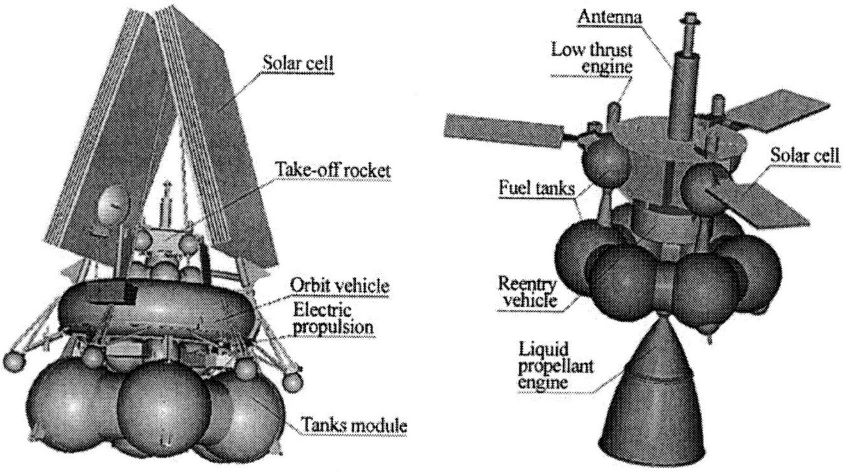

Figure 13. General PSR vehicle configuration (left) and mock up (right) (from Marov et al., 2004).

Even if Mars Sample Return is now planned for the next decade, sample return from Mars stays one of the major components for the understanding of the planet, its evolution and whether extinct or extant life exist on this planet. From the 2006 MEPAG and by using actual available Earth instrument, MSR could provide answer to these questions:

- Precise age of fundamental Mars crustal units
- Nature of the interaction between Mars' atmosphere/hydrosphere and regolith
- Temperatures and chemical evolution of water deposited chemical sediments
- Evolution of mars' atmosphere
- Confirmation or refutation of the past or present existence of life on Mars
- Habitability
- Prepare for human exploration

However, in addition to the fact that the probability of success for a sample return mission is more low-level than for in situ mission (respectively 0.68 against 0.81 (Sims et al., 2002)) many other problems are induced by the sample return: analysis procedure, contamination, public perception, strategy, architecture, financing… All these parameters have to be carefully considered.

First of all, the cost of the future Mars sample return missions may limit the number of geological site of drilling and limit the number of samples. Moreover, the cost and space constraint may limit the volume, the size and the weight of return sample. Thus, the analysis of future sample may be limited at few laboratories able to work with analytical processes that do not need a large volume of sample and with methods compatible with other investigation techniques. For example it may be required that physical methods are compatible with chemical analysis procedures (or conversely) in order to minimize the amount of sample used. Then some efficient techniques should be eliminated.

But, the major limitation concerning the sample return analysis is the contamination. All the past Mars missions tried to fulfil with the planetary protection requirements which were based on avoiding contamination of Mars with Earth organisms. Sample return should take in account two major new requirements: first requirement is to avoid contamination of Earth with possible Mars organisms. Second is to keep clean the Mars sample from round-trip Earth contamination.

A big issue which has to be taken into consideration is to know if it will be safe to release the Mars sample toward the science team. Indeed, to be sure of the non dangerousness of this sample, it will be essential to ensure that no martian organism, if any exists, contaminates the Earth. Earth is meaning as Human and biosphere. Then biological investigation should be done. In order to protect the Earth biosphere from possible Mars biological contamination, it has been proposed to define a probability number of possible release single particle of size superior to 0.2 μm. A number of 10^{-6} is under consideration (Cazeaux et al., 2004). But, biological issue exists. Indeed, as we know on earth, only 1-10% of micro-organisms are cultivable and if life exists and uses another form of biochemistry then perhaps we will not able to detected a bio-harzard (Sims et al., 2002).

In addition, martian sample which returns on Earth must be exempt of all contamination. This requirement is one of the major needs if we don't want to meet the problem encountered with results which came from martian meteorite. Indeed, contamination of Murchinson or Tagish lake have induced several polemics about the presence of biotic/abiotic organisms and Earth contamination such as atmospheric hydrocarbon (Glavin et al. 2004; Brown et al., 2000). In the same way it has been reported that due to the presence of contamination the Aib (amino isobutyric acid) detection from lunar sample was limited about 0.1 ppb (Brinton et al., 1996).

Moreover, we search for both extinct and extant martian life, then, erasing living organism before launch from Earth is an important challenge. But cleaning spacecraft from all the dead bodies is another big challenge (Sims et al., 2002). With these goals, a cleaning procedure must be developed and the detection procedure associated. Then in order to answer whether or not life on Mars exists or existed, Earth contamination problem must be solved. And to express this requirement a probability number of less than 10^{-2} of having a single Earth organism contamination is under discussion.

Another requirement in the field of the sample contamination has been defined in order to preserve the properties of the extraterrestrial sample. Goal is to avoid false results concerning presence or absence of biohazard materials. False-positive and false-negative would imply false interpretation concerning life presence on Mars but would also induce dramatic change in the planetary protection policy (Debus, 2006).

Other requirements, in relation with the contamination problem, have to be taken in account:

- → Avoid contamination of the container with martian materials. Heritage of other mission such Apollo, stardust or genesis will be used to design the future entry probe.
- → Avoid meteoritic impact which could contaminate the probe (Gershman et al., 2004; Cazeaux et al., 2004).

It must be noticed that against the need of precaution with Mars sample return some authors argue that Earth is impacted by martian meteorites periodically ejected from Mars (Gladman, et al., 1996; Gladman et al., 1997; Zubrin et al., 2000). They conclude that back contamination can be considered as a minor concern or even as not at all concern.

Nevertheless, percentage of infall meteorites on Earth identified as martian meteorites represents only 0.1 % of the total infall and it seems difficult to conclude from a so low cross-section a real frequency of infall (Clark, 2002). Moreover, infall meteorites are between 500 to 20 million years old. Then, it seems again difficult to be sure that they have never contained hazardous materials which could be the cause of some extinction or decline in the life history of the Earth. In addition, even if some meteorites sheltered organisms then probability that space conditions such as radiation may destroy them is enormous (Clark, 2001). But in the case of sample return organism could survive. Thus the contamination problem is not yet solved.

This issue is not only a scientific problem, it is also a public perception problem (Lofstedt et al., 2003). In regard of the expense and possible Earth contamination Public could not blind accept the Mars sample return. Another argument in favour the planetary protection is this public perception. It's a first importance to develop a planetary program for sample return and take the public informed about the risk (Race, 1998; Billings, 2006).

Landing Site for Future In Situ Experiment or Future Sample Return

Whatever the mission will be, choice of a prime candidate site for the future in situ exploration or sample return mission should be based on actual knowledge of the martian geology, paleohydrology, paleoclimatology and exobiology.

Exobiolological site of interest should be selected in function of the hydrological cycle on Mars and in function of the relation between the putative geological and paleohydrological histories and the possible biological evolution of Mars. From data of past and present Mars missions three prime possible sites could be taken in consideration (Fairèn et al., 2005).

First candidate site is located in Meridiani Planum. In this place basaltic rock material with 10-15 % of hematite deposit has been detected by THEMIS (Christensen et al., 2001). Hematite may have been precipitated from the circulation of fluids or by precipitation of nano-crystallin iron oxyhydroxides from solution via biologically driven system (Banfield et al., 2001). Futhermore, from the recent MER and THEMIS analysis original surface which contains hematite is probably early to middle Noachian age (Lane et al., 2003; Hynek 2004). Regarding the Rion Tinto earth place where goethite and hematite in Pleistocene and Holocene terraces have preserved bacterial, microalgal and fungal microfossils (Fernandez-Remolar et al., 2005) meridiani planum is one of the first place where next martian missions have to focus his research for in situ or sample return.

Second site is the Mangala Valles located in the northwestern slope valleys region considered as being late Hesperian to early Amazonian age. At this place hydrothermal environment noted by the high density of dark slope streaks has probably existed and/or may still exist and then could be a primordial exobiological interesting site for the development and sustenance of life (Ferris et al., 2002).

Third site is the Orcus Patera located between the Elysium and Amazonis Planitias. Orcus Patera is a modern age site with an absence of extensive deformation (Berman and Hartmann, 2002). Moreover hydrothermal activity may have occurred since the early Amazonian and could have persisted to date (Plescia, 2003). Then, possible fossil record may be preserved.

Conclusion

The primary goal of exobiological research is to reach a better understanding of the processes leading to the origin, evolution and distribution of life on Earth or elsewhere in the universe. In this endeavour, scientists from a wide variety of disciplines are involved, such as astronomy, planetary research, organic chemistry, palaeontology and biology.

Space technology plays an important part by offering the opportunity for exploring our solar system. Exobiological activities include comparison of the overall pattern of chemical evolution of potential precursors of life, in the interstellar medium, and on the planets and small bodies of our solar system; tracing the history of life on Earth back to its roots; searching for other planetary systems in our Galaxy and for signals of extraterrestrial civilizations; testing the impact of space environment on survivability of resistant life forms.

In the solar system, Mars is the planet which has the surface environment which is most like Earth. Due to the strong connection between life and organics, it clearly appears that the research for organics of Mars is of primary interest to discover a potential martian life form. The detection of organics on Mars remains an important scientific objective and represents a key point in the search of the origin of life.

With evidence of presence of liquid water on Mars and with the development of new high sensitivity spaceflight instruments, chance to answer to the question of the life presence on Mars are close.

Acronyms

CHEMCAM	CHEMistry CAMera
CE	Capillary Electrophoresis
CRISM	Compact Reconnaissance Imaging Spectrometer for Mars
DAN	Dynamic of Albedo Neutrons
DMF	Dimethylformamide
DMF-DMA	Dimethylformamide dimethylacetal
DNA	DeoxyriboNucleic Acid
DSC	Differential Scanner Calorimeter
ESA	European Space Agency
GAP	Gas Analysis Package
GC-MS	Gas Chromatography – Mass Spectrometry
GEx	Gas Exchange
GSR	Gamma Ray Spectrometer
HCP	high-calcium pyroxenes
HPLC	High Performance Liquid Chromatography
HRSC	High Resolution Stereo Camera
IRIS	Infrared Intersferometer Spectrometer.
IRR	Infrared radiometer
LCP	low calcium pyroxene
LR	Labelled Release
MAG	MagnetoMeter

MAOS	Mars AtmOspheric Sensor
MARIE	Martian Radiation Experiment
MaRS	Mars Radio Science
MARSIS	Mars Advanced Radar for Subsurface and Ionospheric Sounding
MECA	Microscopy, electrochemistry and conductivity analyser
MEPAG	Mars Exploration Program Analysis Group
MER	Mars Exploration Rover
Mini-TES	Miniature-Thermal Emission Spectrometer
MGS	Mars Global Surveyor
MOMA	Molecular Organic Molecule Analyzer
MOC	Mars Orbiter Camera
MOD	Mars Organic Detector
MOI	Mars Oxidant Instrument
MOx	for Mars Oxidant experiment
MOLA	Mars Laser Altimeter
MSP	Mars Surveyor Program
MRO	Mars Reconnaissance Orbiter
MS	Mass Spectrometry
MSL	Mars Sample Laboratory
MSR	Mars Sample Return
MTBSTFA	N-(t-butyldimethylsilyl)-N-methyltrifluoroacetamide
MUSE	Mars Ultraviolet Science Experiment
NASA	National Aeronautics and Space Administration
OMEGA	Observatoire pour la Minéralogie, l'Eau, les Glaces, et l'Activité
PCR	Polymerase Chain Reaction
PFS	Planetary Fourrier Spectrometer
PSR	Phobos Sample Return
QMS	Quadrupole Mass Spectrometer
RA	Robotic Arm
RISR	Rapid Impactor Sample Return
RNA	RiboNucleic Acid
SAM	Sample Analysis at Mars
SCWE	SubCritical Water Extraction
SHARAD	SHAllow Subsurface Soundind RADar
SMILE	Specific Molecular Identification of Life
SPICAM	Spectroscopy for Investigation of Characteristics of the Atmosphere of Mars
TEGA	Thermal evolved gas analyzer
TES	Thermal Emission Spectrometer
THEMIS	Thermal Emission Imaging System
TLS	Tunable Laser Spectrometer
UV	Ultra-Violet
WCL	Wet Chemistry Lab

REFERENCES

Allamandola, L.J.; Bernstein, M.P.; Sandford, S.A. et al. (1999). Evolution of interstellar ices. *Space Science Reviews, 90*, 219-232.

Amblès, A.; Jambu, P.; Parlanti, E.; Joffre, J. and Riffe C. (1994). Incorporation of natural monoacids from plant residues into an hydromorphic forest podzol. *European Journal of Soil, Science, 45*, 175–182.

Bada, J.L. and McDonald, G.D. (1995). Amino acid racemization on Mars: implications for the preservation of biomolecules from an extinct Martian biota, *Icarus, 114,* 139-143.

Bada, J. L. and McDonald, G.D. (1996). Detecting amino acids on Mars, *Analytical Chem*istry, *68*, 668A-673A.

Baker, V.R.; Strom, R.G.; Gulick, V.C.; Kargel, J.S.; Komatsu, G. and Kale, V.S. (1991). Ancient oceans, ice sheets and the hydrological cycle on Mars, *Nature*, *352*, 589–594.

Banfield, J.F.; Moreau, J.W.; Chan, C.S.; Welsh, S.A. (2001). Mineralogical Biosignatures and the Search for Life on Mars, *Astrobiology,* 1, 447-465.

Barbier, B.; Henin, O.; Boillot, F. et al. (2002). Exposure of amino acids and derivatives in the Earth orbit, *Planetary Space Science*, *50*, 353-359.

Becker, L.; Glavin, D.P. and Bada, J.L. (1997). Polycyclic aromatic hydrocarbons (PAHs) in Antarctic Martian meteorites, carbonaceous chondrites, and polar ice, *Geochimica et Cosmochimica Acta*, *61,* 475-481.

Benner, S.A.; Devine, K.G.; Matveeva, L.N. and Powell, D.H. (2000). The missing organic molecules on Mars. *Proc. Natl. Acad. Sci. USA,* 97, 2425–2430.

Berman, D.C. and Hartmann, W.K. (2002). Recent fluvial, volcanic and tectonic activity on the Cerberus plains of Mars. *Icarus*, 159, 1-17.

Bernstein, M. (2006). Prebiotic materials from on and off the early Earth Philosophical, *Transactions of the Royal Society B-Biological Sciences, 361*, 1689-1700.

Bibring, J.P.; Langevin, Y.; Poulet, F.; Gendrin, A.; Gondet, B.; Berthé, M.; Soufflot, A.; Drossart, P.; Combes, M.; Bellucci, G.; Moroz, V.; Mangold, N.; Schmitt, B. (2004). Perennial water ice identified in the south polar cap of Mars, *Nature, 428*, 627-630.

Bibring, J.P.; Langevin, Y.; Mustard, J.F.; Poulet, F.; Arvidson, R.; Gendrin, A.; Gondet, B, Mangold, N.; Pinet, P.; Forget, F. (2006). Global mineralogical and aqueous mars history derived from OMEGA/Mars Express data, *Science, 312*, 400–404.

Biemann, K.; Oro, J., Toulmin III, P.; L.E. Orgel, A.O.; Nier, D.M.; Anderson, P.G.; Simmonds, D.; Flory, A.V.; Diaz, D.R.; Rushneck, J.E.; Biller, A.L.; LaFleur, J. (1977). The search of organics substances and inorganic volatiles compounds in the surface of Mars, *Journal Geophysical Research, 82*, 4641–4658.

Billings, L. (2006). Public communication strategy for NASA's planetary protection program: Expanding the dialogue. *Advances in Space Research, 38*, 2225-2231.

Bland, P.A. and Smith, T.B. (2000). Meteorite accumulations on Mars. *Icarus, 144*, 21–26.

Bonner, W.A. (1991). Terrestrial and extraterrestrial sources of molecular homochirality, *Origins of Life and Evolution of Biospheres, 21*, 407-420.

Boynton, W.V.; Feldman, W.C.; Squyres, S.W.; Prettyman, T.H.; Brückner, J.; Evans, L.G.; Reedy, R.C.; Starr, R..; Arnold, J.R.; Drake, D.M.; Englert, P.A.J.; Metzger, A.E.; Mitrofanov, I.; Trombka, J.I.; d'Uston, C.; Wänke, H.; Gasnault, O.; Hamara, D.K.; Janes, D.M.; Marcialis, R.L.; Maurice, S.; Mikheeva, I.; Taylor, G.J.; Tokar, R.;

Shinohara, C. (2002). Distribution of hydrogen in the near Surface of Mars: evidence for subsurface ice deposits, *Science, 297*, 81-85.

Brack, A. (1996). Why exobiology on Mars? *Planetary and Space Science, 44,* 1435-1440

Brack, A. (1997). Life on Mars: a clue to life on Earth? *Chemistry & Biology*, *4,* 9-12.

Brinton, K.L.F and Bada, J.L. (1996). A reexamination of amino acids in lunar soils: implications for the survival of exogenous organic material during impact delivery. *Geochimica et Cosmochimica Acta*, 60, 349-354.

Brocks, J.J.; Logan, G.A.; Buick, R. and Summons, R.E. (1999). Archean molecular fossils and the early rise of eukaryotes. *Science, 285,* 1033–1036.

Brown, P.G.; Hildebrand, A.R.; Zolensky, M.E.; Grady, M.; Clayton, R.N.; Mayeda, T.K.; Tagliaferri, E.; Spalding, R.; MacRae, N.D.; Ho:man, E.L.; Mittlefehldt, D.W., Wacker, J.F., Bird, J.A., Campbell, M.D., Carpenter, R., Gingerich, H.; Glatiotis, M.; Greiner, E.; Mazur, M.J.; McCausland, P.J.; Plotkin, H.; Mazur, T.R. (2000). The fall, recovery, orbit, and composition of the Tagish Lake meteorite: a new type of carbonaceous chondrite. *Science, 290*, 320–325.

Buch, A.; Sternberg, R.; Meunier, D.; Rodier, C.; Laurent, C.; Raulin, F. and Vidal-Madjar, C. (2003). Solvent extraction of organic molecules of exobiological interest for in situ analysis of the Martian soil, *Journal of Chromatography A, 999, 165-174.*

Buch, A.; Glavin, D. P.; Sternberg, R.; Szopa, C.; Rodier, C.; Navarro-González, R.; Raulin, F.; Cabane, M. and Mahaffy, P. R. (2006). A new extraction technique for *in situ* analyses of amino and carboxylic acids on Mars by gas chromatography mass spectrometry, *Planetary Space Science*, *54,*1592-1599.

Burns R.G. (1987) Ferric sulfates on Mars, *Journal of Geophysical Research, Solid Earth Planets* 92, E570–E574.

Cabane, M., Coll, P., Rodier, C., et al. (2001). In situ inorganic and organic analysis (Pyr/CD-GC/MS) of the martian soil, on Mars 2005 mission, *Planetary Space Science*, 49, 523-531.

Cabane, M.; Coll, P.; Israël, G.; Raulin, F.; Niemann, H.; Mahaffy, P.; Sternberg, R. (et al.). (2002). Organic and inorganic signatures in Mars ground and underground, one of the goals for "SAM" (sample analysis at Mars), *European Space Agency, (Special Publication) ESA SP*, 518, 323-326.

Cabane, M.; Coll, P.; Szopa, C.; Israël, G.; Raulin, F.; Sternberg, R; Mahaffy, P.; Person, A.; Rodier, C.; Navarro-Gonzàlez, R. *et a.l* (2004) Did life exist on Mars? Search for organic and inorganic signatures, one of the goals for "SAM" (sample analysis at Mars), *Advances in Space Research, 33,2240-2245.*

Carlson, A. and Gyllenhaal, O. (1990). Separation of carboxylic acid enantiomers by gas chromatography after rapid dramatisation with (*R*)- or (*S*)-1-phenylethylamine after activation by ethyl chloroformate, *Journal of Chromatography 508*, 333.

Carr, M.H. (1996). *Water on Mars*, Oxford University Press, Oxford.

Cazeaux C., Naderi F., Whetsel C., Beaty D., Gershman B., Kornfeld R., Micheltree B., Sackheim B. (2004) The Nasa/CNES Mars sample return – a status report. *Acta Astronaumica* 54 601-617.

Challinor, J.M. (2001). Review: the development and applications of thermally assisted hydrolysis and methylation reactions. *Journal of Analitytical Applied Pyrolysis. 61*, 3-34.

Charbonneau, D., Brown, T.M., Noyes, R.W., et al. (2002). Detection of an extrasolar planet atmosphere. *Astrophysical Journal, 568*, 377-384.

Christensen P.R., Morris R.V., Lane M.D., Bandfield J.L., Malin M.C. (2001) Global mapping of martian hematite mineral deposits: remnants of water-driven processes on early Mars. *Journal of Geophysical Research, 106, 23,873–23,885.*

Christensen P.R., Bandfield J.L., Hamilton V.E., Ruff S.W., Kieffer H.H., Titus T.N., Malin M.C., Morris R.V., Lane M.D., Clark R.L., Jakosky B.M., Mellon M.T., Pearl J.C., Conrath B.J., Smith M.D., Clancy R.T., Kuzmin R.O., Roush T., Mehall G.L., Gorelick N., Bender K., Murray K., Dason S., Greene E., Silverman S., Greenfield M.. (2001b) Mars Global Surveyor Thermal Emission Spectrometer experiment: Investigation description and surface science results. *Journal of Geophysical Research, 106*, 23,823-871.

Christensen, P.R.; Mehall, G.L.; Silverman, S.H.; Anwar, S.; Cannon, G.; Gorelick, N.; Keehn, R.; Tourville, T.; Bates, D.; Ferry, S.; Fortuna, T.; Jeffryes, J.; O'Donnell, W.; Peralta, R., Wolverton, T.; Blaney, D.; Denise, R.; Rademacher, J.; Morris, R.V. and Squyres, S. (2003). The Miniature Thermal Emission Spectrometer for the Mars Exploration Rovers. *Journal of Geophysical Research, 108*, 8064.

Christensen, P.R.; Ruff, S.W.; Fergason, R.L.; Knudson, A.T.; Anwar, S.; Arvidson, R.E.; Bandfield, J.L.; Blaney, D.L.; Budney, C.; Calvin, W.M.; Glotch, T.D.; Golombek, M.P.; Gorelick, N.; Graff, T.G.; Hamilton, V.E.; Hayes, A.; Johnson, J.R.; Mcsween Jr., H.Y.; Mehall, G.L.; Mehall, L.K.; Moersch, J.E.; Morris, R.V.; Rogers, A.D.; Smith, M.D.; Squyres, S.W.; Wolff, M.J.; Wyatt, M.B. (2004). Initial results from Mini-TES experiment in Gusev Crater from Spirit Rover. *Science, 305*, 837-841.

Chyba, C.F.; Thomas, P.J.; Brookshaw, L. and Sagan, C. (1990). Cometary delivery of organic molecules to the early Earth. *Science, 249*, 366-373.

Chyba, C.F. and Sagan, C. (1992). Endogenous production, exogenous delivery and impact-shock synthesis of organic molecules: An inventory for the origin of life. *Nature, 355*, 125-132.

Clancy, R.T.; Sandor, B.J.; Moriarty-Schieven, G.H. (2004). A measurement of the 362 GHz absorption line of Mars atmospheric H2O2. *Icarus, 168*, 116-121.

Clark B.C. (2001) Planetary interchange of bioactive materials: probability factors and implications. *Origin of Life and Evolution of Biosphere, 31,* 185-197.

Clark, B.C. (2002) Martian meteorites do not eliminate the need for back contamination precautions on sample return missions. *Advances in Space Research, 30*, 1593-1600.

Clark, B.C. and Baird, A.K. (1979). Is the Martian lithosphere sulphur rich? *Journal of Geophysical Research, 84*, 8395-8403.

Clemett, S.J., Dulay, M.T., Gillette, J.S., Chillier, S.D.F., Mahajan, T.B and Zare, R.N., (1998). Evidence for the extraterrestrial origin of polycyclic aromatic hydrocarbons in the Martian meteorite ALH84001. *Faraday Discussion, 109*, 417–436.

Cottin, H.; Gazeau, M.-C.; Raulin, F. (1999). Cometary organic chemistry: a review from observations, numerical and experimental simulations, *Planetary Space Science, 47*, 1141-1162.

Cottin, H., Szopa, C.Moore, M.H. (2001). Production of hexamethylenetetramine in photolyzed and irradiated interstellar cometary ice analogs, *Astrophysical Journal, 561*, L139-L142.

Cronin, J. R. and Chang, S. (1993) 'Organic Matter in Meteorites: Molecular and Isotopic Analyses of the Murchison Meteorites', in J. M. Greenberg, C. X. Mendoza-Gomez & V.

Pirronello (eds.), *The Chemistry of Life's Origin*, Kluwer Academic Publishers, The Netherlands, pp. 209-258.

Cronin, J.R. and Pizzarello, S. (1983). Amino acids in meteorite. *Advances in Space Research, 3*, 5-18.

Cronin, J.R. and Pizzarello, S. Organic matter in carbonaceous chondrites, planetary satellites, asteroids and comets in Kerridge, J.F.; Matthews, M.S.(Eds.), *Meteorites and the early solar system.* Univ. of Arizona Press, (1988), 819-857.

Cronin, J.R. and Pizzarello, S. (1997). Enantiomeric excesses in meteoritic amino acids, *Science, 275*, 951-955.

Cronin, J.R.and Pizzarello, S. (1999). Amino acid enantiomer excesses in meteorites: Origin and significance, *Advances in Space Research, 23, 293-299*

Davies, P. Are we alone?. In: J. Chela-Flores and F. Raulin, Editors, *Exobology: matter, energy, and information in the origin and evolution of life in the universe*, Kluwer Academic Publishers, Dorcrecht, Nertherland (1998), 61–68.

Deamer, D. W. (1985). Boundary structures are formed by organiccomponents of the Murchison carbonaceous chondrite. *Nature, 317*, 792-794.

Debus A. (2006) The European standard on planetary protection requirements. *Research in Microbiology, 157,* 13-18.

Douté S., Schmitt B., Langevin Y., Bibring J.P., Altieri F., Belluci G., Gondet B., Poulet F. (2007). The MEX OMEGA Team. South pole of Mars: nature and composition of the icy terrains from Mars Express OMEGA observations. *Planetary and Space Science, 55,* 113-133.

Eckhard, F.; Arnoult, S.; van Driel, C.; Hoogakker, G.; Kiewiet, P.; van Loon, D.; Heideman, R.; Mateman, R.; Prak, A.; Wehrmeijer, M.; van den Assem, D.; van Benthem, R. (2003). MUSE—Lab on Chip for chemical analysis in situ. Presented at *Fourth Round Table on Micro/Nano Technologies for Space* 20th–22nd May 2003, Noordwijk, The Netherlands.

Eglinton, G. and Hamilton, R.J. (1967). Leaf epicuticular waxes, *Science 156*, 1322–1335.

Ehrenfreund, P. and Charnley, S.B. (2000). Organic molecules in the ISM, comets and meteorite: A voyage from dark clouds to the early Earth, *Annual Review of Astronomy and Astrophysics 38* 427–483.

Ehrenfreund, P.; Ruiterkamp,R.; Peeters,Z.; Foing, B.; Salama, F. and Martins, Z. (2007). The ORGANICS experiment on BIOPAN V: UV and space exposure of aromatic compounds. *Planetary and Space Science, 55,* 383-400.

Encrenaz, T. (2005). Neutral atmospheres of the giant planets: an overview of composition measurements. *Space Science Reviews, 116*, 99-119.

Fairén, A.G.; Dohm, J.M.; Uceda, E.R.; Rodriguez, A.P.; Baker, V.R.; Fernandez-Remolar, D.; Schulze-Makuch, D.; Amils, R. (2005). Prime candidate sites for astrobiological exploration through the hydrogeological history of Mars, *Planetary and Space Science,* 53, 1355-1375.

Farmer, J.D. and Desmarais, D. J.(1994). Exopaleontology and the search for a fossil record on Mars. In Lunar and Planetary Inst., *The Twenty-Fifth Lunar and Planetary Science Conference*. Part 1: A-G p 367-368.

Feldman, W.C.; Prettyman, T.H.; Maurice, S.; Nelli S.; Elphic, R.; Funsten, H.O.; Gasnault, O.; Lawrence, D.J.; Murphy, J.R.; Tokar, R.L.; Vaniman, D.T. (2005) Topographic control of hydrogen deposits at low latitudes to midlatitudes of Mars. *Journal of Geophysical Research 110,* 1-15.

Ferris, J.C.; Dohm, J.M.; Baker, V.R. and Maddock III, T. (2002). Dark slope streaks on Mars: Are aqueous processes involved? *Geophyical. Research Letters,* 29, 1490.

Fernandez-Remolar, D.C.; Morris, R.; Gruener, J. E.; Amils, R. and Knoll, A.H. (2005). The Rio Tinto Basin, Spain: Mineralogy, sedimentary geobiology, and implications for interpretation of out- crop rocks at Meridiani Planum, Mars, *Earth Planetary Science Letters, 240*, 149–167.

Flynn, G.J. (1996). The delivery of organic matter from asteroids and comets to the early surface of Mars. *Earth Moon Planets,* 72, 469–474.

Formisano V.; Grassi D.; Ignatiev, N.I., Zasova L. (2001) IRIS Mariner 9 data revisited: water and dust daily cycles. *Planetary and Space Science, 49*, 1331-1346.

Formisano,V.; Atreya, S.; Encrenaz, T.; Ignatiev, N.; Giuranna, M. (2004). Detection of Methane in the Atmosphere of Mars, *Science*, 306, 1758-1761.

Gendrin, A.; Bibring, J.P.; Langevin, Y.; Gondet, B.; Poulet, F.; Mangold, N.; Bonello, G.; Quantin, C.; Mustard, J.; Arvidson, R.; LeMouelic, S. (2005). Sulfates in martian layered terrains: the OMEGA/Mars Express view. *Science, 307*, 1587-1591.

Gershman R., Adams M., Mattingly R., Rohatgi N., Corliss J., Dillman R., Fragola J., Minarick J. (2004). Planetary protection for Mars sample return, *Advances in Space Research, 34,* 2328-2337.

Gladman B.J. and Burns, J.A. (1996). Mars meteorite transfer: simulation, *Science, 274*, 161-162.

Gladman B.J. (1997). Destination Earth: martian meteorite delivery, *Icarus, 130*, 228-246.

Glavin, D.P.; Schubert, M.; Botta, O.; Kminek, G. and Bada, J.L. (2001). Detecting pyrolysis products from bacteria on Mars. *Earth Planeary Science Letters, 185*, 1–5.

Glavin, D. P.; Matrajt, G.; Bada, J. L. (2004). Re-examination of amino acids in Antarctic micrometeorites, *Advances in Space Research, 33,* 106-113.

Goetz, W.; Bertelsen, P.; Binau, C.S.; Gunnlaugsson, H.P.; Hviid, S.F.; Kinch, K.M.; Madsen, D.E.; Madsen, M.B.; Olszen, M.; Gellert, R.; Klingehöfer, G.; Ming, D.W.; Morris, R.V.; Rieder, R.; Rodionov, D.S.; de Souza Jr., P.A.; Schröder, C.; Squyres, S.; Wdowiak, T.; Yen, A. (2005). Indication of drier periods on Mars from the chemistry and mineralogy of atmospheric dust. *Nature, 436*, 62-65.

Grady, M.M.; Wright, I.P.; Douglas, C. and Pillinger, C.T. (1994). Carbon and nitrogen in ALH84001. *Meteoritics, 29*, 469.

Graf, J.E.; Zurek, R.W.; Eisen, H.J.; Jai, B.; Johnston, M.D.; DePaula, R. (2005). Mars Reconnaissance Orbiter Mission. *Acta Astronautica, 57*, 566-578.

Hansen G.; Giuranna, M.; Formisano, V.; Fonti, S.; Grassi, D.; Hirsh, H.; Ignatiev, N.; Maturill,i A.; Oleanski, P.; Piccioni, G.; Rataf, M.; Saggin, B.; Zasova, L. (2005). PFS-MEX observation in the resudual south polar cap of Mars, *Planetary and Space Science, 53,* 1089-1095.

Helfer, H. L. (1990). Of Martian atmospheres, oceans, and fossils, *Icarus*, *87*, 228-235.

Horneck, G. (1995). Exobiology, the study of the origin, evolution and distribution of life within the context of cosmic evolution: a review. *Planetary and Space Science*, *43*, 189-217.

Hunten, D.M. (1979). Possible oxidant sources in the atmosphere and surface of Mars, *Journal of Molecular Evolution, 14*, 71-78.

Huten, D.M. (1974). Aeronomy of the lower atmosphere of Mars. *Review of Geophysical Space Phys. 12*, 529-535.

Hutt, L.D.; Glavin, D.P.; Bada, J.L.; Mathies, R.A. (1999). Microfabricated capillary electrophoresis amino acid chirality analyzer for extraterrestrial exploration, *Analytical Chemistry, 71*, 4000-4006.

Hynek, B.M. (2004) Implications for hydrologic processes on Mars from extensive bedrock outcrops throughout terra Meridiani. *Nature*, 431, 156-159.

Imbus, S.W. and McKirdy, D.M. (1993). Organic geochemistry of Precambrian sedimentary rocks. In: M.H. Engel and S.A. Macko, Editors, *Organic Geochemistry*, Plenum, 657–684.

Kanavarioti, A. and Mancinelli, R.L (1990). Could organic matter have been preserved on Mars for 3.5 billion years? *Icarus*, *84,* 196-202.

Kerr, R.A. (2004). Methane means Martians? *Science, 324*,1.

Kissel, J. and Krueger, F.R. (1987). The organic component in dust from comet Halley as measured by the PUMA mass spectrometer on board Vega 1. *Nature, 326*, 755–757.

Klein, H.P. (1978). The Viking biological experiments on Mars, *Icarus, 34*, 666–674.

Klein, H.P. (1979). The Viking mission and the search for life on Mars. *Review of Geophysical Space Phys. 17*, 1655–1662.

Klingelhoefer, G.; Morris, R.V.; Bernhardt, B.; Schroeder, C.; Rodionov, D.S.; de Souza Jr., P.A.; Yen, A.; Gellert, R.; Evlanov, E.N.; Zubkov, B.; Foh, J.; Bonnes, U.; Kankeleit, E.; Gutlich, P.; Ming, D.W.; Renz, F.; Wdowiak, T.; Squyres, S.W.; Arvidson, R.E. (2004). Jarosite and hematite at Meridiani Planum from Opportunity's Mössbauer spectrometer, *Science, 306*, 1740–1745.

Kminek, G.and Bada, J.L. (2006).The effect of ionizing radiation on the preservation of amino acids on Mars, *Earth Planetary Science Letters, 245*, 1-5.

Knoll, A.H.; Carr, M.; Clark, B.; Des Marais, D.J.; Farmer, J.D.; Fischer, W.; Grotzinger, J.P.; McLennan, S.M.; Malin, M.; Schröder, C. *et al.* (2005). An astrobiological perspective on Meridiani Planum. *Earth and Planetary Science Letters, 240, 179-189.*

Koenig, W.A. and Benecke, I. (1980), Gas chromatographic separation of chiral 2-hydroxy acids and 2-alkyl-substituted carboxylic acids, *Journal of Chromatography*, 195, 292.

Krasnopolsky, V.A.; Maillard, J.P.; Owen, T.C. (2004). Detection of methane in the martian atmosphere: evidence for life*? Icarus, 172*, 537-547.

Kuan, Y.-J.; Charnley, S.B.; Huang, H.-C.;Tseng, W.-L.; Kisiel, Z. (2003). "Interstellar Glycine", *Astrophysical Journal, 593*, 848.

Kvenvolden, K.; Lawless, J.; Pering, K.; Peterson, E.; Flores, J.; Ponnamperuma, C.; Kaplan, I.R.; and Moore, C. (1970). *Nature, 228*, 923–925.

Lane, M.D.; Christensen, P.R. and Hartmann, W.K. (2003) Utilization of the THEMIS visible and infrared imaging data for crater population studies of the meridiani planum landing site. *Geophysical Research Letters,* 30, 1770.

Langevin, Y.; Poulet, F.; Bibring, J.P.; Gondet, B. (2005). Sulfates in the north polar region of Mars detected by OMEGA/Mars express. *Science, 307,* 1584-1586.

Larralde, R.; Robertson, M.P. and Miller, S.L. (1995). Rates of decomposition of ribose and other sugars: implications for chemical evolution, *Proc. Natl. Acad. Sci. USA,* 92, 8158–8160.

Lebonnois, S.; Bakes, E.L.O.; McKay, C.P. (2002). Transition from gaseous compounds to aerosols in Titan's atmosphere. *Icarus*, 159, 505-517.

Leighton, R.B.; Murray, B.C.; Sharp, R.P.; Allen, J.D. and Sloan, R.K. (1965). Mariner IV photography of Mars: initial results. *Science*, 149, 627-630.

De Leeuw, J.W. and Largeau, C. (1993). A review of macromolecular organic compounds that comprise living organisms and their role in kerogen, coal and petroleum formation. In: M.H. Engel and S.A. Macko, Editors, *Organic Geochemistry*, Plenum, 23–71.

Levin, G.H. and Levin, R.L. (1998). Liquid water and life on Mars. *SPIE, 3441*, 30–41.

Levin, G.V. (1997). The Viking labelled-release experiment and life on Mars, *SPIE, 3111*, 146–157.

Lindsay, J.F.; Brasier, M.D.; McLoughlin, N.; Green, O.R.; Fogel, M.; Steele, A. and Mertzman, S.A. (2005).The problem of deep carbon—An Archean paradox. *Precambrian Research, 143,* 1-22.

Lofstedt R.E. (2003). Public perception of the Mars sample return program, *Space Policy, 19* 283-292.

Maurette, M. (1998). Carbonaceous micrometeorites and the origin of life, *Origins Life Evol. Biosphere,* 28, 385–412.

Margulis, L.; Mazur, P.; Barghoorn, E. S.; Halvorson, H. O.; Jukes, T. H.; Kaplan, I. R. (1979). The Viking Mission: Implications for Life on Mars. *Journal of Molecular Evolution, 14*, 223-232.

Marov M.Y., Avduevsky V.S., Akim E.L., Eneev T.M., Kremnev R.S., Kulikov S.D., Pichkhadze K.M., Popov G.A., Rogovsky G.N. (2004). *Advances in Space Research, 33,* 2276-2280.

McCollom T.M. (2003). Formation of meteorite hydrocarbons from thermal decomposition of siderite ($FeCO_3$). *Geochimica et Cosmochimica Acta, 67,* 311-317.

McKay, C. P. and Stoker, C. R. (1989). The Early Environment and its Evolution on Mars: Implications for Life. *Reviews of Geophysics, 27*, 189-214.

McKay, C. P. and Stoker, C. R. (1990). Could the early environment of Mars have supported the development of life? *Earth in Space, 2,* 10-12.

McKay, C.P. and Davis, W.L. (1991). Duration of liquid water habitats on early Mars. *Icarus, 90,* 214-221.

McKay, C.P. (1996). The search for life on Mars, *Origins of life and evolution of the biosphere*, 27, Kluwer Academic Publishers, Dorcrecht, Nertherland, 263–289.

McKay, D.S.; Gibson Jr, E.K.; Thomas-Keprta, K.; Vali, H.; Romanek, C.S. and Clemett S.J. *et al.*, (1996). Search for past life on Mars: possible relic biogenic activity in Martian meteorite ALH84001. *Science,* 273, 924–930.

McKay, C. P;. Grunthaner, F.J;. Lane, A.L.; Herring, M.; Bartman, R.K.; Ksendzov, A.; Manning, Lamb, C.M.; J.L.; Williams, R.M.; Ricco, A.J.; Butler, M.A.; Murray, B.C.; Quinn, R.C.; Zent, A.P.; Klein, H. P.; Levin, G.V. (1998). The mars oxidant experiment (MOx) for Mars '96. *Planetary Space Science, 46*, 769-777.

Meierhenrich, U. J; Munoz Caro, G. M.; Bredehoft, J. H.; Jessberger, E. K.; Thiemann, W. H. (2004). Identification of diamino acids in the Murchison meteorite. *Proc. Natl. Acad. Sci. USA 101*,9182-9186.

Meierhenrich,U.; Wolfram H. -P. Thiemann, W.H.P and Rosenbauer,H. (2001). Pyrolytic methylation assisted enantioseparation of chiral hydroxycarboxylic acids, *Journal of Analytical and Applied Pyrolysis, 60, 13-26.*

Melosh, H.J. (1988). The rocky road to panspermia. *Nature,* 332, 687–688.

Miller S. L. and Urey H. C. (1959). Organic compound synthesis on the primitive Earth. *Science*, 130 245-251.

Mitrofanov I., Anfimov D., Kozyrev A., Litvak M., Sanin A., Tret'yakov V., Krylov A., Shvetsov V., Boynton W., Shinohara C., Hamara D., Saunders R.S. (2002). Maps of subsurface hydrogen from the high energy neutron detector, Mars Odyssey. *Science, 297,* 78-81.

Mitrofanov G., Zuber M.T., Litvak M.L., Boynton W.V., Smith D.E., Drake D., Hamara D., Kozyrev A.S., Sanin A.B., Shinohara C., Saunders R.S., Tretyakov V. (2003) CO_2 snow depth and subsurface water-ice abundance in the northern hemisphere of mars. *Science, 300,* 2081-2084.

Mongenot, T.; Riboulleau, A.; Garcette-Lepecq, A.; Derenne, S.; Pouet, Y.; Baudin F.; and Largeau, C.(2001). Occurrence of proteinaceous moieties in S- and O-rich Late Tithonian kerogen (Kashpir oil Shales, Russia), *Organic Geochemistry*, *32*, 199-203.

Munoz-Caro, G.M., Meierhenrich, U.J., Schutte, W., et al. Amino acids from ultraviolet irradiation of interstellar ice analogues *Nature* 416, 403-406, 2002.

Navarro-Gonzalez, R.; Navarro, K. F.; de La Rosa, J.; Molina, P.; Iñiguez, E.; Miranda, L. D.; Morales, P.; Cienfuegos, E.; Coll, P.; Raulin, F.; Amils, R.; McKay, C. P. (2006). The limitations on organic detection in Mars-like soils by thermal volatilization–gas chromatography–MS and their implications for the Viking results. *PNAS*, 103, 16089-16094.

Novotny, M.; Hayes, J.M.; Bruner, F. (1975). Gas Chromatographic Column for the Viking 1975 Molecular Analysis Experiment, *Science*, 189, 215-216.

Orgel, L.E. (1998). The origin of life—a review of facts and speculations, *Trends in Biochemical Sciences*, *23*, 491-495.

Oró, J. (1961). Comets and the formation of biochemical compounds on the primitive Earth. *Nature, 190*, 389–390.

Oyama, V.I. (1963). Use of Gas Chromatography for the Detection of Life on Mars. *Nature, 200*, 1058-1059.

Oze, C. and Sharma, M. (2005). Have olivine, will gas: Serpentinization and the abiogenic production of methane on Mars, *Geophysical Research Letters,* 32, L10203.

Parker, T.J.; Saunders, R.S. and Schneeberger, D.M. (1989). Transitional morphology in West Deuteronilus Mensae, Mars: Implications for modification of the lowland/upland boundary. *Icarus*, *82,* 111-145.

Parker, T.J.; Gorsline, D.S.; Saunders, R.S.; Pieri, D.C. and Schneeberger, D.M. (1993). Coastal morphology of the Martian Northern Plains *Journal of Geophysic Research, 98,* 11061–11078.

Patel, M.R.; Berces, A.; Kolb, C.; et al. (2003). Seasonal and diurnal variations in Martian surface ultraviolet irradiation: biological and chemical implications for the Martian regolith. *International Journal of Astrobiology, 2*, 21-34.

Pierazzo, E. and Chyba, C.F. (1999). Amino acid survival in large cometary impacts - *Meteoritics Planetary Science, 34*, 909-918.

Pietrogrande, M.C.; Zampolli, M.G.; Dondi, F. Szopa, C. Sternberg, R. Buch, A. and Raulin, F. (2005). In situ analysis of the Martian soil by gas chromatography: Decoding of complex chromatograms of organic molecules of exobiological interest. *Journal of Chromatography A*, *1071,* 255-261.

Plescia, J.B. (2003). Cerberus Fossae, Elysium, Mars: a source for lava and water. *Icarus*, 164, 79-95.

Pollack, J.B.; Kasting, J.F.; Richardson, S.M. and Poliakoff, K. (1987). The case for a wet, warm climate on early Mars. *Icarus, 71*, 203–224.

Poulet F., Bibring J.P., Mustard J.F., Gendrin A., Mangold N., Langevin Y., Arvidson R.E., Gondet B., Gomez C. (2005) Phyllosilicates on Mars and implications for early martian climate. *Nature*, 438, 623-625.

Powers,L.; Ciraolo, S.T.; Agarwal, K.C.; Kumar, A.; Bomont, C.; Soloviev, M.V.; David, F.; Desrochers, S. and Brunengraber, H. (1994), Assay of the enantiomers of 1,2-propanediol, 1,3-butanediol, 1,3-pentanediol, and the corresponding hydroxyacids by gas chromatography–mass spectrometry, *Analytical Biochemistry, 221,* 323.

Race M.S. (1998). Mars sample return and planetary protection in a public context. *Advances in Space research, 22*, 391-399.

Remusat, L.; Derenne, S.; Robert, F. and Knicker, H. (2005). New pyrolytic and spectroscopic data on Orgueil and Murchison insoluble organic matter: A different origin than soluble? *Geochimica et Cosmochimica Acta, 69,* 3919-3932.

Rieder, R.; Gellert, R.; Brockner, J.; Klingelhofer, G.; Dreibus, G.; Yen, A.; Squyres, S.W. (2003). The new Athena alpha particle X-ray spectrometer for the Mars Exploration Rovers, *Journal of Geophysical Research*, 108, 1-7.

Rushneck, D.R.; Diaz, A.V.; Howarth, D.W.; Rampacek, J.; Olson, K.W.; Dencker, W.D.; Smith, P.; McDavid, L.; Tomassian, A.; Harris, M.; Bulota, K.; Biemann, K. LaFleur, A.L.; Biller, J.E.; Own, T. (1978). Viking gas chromatograph–mass spectrometer. *Review of Scientific Instruments, 49*, 817-834.

Rodier, C.; Vandenabeele-Trambouze, O.; Sternberg, R.; Coscia, D.; Coll, P.; Szopa, C.; Raulin, F.; Vidal-Madjar, C.; Cabane, M.; Israel, G.; Grenier-Loustalot, M.F.; Dobrijevic, D. and Despois, D. (2001). Detection of martian amino acids by chemical derivatization coupled to gas chromatography: in situ and laboratory analysis. *Advances in Space Research,* 27, 195-199.

Rodier, C.; Laurent, C.; Szopa, C.; Sternberg, R.; Raulin, F. (2002) Chirality and the origin of life: in situ enantiomeric separation for future space missions. *Chirality, 14*, 527-532.

Rodier, C.; Sternberg, R.; Szopa, C.; Buch, A.; Cabane, M. and Raulin, F. (2005). Search for organics in extraterrestrial environments by in situ gas chromatography analysis. *Advances in Space Research,36,* 195-200.

Sandford, S.A.; Aleon, J.; Alexander, C.M.O.; et al. (2006). Organics captured from comet 81P/Wild 2 by the Stardust spacecraft *Science* 314, 1720-1724.

Schutte, W.A., Allamandola, L.J.Sandford, S.A. (1993). An Experimental Study of the Organic Molecules Produced in Cometary and Interstellar Ice Analogs by Thermal Formaldehyde Reactions *Icarus* 104, 118-137.

Schurig, V. (2001). Separation of enantiomers by gas chromatography. *Journal of Chromatography A* 906, 275-299.

Sephton, M.A. (2002). Organic compounds in carbonaceous meteorites. *Nat. Prod. Rep.* 19, 292.-311.

Shotwell, R. (2005). Phoenix--the first Mars Scout mission.,*Acta Astronautica,57*, 121-134.

Sims, M.R;. Pillinger, C.T.; Wright, I.P.; Dowson, J.; Whitehead, S.; Wells, A.; Spragg, J.E.; Fraser, G.; Richter, L.; Hamacher, H.; Johnstone, A.; Meredith, N.P.; de la Nougerede, C.; Hancock, B.; Turner, R.; Peskett, S.; Brack, A.; Hobbs, J.; Newns, M.; Senior, A.; Humphries, M.; Keller, H.U.; Thomas, N.; Lingard, J.S.; Underwood, J.C.; Sale, N.N.; Neal, M.F.; Klingelhofer, G.; Ng., T.C. (1999) Beagle 2: The exobiology lander on

ESA's 2003 Mars Express Mission. SPI Conference on Instruments, Methods and Mission for Astrobiology II, *SPIE, 3755*, 10-23.

Sims M.R.; Pullan D.; Pillinger C.T.; Wright I.P. (2002) An evaluation of in situ analysis and sample return in the exploration of Mars; *Planetary and Space Science, 50*, 657-668.

Sims M.R., Cullen D.C., Bannister N.P., Grant W.D., Henry O., Jones R., McKnight D., Thompson D.P., Wilson P.K. (2005) The specific molecular identification of life experiment (SMILE). *Planetary Space Science, 53*, 781-791;

Skelley, A.M.; Grunthaner, F.J.; Bada, J.L.; Mathies, R.A. (2003) Mars organic detector III: A versatile instrument for detection of bio-organic signature on Mars. SPIE – *The international society for optical engineering proceedings, first Jet Propulsion Laboratory in situ instruments workshop*, june 11-13, 2002, Pasadena, CA, G.H. Beaman and P.M. Beauchamp, Editors, 4878, 59-67.

Skelley, A.M.; Scherer, J.R.; Aubrey, A.D.; Grover, W.H.; Ivester, R.H.C.; Ehrenfreund, P.,; Grunthaner, F.J.; Bada, J.L.; Mathies, R.A. (2005) Development and evalutation of a microdevice for amino acid biomarker detection and analysis on Mars, *PNAS, 102*, 1041-1046.

Skelley, A.M.; Mathies, R.A.; Aubrey, A.D.; Bada, J.L.; Willis, P.; Amashukeli, X.; Grunthaner F.J.; Ehrenfreund, P. (2006). Detection of trace biomarkers in the Atacama desert with the UREY *In Situ* Organic Compound Analysis Instrument. *Astrobiology*, 6, 138.

Soffen, G.A. (1976). Scientific Results of the Viking Missions, *Science, 94*, 1274 – 1276.

Squyres, S.; Arvidson, R.E.; Bell, R.F.; Brückner, J.; Cabrol, N.A;. Calvin, W.; Carr, M.H.; Christensen, P.R.; Clark, B.C.; Crumpler, L. et al. (2004).The Opportunity rover's Athena science investigation at Meridiani Planum Mars, *Science, 306*, 698–1703.

Stalport, F.; Coll, P.; Szopa, C.; et al. Search for organic molecules at the Mars surface: The "Martian Organic Material Irradiation and Evolution" (MOMIE) project *Adv. Space Res.* Under review.

Stoker, C. and Bullock, M.A. (1997). Organic degradation under simulated Martian conditions. *Journal of Geophysical Research, 102*, 10881-10888.

Strazzulla, G. and Palumbo, M.E. (2001). Organics produced by ion irradiation of ices: Some recent results. *Advances in Space Research, 27,* 237-243.

Summons, R.E; Jahnke, L.L.; Hope, J.M. and Logan, G.A. (1999). 2-Methylhopanoids as biomarkers for cyanobacterial oxygenic photosynthesis. *Nature, 400*, 554-557.

Szopa, C.; Sternberg, R.; Rodier, C.; Coscia, D.; Raulin, F. (2001). Development and analytical aspects of gas chromatography for space exploration. *LC-GC, 14*, 114-120.

Szopa, C.; Buch, A.; Cabane, M.;Coll, P.; Mahaffy, P.; Brinckerhoff, W.; Israel., G. et al. (2004).The sample analysis at mars (SAM) experiment and the search for organic molecules at the martian surface, *European Space Agency, (Special Publication) ESA SP*,545, 105-108

Tsapin, A.I.; Goldfeld, M.G.; McDonald, G.D. et al. (2000). Iron (IV): hypothetical candidate for the martian oxidant. *Icarus, 147*, 68-78.

Tielens,A.G.G.M.; Hony, S.; van Kerckhoven, C. and Peeters, E. (1999). Interstellar and circumstellar PAHs. In: The Universe as seen by ISO, *ESA SP 427*, 579–588.

Vago, J.; Gardini, B.; Kminek, G. et al. (2006). ExoMars - Searching for life on the red planet *Esa Bulletin*-European Space Agency 16-23.

West, R.A.; Strobel, D.F.; Tomasko, M.G. (1986). Clouds, aerosols, and photochemistry in the Jovian atmosphere. *Icarus, 65*, 161-217.

Westall, F. (1999). The nature of fossil bacteria: a guide to the search for extraterrestrial life. *Journal of Geophysical Research* 104, 16437–16451.

Winnewisser, G. (1997). Interstellar and laboratory spectroscopy in the terahertz region, *Journal of Molecular Structure, 408-409, 1-10.*

Wright, P.; Grady, M.M. and Pillinger, C.T. (1989). Organic materials in a martian meteorite. *Nature,* 340, 220–222

Yan, L.; Chary, R.; Armus, L.; Teplitz, H.; Helou, G.; Frayer, D.; Fadda, D.; Surace, J. and Choi, P. (2005). Spitzer detection of polycyclic aromatic hydrocarbon and silicate dust features in the mid-infrared spectra of $z\sim2$ ultraluminous infrared galaxies, *Astrophysical Journal 628*, 604–610.

Yen, A.S.; Kim, S.S.;, Hecht, M.H. et al.(2000). Evidence that the reactivity of the martian soil is due to superoxide ions, *Science, 289*, 1909-1912.

Zampolli, M.G.; Basaglia, G., Dondi, F.;Sternberg, R.; Szopa, C. and Pietrogrande, M.C.(2006). Gas chromatography–mass spectrometry analysis of amino acid enantiomers as methyl chloroformate derivatives: Application to space analysis, *Journal of Chromatography A, In Press.*

Zasova, L.; Formisano, V.; Grassi, D.; Ignatiev, N. and Maturilli, A. (2002). Martian winter atmosphere at north high latitudes: Mariner 9 IRIS data revisited, *Advances in Space Research, 29,* 151-156.

Zent, A.P. (1998). On the thickness of the oxidized layer of the Martian regolith. *Journal of Geophysical Research, 103*, 31491-31498.

Zubrin R. (2000) Contamination from Mars: no threat. Planetary Report, Ed. *The Planetary Society* July/Aug. 4-5.

In: Planet Mars Research Focus
Editor: L. A. Costas, pp. 93-103
ISBN: 978-1-60021-826-2

Chapter 3

THE ODDS FOR LIFE ON MARS

Mark K. Moran[1,]* and Kevin M. Shaw[2]
[1]Mars Society, 606 S Alder Avenue, Sterling VA 20164, USA
[2] 701 Rock Elm Court, Odenton, MD 21113, USA

ABSTRACT

In the solar system and elsewhere, zones of habitability suitable for life are likely to be sparse, given the particularity of such life as science has encountered to date. Before any direct observation of present or past life on Mars is available, a forecast of the prospects for finding such a specimen must overcome many weaknesses in the existing body of knowledge. Can we be sure even whether natural or NASA-type transportation mechanisms could have already introduced terrestrial species to Mars, casting a "discovery" of the future in a notably less dramatic light? It can be shown that terrestrial extremophiles are capable of surviving across a range of Mars conditions. In this chapter, we will undertake a conceptual and ultimately statistical analysis of the probability for tracking down a rare but postulated existing lower-order life form on Mars. The Ward-Brownlee hypothesis, that microbial life forms are common in the universe but higher-order life forms are rare, is accepted as credible.

INTRODUCTION

In our solar system and elsewhere, zones of habitability suitable for life are likely to be sparse. Although the solar system in which we find ourselves bears one spectacularly fruitful biosphere (Earth), no sign of a second teeming biosphere in that same solar system is evident. At the same time, even the bottom rung of recently discovered terrestrial extremophiles (e.g., nanobacteria) blurs the line between habitability and uninhabitability. Having scoured the proliferating body of literature, Peter Ward and Donald Brownlee theorize that low-order life forms are commonplace within scattered habitable zones throughout the universe but habitable zones bearing higher-order life forms, on the other hand, are exceedingly rare.[1] If preconditions for life are indeed widely divergent between microbial and more sophisticated orders of life, there would likely be no grandiose zone of habitability across the spectrum of life but rather at least two sometimes overlapping zones of habitability, if not more: one corresponding to microbial life and another corresponding to higher-order life.

In either case, this chapter aims to investigate the odds for life on Mars. At minimum, we aim to outline a method capable of integrating the available data and estimating these odds. Although by the end of the chapter we will have calculated odds, it will also be clear that the odds forecast is no more reliable than the data from which it will be derived, and a number of weaknesses apply to the data. First, what is the probability that life on Mars exists given the present state of knowledge about the planet? Second, what is the probability that we can find life, given that it exists, under realistic scenarios? Our answers to these questions may be less important than our formulation of a prototype method capable of future elaboration and improvement.

An image of Mars from Hubble taken in 2003.
[Image: Courtesy NASA/JPL-Caltech]

In order to explore both questions we must first attempt to outline some working definition of life. Even a tentative and flawed definition will serve for the purposes of this chapter better than none. Is a "sample of one" example of a life-sustaining biosphere (Earth) sufficient to construct a definition capable of transcending the one planet? Only if what we have is indeed a sample of one. Recent findings argue that the envelope of data has already been pushed outward.

The meteorite ALH84001, containing carbonate globules reminiscent of fossilized microbial life, has been touted since the mid 1990's as biogenic evidence that microbial or sub-microbial extraterrestrial life of some kind has existed and been transported via this particular meteorite from the planet Mars to the planet Earth.[2] This discovery elated many of those who followed the search for life on Mars. Early critics rallied around the multitude of ways the sample could have been contaminated with terrestrial agents. More recent critics find evidence that somewhat similar features can be detected even in meteorites (or at least the fewer than 5% of all chondritic meteorites that are carbonaceous). The discovery of the feature in the barren, presumably sterile setting of meteoroids has dampened the enthusiasm of many and convinced some scholars that non-

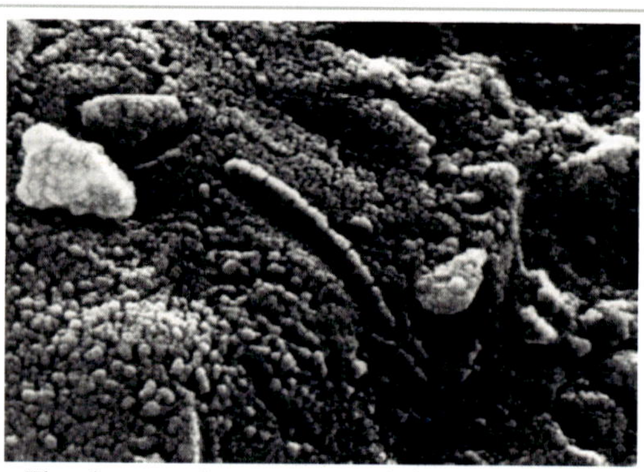

The electron microscope revealed an unusual tube-like structure in meteorite ALH84001.
[Image: Courtesy NASA/JPL-Caltech]

biogenic processes may account better or at least more simply for it than biogenic ones.

Researchers such as Richard Hoover of NASA have found unambiguous microfossils in the carbonaceous chondrites.[3] Even though the ALH84001 specimens exhibit several features different from specimens found in carbonaceous chondrites, the biological interpretation of the data – i.e., that both the ALH84001 martian-origin meteorite and the carbonaceous chondrites exhibit extraterrestrial biomarkers – actually coheres strongly with the Ward-Brownlee theory that microbial life forms are common on a cosmologic scale, while higher order life forms (mainly multicellular animalian) are comparatively rare on a cosmologic scale. An interpretation of the facts consistent with Ward-Brownlee implies that, in fact, the sample of biospheres has grown from one (Earth alone) to at least three (Earth, Mars, and meteoroids).

Since ALH84001, further evidence of possible biofossils in a martian meteorite has come to light in regard to the Nakhla meteorite, remains of which descended to Earth over Egypt in 1911.[4] Although controversy continues to wax and wane around the complicated topic of terrestrial contamination in both meteorites, arguments favoring certain mechanisms of contamination are beginning to be overcome.[5] This trend again bodes well for the Ward-Brownlee theory.

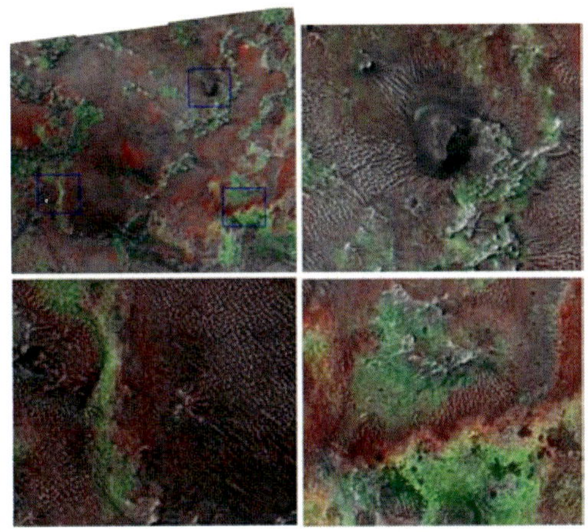

Is there life among the clays at Nili Fossae? [Image: Courtesy NASA/JPL-Caltech]

Definition of Life

How do we define life? The order within an unopened pack of playing cards has been compared to low entropy, while the "order" within the same pack of cards after it has been thoroughly and honestly shuffled has been compared to high entropy. Conversely the "information content" of such decks of cards is said to be "high" when the entropy is low (packaged deck) and "low" when the entropy is high (shuffled deck). Throughout the universe, the ubiquitous trend is for lower-entropies/higher information content entities to deteriorate into higher-entropies/lower information content entities. This phenomenon is even true for life forms when they die. On the other hand, life forms when they are "alive" reverse, nullify, or at least quash, this pattern: "consuming" information content and inhibiting if not entirely suspending entropy.

A cursory search of the literature reveals many competing definitions of life. Based on the one or as many as three biospheres in the realm of experience, there appear to be more or less five minimal features of life beyond the superficial assumption that life is a phenomenon with respect to *distinct entities* – that is, entities that:

1. actively work more or less continuously so as to sustain themselves (in status-keeping or *homeostasis* of some kind paired with *metabolism* or the intake of sustenance),
2. as a community *reproduce their own kind* regularly in accord with a pattern extending across successive generations,
3. temporarily appear to *successfully null out entropy* while alive*,
4. embody some kind of day-long (circadian) *rhythm of activity* (such as sleep/wake), and
5. adapt themselves toward their environment or, in some cases, outwardly adapt their environment to suit themselves (the latter applies mainly to the highest order species, notably humankind).

*Within this 5-part tentative definition, the only terminology that risks circularity is the use of the term "alive." By this term, one simply means that features (1), (4), and (5) are operational. Features (1), (4), and (5) tend to hold together or fall apart as a unit, and when they fail (3) inevitably fails, too, and death ensues.

Our first test of the definition takes up simple cases:

Consider our solar system's sun as a candidate, albeit an unlikely one, for life form. Although the sun, for example: sustains itself over millennia; represents one example of numerous entities called stars; theoretically could, under freak circumstances – perhaps stellar-scale collision – split into two "descendants"; nevertheless the sun is not thought to "reproduce" in accord with any known pattern. The sun does not self-organize nor has ever been characterized as resisting entropy, even temporarily, but rather expends its thermonuclear career according to the principles of entropy. While the sun has its own rhythm called the solar cycle, there exists no evidence that it adapts to its environment purposefully. Therefore, the sun, as one would expect, is *not* a life form by the proposed definition.

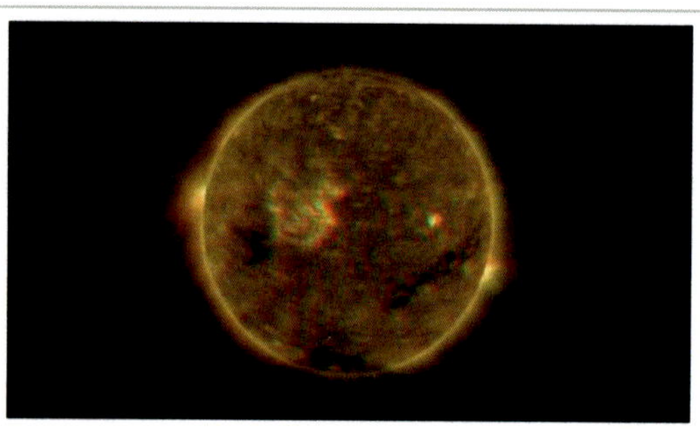

A full disk image of the sun.
[Image: Courtesy NASA/JPL-Caltech]

On the other hand, even the most rudimentary cyanobacteria (simplest of the bacteria) sustain themselves insofar as possible, reproduce themselves across successive generations, self-organize and therefore counter entropy while living, possess circadian rhythm,[6] and adapt to their environment. Therefore, cyanobacteria clearly qualify as life forms.

The case of terrestrial viruses has been debated even in the biology textbooks as a topic at the borderline between life and non-life. As one might expect, our definition does not conveniently trump the longstanding debate. There is a sense in which terrestrial viruses seem to sustain themselves, as in requirement (1), but whether terrestrial viruses *actively work* to sustain themselves is much more problematic except on such occasions when they happen to be in proximity to a host cell. It is unclear if "homeostasis" applies to a terrestrial virus, much less a prion. For example, a terrestrial virus can be crystallized to remain dormant in a block of ice for centuries. Upon thawing, the same terrestrial virus returns to its default "functional" status without the slightest sign of impairment. Virtually all other known life would at least be challenged, if not permanently impaired, by even the most delicately-supervised process of being frozen into crystals and then thawed. Viruses cannot reproduce themselves without the cooperation (or hijacking) of host cells, qualifying them to that attribute in the definition, if at all, only in the most strained way. Although a terrestrial host to a terrestrial virus always seems to embody circadian rhythms in some fashion, whether terrestrial viruses themselves enact circadian rhythm remains open to investigation but biologists have left the question largely unexplored because none expect viral "life" to emdoy circadian rhythm. It appears that viruses, then, sit just outside the boundaries of life as we have found them, but not far outside them.

For purposes of this chapter, life is defined as a phenomenon in which multiple entities (1) self-sustain themselves, (2) reproduce, not as an individual itself, but among a community of fellow life forms whether or not specific members of the community reproduce, (3) over calendars of time (not just minutes or hours, but at least on the scale of magnitude noted under the following item listed next), neutralize entropy (i.e., self-organize, fight impairment or heal, as opposed to giving in to the unrelenting disintegration ubiquitous to all non-living phenomena, including former life forms from the moment they die), (4) embody periodicity on a macroscopic time scale, and (5) adapt.

This working definition of life largely obscures, rather than implies, how one might undertake measurements in an effort to detect life. We emphasize the existence of two distinct probabilities in question. They are not the same. One is the probability that life on Mars exists; the other is the probability that life on Mars can be detected, given that life exists in the first place on Mars. The latter probability depends on the distribution of life on Mars, the distribution of impediments to the detection of life, the state-of-the-art in detection instrumentation, the mobilization of resources (people, robots, time, money, etc.) in order to mount the campaign to detect it, and finally the reliability of the combined measures employed. It is the final item listed in this latter probability that most depends on the testability of one's definition of life. That issue will be bracketed until the last part of this chapter.

Probability of Life Forms on Mars

In the meantime, we must directly consider the probability that life exists on Mars – with detection of it, again, bracketed and set aside. In this case, it should be clear that the five attributes of life proposed above can each, for the most part, exist in greater or lesser degrees. One life form may work forcefully at sustaining itself while another life form, under virtually identical conditions, might work minimally. (We'll briefly postpone the attribute of reproduction.) One life form may appear to reverse entropy through healing itself aggressively, while another might even succumb to entropy through, for example, the problem of cancer. Circadian rhythm may play a pronounced foreground role in one life form, but play a subdued background role in another. Obviously some life forms are better capable of adapting to given environmental challenges, others less capable. Different environmental challenges could even reverse these situations. In all these cases lesser and greater degrees are possible. Only in the case of reproduction do we seem to have no room, or next to no room, for gradation between reproducing and non-reproducing. The virus is perhaps the best example of a peculiar middle ground: it can reproduce itself provided the machinery of reproduction of a host life form is readily available to it, often at its host's expense. We have already argued that a virus is (apparently) on the non-living side of the line.

The important point to recognize is that an entity characterized with an abundance of four of the five attributes but none of the fifth attribute is not alive at all, nor has it ever been alive. All five attributes must together apply to the same entity in order for it to qualify as "living." How must one take account, then, of degrees of each attribute? The authors would argue that life forms may differ superficially in the degrees of some or all their five attributes, but in any case where an attribute sits below some threshold the entity cannot be classified as a life form. There are five attributes, and each attribute inherits a threshold below which the cluster of five attributes cease to signify life at all. Reproduction, on the community level, comes closest of all five attributes to a binary/all-or-nothing rather than graded or shaded variable.

Let A_1 = working to self-sustain, A_2 = reproducing as a community, A_3 = calendar of entropy nullification, A_4 = circadian periodicity, A_5 = adapting. Further, let T_1 be the threshold of self-sustaining work inherent to a life form, T_2 the threshold of reproduction, T_3 the threshold of calendric entropy nullification, T_4 the threshold of periodicity, and T_5 the threshold of adaptation. If and only if for an entity ($A_1>T_1$, $A_2>T_2$, $A_3>T_3$, $A_4>T_4$, $\cap$ $A_5>T_5$), is then the entity a life form. More compactly, the vector $\{A_i\} > \{T_i\}$ for all $i \in \{1,2,3,4,5\}$. For example, on earth, if the T's are nearly zero then viruses qualify as life forms along with tens of millions of other species. Raise them just slightly and viruses drop off the radar screen of life forms. The bar can be raised thereafter at will through a considerable distance without affecting the remaining tens of millions of terrestrial life forms. From then on, entities are simply alive or not alive.

Based on our attributes of life and our thresholds, what are the bounds within which life is constrained to exist? We don't know. What we know is that over a narrow range of conditions on Earth, the five conditions are met and life exists. On the Earth, life is now known to exist from the heights of the stratosphere to nearly a kilometer beneath the bottom of the sea floor. The entire terrestrial biosphere is considerably larger than once thought. If perchance life exists on Earth under an envelope of temperatures, pressures, desiccations, salinities, pH's, etc., that also exists on Mars, then the probability for life to exist on Mars under those conditions should be on a similar scale. The one anomaly to an exact match between Earth and Mars conditions is gravity: Regardless of other conditions, a ratio between Mars gravity and Earth gravity of 1:3 will persist regardless of all other matchings or similarities or differences. In this case, our analysis will simply have to gloss over the mismatch.

A thin coating of water ice on the rocks and soil at Utopia Planitia. [Image: Courtesy NASA/JPL-Caltech]

To be conservative, we will constrain our estimation in three ways.

First, gravity is one example of mismatches between Earth conditions and Mars conditions whose implications for life are unknown. Perhaps other important, but obscure, mismatches are also unaccounted for in this analysis. Given that for all these reasons our calculation could easily overestimate the odds for life on Mars by a thousand, as a first stab at this type of calculation one might estimate the probabilities for Mars when conditions approximate conditions where extremophiles can function on Earth (disregarding gravity) as 0.1 percent the probability of their existing within a comparable region on Earth. Features on Mars, just as in features on Earth, vary by geography. Some regions of Mars may be more promising for the existence of life than others. A sophisticated analysis would identify specific volumetric concentrations from Earth regions, take 0.1 percent that number, assign it to regions of Mars that most closely mimic those conditions, continue to other conditions that match other Mars settings, for the entire planet. This chapter will simply undertake to propose a credible method and then to apply the method to a crude example. Refining the method will be left to further research.

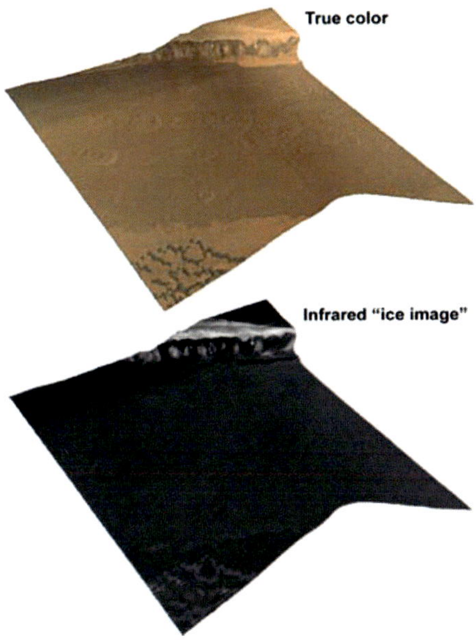

Example of ice from the Chasma Boreale in north polar region.
[Image: Courtesy NASA/JPL-Caltech]

Second, again to be conservative, it would be reasonable to employ water as a constraint. Liquid water has been shown to be stable within the Hellas and Argyre basin and across northern lowlands from about 40° toward the equator.[7] The location with the longest period of stability for liquid H_2O is believed to be in the south-eastern part of Utopia Planitia, where 34% of the year liquid H_2O would remain in that state. However, H_2O anywhere on Mars is likely to contain suspended particles making its stable forms less clear. The general idea is to set to zero the probability of life wherever water quantities are relatively scarce.

Third, given that methane and its oxidized by-product (formaldehyde) may together be associated with life on Mars, we should constrain our hopes for life to the three regions of Mars where methane production overlaps the water-rich region. (One should flag this overlap as one area where measurements expected from the already successfully en route Mars Phoenix spacecraft are likely to update our data relatively soon.) Data from Mars Express restricts these to three regions: Arabia Terra, Elysium Planum, and Arcadia-Memnonia.[8] Given the H_2O and methane constraints together, our calculations will assume a biosphere whose area corresponds to (1) Arabia Terra, (2) Elysium Planitia, and (3) the combined Arcadia-Memnonia regions.

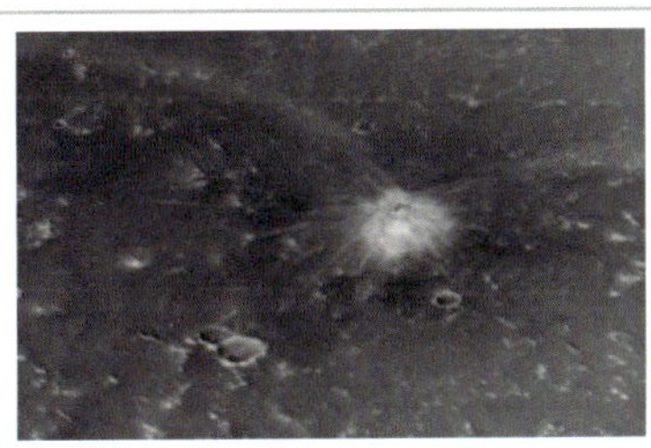
Fresh crater in Arabia Terra.
[Image: Courtesy NASA/JPL-Caltech]

As seen in Table 1, these three regions of H_2O-methane overlap together cover roughly about 8.98% of the surface of Mars.

Table 1. Key Geographic Regions of Mars

Arabia Terra:	7.638860×10^6 km^2
Elysium Planitia:	2.362907×10^6 km^2
Arcadia-Memnia Planitia:	2.751823×10^6 km^2

The above were calculated using a radius of 3396190.0 meters in a Sinusoidal projection.

In order to carry out at least one example calculation, we need terrestrial data. One terrestrial researcher found a sparse population of xerophiles at an estimated population density of 25.0 ± 3.7 extremophiles per 1152 m^2 area.[9] The surface area of Mars is 144 798 465 square kilometers (0.284 Earths). For an equivalent population density on Mars, we would have $3.1423\times10^{12} \pm 4.6506 \times 10^{11}$ life forms. Taking 0.1 percent of our numbers, and applying them to only 8.98% of the Mars surface, we have an hypothesis of $2.8218\times10^8 \pm 4.1763 \times 10^7$ life forms.

However, we began the above calculations (at the top of the preceding paragraph) with terrestrial data on one arbitrarily selected extremophile out of many different candidate species. Depending upon how narrow one's criteria for admitting that a terrestrial extremophile could survive on Mars, there would be at least half a dozen extremophiles on the list to perhaps as many as 30 or more. Instead of this species, for instance, we might have selected cryptoendolithic cyanobacteria, known to grow in the Ross Desert of Antarctica upon cavities in translucent porous rocks. Arid land endolithic communities of terrestrial cyanobacteria account for 14×10^{12} g C of biomass.[10]

An image of the north polar ice cap. [Image: Courtesy NASA/JPL-Caltech]

The selection of any one particular extremophile is problematic in that it is likely to be biased. Each species of extremophile tends to yield its own population density, greater or less than other species of extremophiles. Not only are there a variety of terrestrial species that could survive under the equivalent conditions on Mars, but for all we know there could be a variety of martian species contributing to the total population of indigenous lifeforms, each with its own habitability zone. If we were to chart those population densities we know of and sort them by increasing magnitude, we would obtain something like a bell curve, with certain extremophiles in the left-hand tail tending toward sparser densities and other extremophiles in the right-hand tail tending toward more crowded densities. The certitude of our estimate depends to some degree on the choice of terrestrial model from this empirical distribution. If all the data were available, we could take an average. In the meantime for the sake of convenience and deriving the prototype of a method that future research can improve upon, we will assume a normal form to this distribution and consider the model selected above as, fortuitously, the actual mean population density of life forms on Mars, if they exist. If a *crude* estimate of the standard deviation of this distribution is one-half the mean, and instead of one particular species we now select a random sample of 100 extremophiles across various species, then the 95% confidence interval is

$$(2.8218x10^8 - 4.1763 \text{ x } 10^7) - 1.96\ (2.8218 \text{ x}10^8/2)/\sqrt{100} \text{ (lower limit)} \ldots$$

$$(2.8218x10^8 + 4.1763 \text{ x } 10^7) + 1.96\ (2.8218 \text{ x}10^8/2)/\sqrt{100} \text{ (upper limit)}$$

The resulting 95% confidence interval is (212.76M, 351.60M) life forms. It should be emphasized that these figures are highly speculative. For microbiology, even the greater of the two figures is actually a very miniscule planet-wide population size. On the other hand, Jakosky and Shock calculated energies available to life forms on Mars and speculated that only 5 microbes per cubic centimeter per year could be sustained! [11]

FINDING LIFE GIVEN ITS EXISTENCE

Mars' surface could be characterized as a permafrost similar to the permafrost in Antarctica. The total volume of water ice on Mars is estimated to be 1.6 x 10^6 km^3, equivalent to a global water layer approximately 11 meters thick.[12] The best "fit" for life on Mars is expected among slushes that contain water. How much of the martian year are they stable? We have already seen the figure 34%, even without particulates suspended in the water. Critics point out that Mars water slushes contain peroxides, peroxides that are especially antithetical to life. Hydrogen peroxide has been pigeon-holed as a life-inhibiting substance, a life-inhibiting substance that may be ubiquitous on the surface of Mars. Since the Viking landers, and the controversy over the labeled release experiments to detect life on the planet surface, the presence of agents of this kind have been argued to serve as self-sterilizing factors. On the other hand, phenomena with thriving life forms have been shown to operate wherein increasing doses of hydrogen peroxide can have paradoxical effects, even on known forms of life such as Escherichia coli (e. coli), so that the harm from increasing doses of peroxides flattens and actually decreases to be less harmful to certain microbes than lower doses.[13]

Above, we spoke of five features ubuiquitous to life. Despite advances in computing power and computer intelligence, and although detection of the five features may become programmable at some point in the future, the confirmation of all five features at present would require specimen analysis in a laboratory under the direct supervision of human beings. The good news is that the five features tend either to congregate together as a group, or not be present at all. If this is the case, then detecting only one of the five features should suffice to screen out most or all non life forms. The second step would therefore be to examine the screened candidates at a much more expensive, more sophisticated level of scrutiny, for the other features.

Feature number three, that is the neutralization of entropy, may prove to be the key that unlocks the door to screening for these candidate life forms. Fortes compared biomarkers related to feature three for their expected worth in detecting life on Mars.[14] His analysis concluded that among the two most promising biomarkers was the rate of handedness among organic molecules: that is, densities of homochiral (again, a form of resistance to entropy) as opposed to racemized organics. Tests for homochirality of organics should be promoted as one of the two best methods of exploring Mars in order to detect life. An equally promising biomarker is the hopanoid. Hopanoids, or pentacyclic compounds similar to sterols, are an abundant commodity on Earth and may prove their worth as a second, promising indicator of life form existence. When our sensors can detect H_2O, methane, homochiral organics, and hopanoids, our prospects will be good for screening in or out for feature three and hence detecting any life that may exist, or have existed in the past, on Mars, even in the face of peroxides, UV radiation, and other, even spectacular, hindrances to life.

Conclusion

The Ward-Brownlee theory represents the best available school of thought for approaching the question whether the five co-existent features of life exist, or once existed in the past, on Mars. A method has been outlined for estimating the probability of life on Mars. The estimate forthcoming from the method is only as good as the numbers that feed into it. Improving this method, especially through better data, is a worthwhile goal for Mars research of the coming decade.

Some scientists argue that a vigorous robotics program will suffice to detect life on Mars. They point to the rousing success of Spirit and Opportunity Rovers. Yet, in the entire years that Spirit and Opportunity have been exceeding all expectations, they have explored about as much territory as two or three higher-order specimens of *homo sapiens* equipped with equivalent hardware could have thoroughly investigated in a matter of weeks. A vigorous program that integrates the strengths of human beings with boots on the ground and eyes in the laboratory, on the one hand, accompanied by their robotic technologies, on the other hand, and screening for the presence of water slushes, methane, homochirality, and hopanoids as major predictors of life, will make the most of man's (and machine's) ability to detect life on Mars, should it exist or ever have existed there.

References

[1] Ward, P and Brownlee, D. *Rare Earth: Why Complex Life is Uncommon in the Universe* (Springer-Verlag, New York: 2000).

[2] McKay DS; Gibson Jr EK.; et al. Search for Past Life on Mars: Possible Relic Biogenic Activity in Martian Meteorite ALH84001 by *Science* 16 August 1996: Vol. 273. no. 5277, pp. 924 – 930.

[3] Hoover, R.B. and Rozanov, A Yu., (2005) Microfossils, Biominerals, and Chemical Biomarkers in Meteorites. In *Perspectives in Astrobiology*, Vol. 366 NATO Science Series: Life and Behavioural Sciences (R. B. Hoover, R. Paepe, and A. Yu. Rozanov, Eds.) IOS Press, The Netherlands, pp 1-18.

[4] Science 31 March 2006: Vol. 311. no. 5769, pp. 1858 – 1859 *Lunar And Planetary Science Conference: New Signs of Ancient Life in Another Martian Meteorite?* Richard A. Kerr Science. New Series, Vol. 273, No. 5277 (Aug., 1996), pp. 924-930

[5] Hoover, R.B. and Pikuta, E.V. (2004) Microorganisms on Comets, Europa, and the polar ice caps of Mars, *SPIE*, 5163, 191-202.

[6] *Science* 10 January 1997: Vol. 275. no. 5297, pp. 224 – 227 Circadian Rhythms in Rapidly Dividing Cyanobacteria Takao Kondo, Tetsuya Mori, Nadya V. Lebedeva, Setsuyuki Aoki, Masahiro Ishiura, Susan S. Golden

[7] *Proc Natl Acad Sci U S A.* 2001 February 27; 98(5): 2132–2137. The National Academy of Sciences, Use of spacecraft data to derive regions on Mars where liquid water would be stable by Brad Lobitz, Byron L. Wood, Maurice M. Averner, and Christopher P. McKay

[8] Mumma, MJ; Novak, RE; Hewagama, T; Villanueva GL; Bonev BP; DiSanti MA; Smith, MD; Russo ND. Absolute Abundance of Methane and Water on Mars: Spatial Maps. *37th Meeting of the AAS Division for Planetary Sciences*, 4-9 September 2005. Session 27 Mars IV Oral, Tuesday, September 6, 2005, 4:20-6:00pm.

[9] Nelson SE; Banbury BL; Sosa RA; Powell R; and Parmerlee Jr. *Contemporary Herpetology* 2001 Number 1, 26 January 2001. Natural History of Leiocephalus Semilineatus in Association with Sympatric Leiocephalus Schreibersii and Ameiva Lineolata.

[10] Garcia-Pichel F; Belnap J; Neuer S; and Schanz F. Estimates of global cyanobacterial biomass and its distribution Arch. Hydrobiol. Suppl.bd., Algol. stud. *Symposium of the International Association for Cyanophyte Research (IAC)* No15, Barcelona, ESPAGNE (03/09/2001) 2003, vol. 148 (2 p.3/4), pp. 213-227.

[11] Jakosky. BM., and Shock EL. 1998. The biological potential of Mars, the early Earth, and Europa. *J. Geophys. Res.* 103 E8, 19359-19364.

[12] Plaut JJ, Picardi G, Safaeinili A, Ivanov AB, et al. Science. 2007 Mar 15; [Epub ahead of print] *Subsurface Radar Sounding of the South Polar Layered Deposits of Mars.* Jet Propulsion Laboratory, California Institute of Technology, Pasadena, CA, USA.

[13] Imlay JA; Chin SM; Linn S. Toxic DNA Damage by Hydrogen Peroxide through the Fenton Reaction in vivo and in vitro *Science, New Series*, Vol. 240, No. 4852. (Apr. 29, 1988), pp. 640-642.

[14] Fortes, AD. Dissertation: On the Problem of Ambiguity in Extraterrestrial Biomarkers: Implications for Mars. 2000.

In: Planet Mars Research Focus
Editor: L. A. Costas, pp. 105-115

ISBN: 978-1-60021-826-2

Chapter 4

HOT SPRINGS AND LIFE IN MARS?

Kazue Tazaki[1,*], Masayuki Okuno[1] and Muneyoshi Furumoto[2]
[1]Department of Earth Sciences, Kanazawa University, Kanazawa, Ishikawa, Japan
[2]Graduate school of Environmental Studies, Nagoya University, Nagoya, Japan

ABSTRACT

Biological activity can produce complex patterned structures in accretionary carbonate spheres (pisoliths) from hot springs. Pisoliths from a few millimeters to 50 mm in diameter were collected from a geyser at the high-salinity Arima Hot Springs near Kobe, Japan. High-resolution electron microscope images show that microorganisms produced concentric laminar layers of aragonite alternating with Si- and Fe-rich layers. Thin Fe-bearing layers are included in well-ordered aragonite layers. This Fe-bearing layer is identified as amorphous iron oxides by the diffuse ring pattern of both XRD and electron diffraction TEM. STEM elemental distribution maps show that the cementation of aragonite is associated with NaCl-rich bacterial cells and elevated phosphorous and sulfur concentrations. The filamentous microbes build the concentric framework for mineral laminas with a web-like network of microbial strands. The intricate patterns of mineralogical and bacterial variation in the pisoliths correlate with the change of water chemistry in the hot spring. These patterns could improve our understanding of nano-scale biomineralization. In addition, the terrestrial hot spring pisoliths might be a very counterpart of the tiny spherules recently found on Mars. Liquid water on Mars may persist underground and occasionally burst to the surface, forming pisolith-like spherules in the dry river canyons and craters after volcanic activity wanes. In Eagle Crater's outcrops, Opportunity found millimeter-size spherical grains (blue berries) that may have precipitated out of iron- or salt-bearing water as it evaporated. A geyser at the high-salinity and iron-rich Arima Hot Springs may have had a Mars-like climate. Recent results from Mars rovers and orbiters show that warm, wet conditions may have prevailed on the planet for long periods during its early history. If the eras of Earth-like conditions

* Corresponding author. Phone/Fax + 81-76-264-6512; E-mail address: kazuet@kenroku.kanazawa-u.ac.jp (K.Tazaki)

were frequent and long, life would have had a better chance of evolving on Mars. Terrestrial hot spring pisoliths might be a counterpart of the blue berries on Mars.

Keywords: Biomineralization, pisoliths, hot springs, aragonite, microorganisms, laminar layers

1. INTRODUCTION

Life is ubiquitous at Earth-surface conditions when the necessary water and sufficient elements are present. Hot springs water has ideal conditions for microbiological activity across a wide range of temperatures. It establishes a new temperature maximum of 92 °C for an active biological nitrogen-fixing system, with biotechnological potential; it provides evidence for a new environment in deep-sea hot hydrothermal vent. The biogeochemical importance of nitrogen fixation in hot-vent communities must therefore be evaluated directly (Capone, 2006). Recently, the Mars Global Surveyor discovered present-day fluid flow in gully channels on slopes of two craters, which might be created by a water flowing down the slopes (Malin et al., 2006). This discovery lends support to the existence of liquid water so near the surface, at least in place, that it can spurt out on rare occasions. Water appears to have flowed down two gullies sometime during the past few years, even though liquid water can't long persist on the cold, nearly airless martian surface. And where there's liquid water, there could be life (Kerr, 2006).

Terrestrial hot springs may provide important models for life in past Mars environments because geothermal fluids mineralize a wide variety of natural microbial communities (Tazaki, 2003; Kerr, 2004; Tazaki et al., 2005, 2006). The abundant active volcanoes of Japan are associated with numerous hot springs, which produce a wide variety of bio-minerals (Tazaki, 1999, 2003). Terrestrial hydrothermal systems are of major importance to many environmental and earth science disciplines. They serve as models for understanding processes in the shallowest part of the geothermal systems that produce epithermal mineral deposits (Renaut et al., 2003) and they are a major setting for the study of extremophile microorganisms (5kbach).

Biological activities can produce various patterned structures, though the template mechanisms are still poorly understood (Chan et al., 2004). One of the most attractive structures formed in hot springs is pisoliths, small accretionary carbonate spheres which, as decribed below, are produced with laminar patterns through biomineralogical interaction with a hydrothermal environment. Therefore, if any are discovered on the Mars, they are unvaluable to study not only the life on the planet but also the hydrothermal environment. We here report characteristics of pisoliths sampled from a hot spring in Japan, believing that the information will contribute to the study of the life and related fields on the Mars.

2. MATERIALS AND METHODS

Materials

Spherules were collected from the base of a geyser from Uwanari-sengen, Arima Hot Springs, at Hyogo, Japan (Figure 1). The deep reddish brown spherules from a few cm to 50 mm diameter are composed of complex laminar patterns of carbonate lamina (Figs. 2, 3). The water of the Arima Hot Springs in Hyogo has the highest salt concentration of any hot spring in Japan, with more than twice the salt concentration of seawater. The hot spring water has a pH of 5.8, Eh (electrode potential versus the standard hydrogen electrode) 296 mV, EC (electrical conductivity) 6.07 mS/cm, DO (dissolved oxygen) 4.2 mg/l, temperature 98 °C. The water contains a high content of Na^{+} (10000 mg/kg), K^{+} (2080 mg/kg), Ca^{2+} (1920 mg/kg), Fe^{2+} (102 mg/kg), Cl^{-} (22000 mg/kg), SO_4^{2-} (19 mg/kg) and HCO_3^{-} (127 mg/kg). The HCO_3^{-} is trapped gas bubbles within spherules. The periodic eruption of the geyser associated with the hot spring displaces the pisoliths and encourages the formation of concentric laminations (Figure 3). In addition to the lamina, mineralized filaments cross the concentric layers, suggesting evidence for microbial influence in the construction of a template or framework for the lamina.

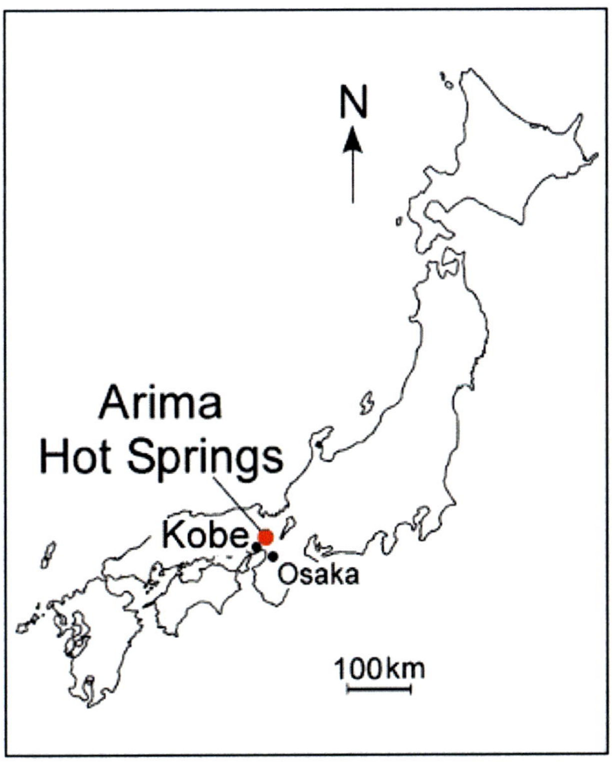

Figure 1. Location map of Arima Hot Springs in Hyogo, Japan.

Figure 2. Field view around geyser at Arima Hot Springs, Hyogo, Japan, shows dark brown spherules.

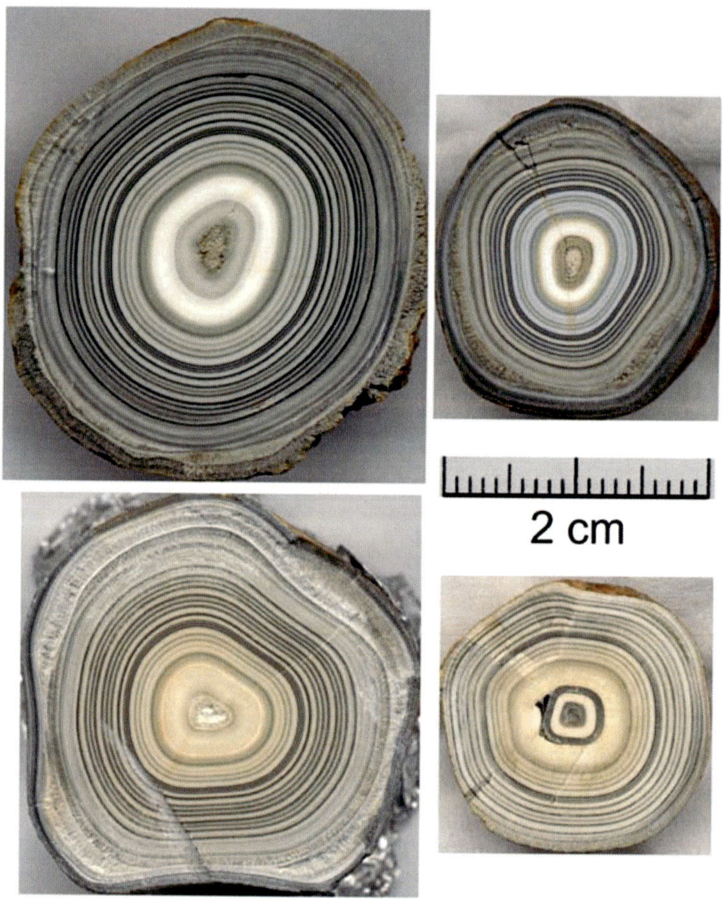

Figure 3. The concentric laminated structure in cross section of dark brown spherules showing few fissured projections. The irregular outer surface may represent an interval of exposure and erosion.

Methods

Chemical composition of hot spring water and the pisoliths were analyzed using the energy-dispersive X-ray fluorescence (ED-XRF) analyzer (JEOL JSX-3201), using Rh Kα, which operated at an accelerating voltage of 30 kV under a vacuum condition. The pisoliths were cut and polished for elemental content mapping. Analysis was carried out by an ED-XRF with equipment for X-ray probe scanning system using Rh Kα (JEOL ED-XRF; JSX-3600 - 50 kV acceleration voltage, 60 s/point analytical time, 19600 analyses). Microbial morphology and the chemistry were observed by scanning electron microscope (SEM; JEOL JSM-5200LV) and transmission electron microscope (TEM; JEOL JEM-2000EX). Additionally, nano-scale observations and elemental distribution analyses of hot spring pisoliths were performed with a transmission electron microscope (JEOL JEM-2010FEF, accelerating voltage 200kV), equipped with a scanning TEM image generator (EM-Z90471) and an energy dispersive X-ray spectrometer (EX-34025JGT). X-ray powder diffraction of pisoliths were done for both white layers and black layers to identify as aragonite, by Rigaku RINT 1200 with Cu Kα radiation.

3. Results

Chemical Analysis

The distribution of chemical components in the pisoliths was analyzed by X-ray fluorescence (ED-XRF). Although ED-XRF chemical analysis of bulk sample shows that the major component is CaO associated with Fe_2O_3, Cl, SiO_2, and K_2O, ED-XRF elemental content maps show the distribution of Fe with concentric striped pattern structure. The core is mainly composed of heavy metals such as Fe, Mn, Ba, and Zn. The lines crossing the lamina are filamentous microorganisms (Figure 4 arrowheads), which eventually form a concentric micron-scale template for the mineralization of a new lamination in the pisolith. EDTA etched sample clearly revealed presence of filamentous microorganisms (arrowheads), transverse cross section through calcified structure, showing thinly laminated nature of Ca cement (Figure 5). The filamentous microorganisms are typically several mm in length and 1μm in diameter, extending outward from the core to the rim of the spherules. SEM photomicrographs of EDTA etched surfaces show thinly laminated Ca-cements encasing microbes.

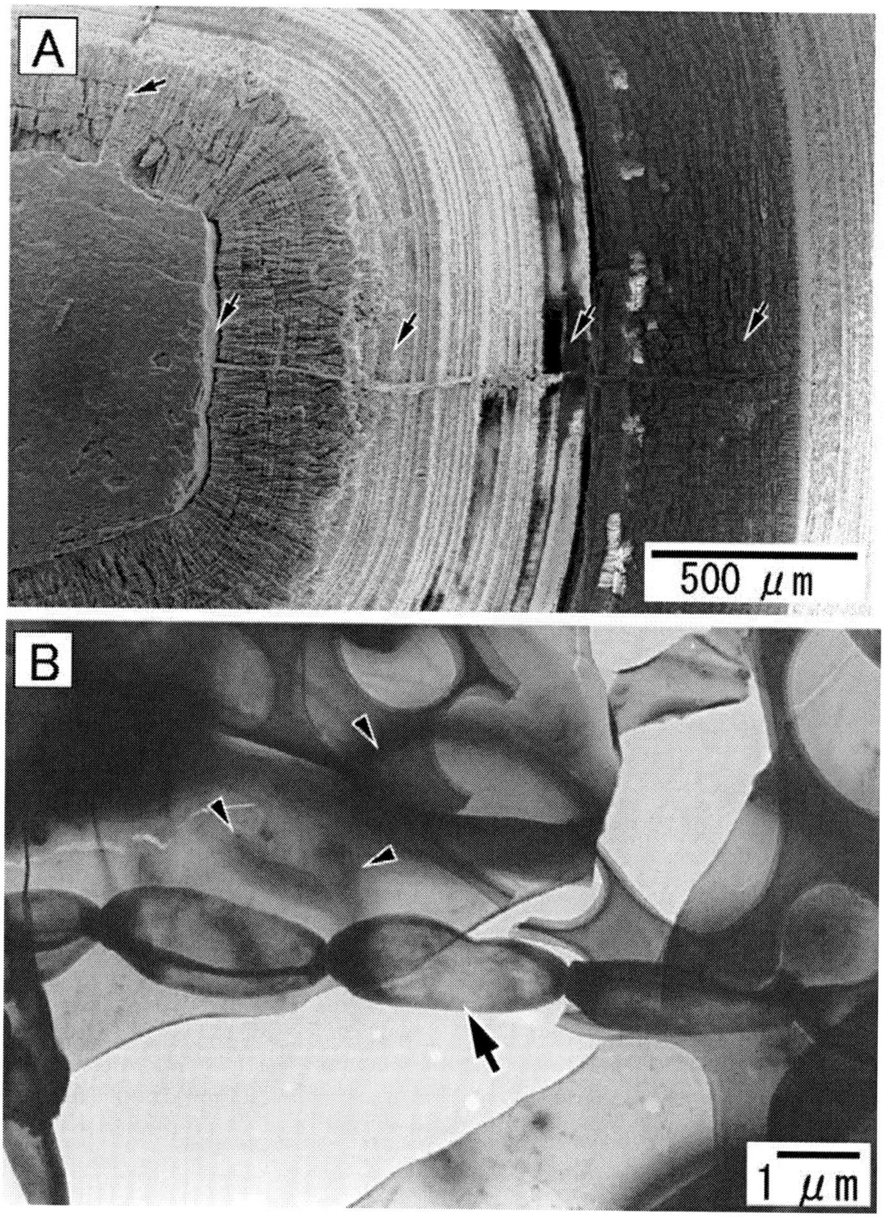

Figure 4. The filamentous microbes build the geometric framework along fissured projections. SEM micrograph of the cross-section of hot spring pisolith (A) showing laminated structure with filamentous microbes (arrowheads in B) by TEM after EDTA etching treatment (B).

Mineralogy and Micromorphology

One pisolith was etched and samples of each laminated layer were examined with scanning electron microscopy (SEM) equipped with EDX and analyzed by X-ray powder diffraction (XRD) for mineralogy, morphology, and elemental composition. The non-mineralized filamentous microorganisms contain large amounts of organic material, associated with Cl and Ca, whereas the mineralized bacteria contain high concentrations of Si and Fe at Fe-bearing layer, with trace amounts of Cl, K, Ca, and Mn. Most filamentous microorganisms were attached to both the Ca-rich layers and the Fe-bearing layers. Arrowheads indicate the location of analytical points of filamentous bacteria.

Transmission electron microscopy (TEM) revealed the micro-morphology of encrusted sheaths of coccus and bacillus type bacteria on micro grids. The mineralized bacteria appear to extend outward from microbial cells, with local precipitation on the surface of the microbial cells at Ca-rich layers. At an advanced stage, most coccus and bacillus type bacteria are cemented by carbonate and/or by amorphous iron hydroxide and SiO_2.

Elemental Maps

Scanning transmission electron microscopy (STEM) was used to carry out semi-quantitative analysis of the hot spring pisoliths after EDTA etching (Figure 5). One drop of the suspension was mounted on the micro grid for imaging. The elemental content mappings illustrate the distribution of the elements in the concentric circular structures. Each bacterial cell was found to contain K, Ca, Mn, and Fe; and the bacterial capsules are surrounded by a region of high Si and Fe, the bacterial cells are cemented together by aragonite, $CaCO_3$. TEM image shows bacterial exopolymers covered with very thin crystals of poorly ordered iron oxides on the left side of bacterial cell wall and within the extracellular exopolymers. Area analysis using ED indicates that the bacterial colony is associated with high levels of Na, Cl, K, P, Ca, and S. The results suggest that the filamentous microbes produce the extraordinary laminated geometric framework that leads to the complex layering in the pisolith. The bacteria associated with the laminations are morphologically all similar species; the only difference is in the cementing minerals bound onto the microbe cell capsule. The mineral composition of both the Ca-rich and Fe-bearing layers is well-ordered aragonite based on bulk XRD analysis. The Fe-bearing layer is identified as amorphous iron oxides by the diffuse ring pattern of both XRD and electron diffraction TEM. The results indicate that only STEM mappings make it possible to discriminate non-mineralized cells from mixtures of Mn-, Fe-, Ca-mineralized matter attached to the bacterial cell capsule. Based on these analytical results, these hot spring pisoliths include a significant portion of organic material derived from both filamentous microorganisms and aragonite-mineralized bacteria.

4. Discussion

Although the occurrence of biogenic minerals in natural environments is well documented, the exact formation mechanisms are still poorly understood. Insight into how biological organisms can produce complex inorganic structures has been provided by Chan et al., (2004), who studied bacteria from a flooded mine. They found that pseudo-single crystals of akaganeite (FeOOH) with incredibly high-aspect ratios had formed on polysaccharide templates extruded by bacteria. In this study, each bacterial cell was found to contain K, Ca, and Fe; and the bacterial capsules are surrounded by a region of high Si and Fe at the bacterial cell which are cemented together by aragonite, $CaCO_3$. The filamentous microbes produce the extraordinary laminated geometric framework that leads to the complex layering in the pisolith. The intricate pattern of filamentous microbes related to the cyclic formation of lamina with differing chemistry. The bacteria associated with the laminations are morphologically all similar species; the only difference is in the cementing minerals bound onto the microbe cell capsule. The amorphous iron (FeOOH) and/or $CaCO_3$ in pisoliths appear to direct the cementation around bacterial cells. The ripening of amorphous iron oxides and/or $CaCO_3$ to aragonite, which precipitated as outer coatings on the cell, has been observed (Figure 5). The bacterial exopolymers are covered with poorly ordered iron oxides. The very thin crystals of iron oxides surround the cell wall. These mineralized materials could also be synthesized in the hot springs by immersing polysaccharides on the cell wall. But the concentric layered structures appear to be controlled by microorganisms. Biogenic minerals are generally those formed in the presence of biological cells and structures outside cells.

A previous study (Lowe and Braunstein, 2003) has suggested the role of microbial and biofilm processes in developing the microstructure of high-temperature (> 73 ^{o}C) siliceous sinter deposited around hot springs and geysers at Yellowstone National Park. Siliceous pisoliths that are permanently sub-aqueous in Bead Geyser were uniformly spherical and coated with fine, hard, granular sinter. In thin section, the pisoliths were found to be composed of finely laminated, concentric, uniform layers of silica. Silica-coated microbial filaments of 0.5-2.0 μm thick and 3-50 μm long are occasionally visible in small void spaces within pisoliths. The characteristics of the laminations in that case while having some similarity to those in the spherules have been examined. In our study, the laminations of the hot spring pisoliths are formed sub-aqueous through the activity of microorganisms. The bacteria do profit from encrusting themselves with various minerals, such as siliceous pisoliths or carbonate pisoliths, to gain useful energy at high temperature. The results suggest that the spherules form in constantly moving waters that are suitable habitats for the microorganisms involved. The use of high spatial resolution tools of STEM (Figure 5) and powerful mineralogical analysis to characterize pisoliths containing cemented organic structures with concentric laminar layers of aragonite should improve our understanding of biomineralization mechanisms.

In addition, a clear demonstration that bacteria can template mineral crystallization is also crucial because it might lead to the development of new tools in the search for evidence of past life on Earth and other planets (Fortin, 2004). Recent results from Mars rovers and orbiters show that warm, wet conditions may have prevailed on the planet for long periods during its early histry. A recent study reports the first physical evidence that liquid water exists on Mars today (Malin et al., 2006). The water appears to have come from beneath the surface, perhaps from an outcropping aquifer normally sealed by ice (Kerr, 2006). If the eras of Earth-like conditions were frequent and long, life would have had a better chance of evolving. In Eagle Crater's outcrops, Opportunity found millimeter-size spherical grains (nicknamed blueberries) that may have precipitated out of iron- or salt-bearing water as it evaporated. Mars may have had an Earth-like climate for as much as a third of the planet's history. We believe that the blueberries are what geologists call concretions, grains that precipitate out of iron- or salt-bearing water as it evaporates. On Earth, some concretions grow to the size of marbles or ping-pong balls, suggesting possible variations in the duration of the watery environment or the rate of the water's evaporation (Bell, 2006).

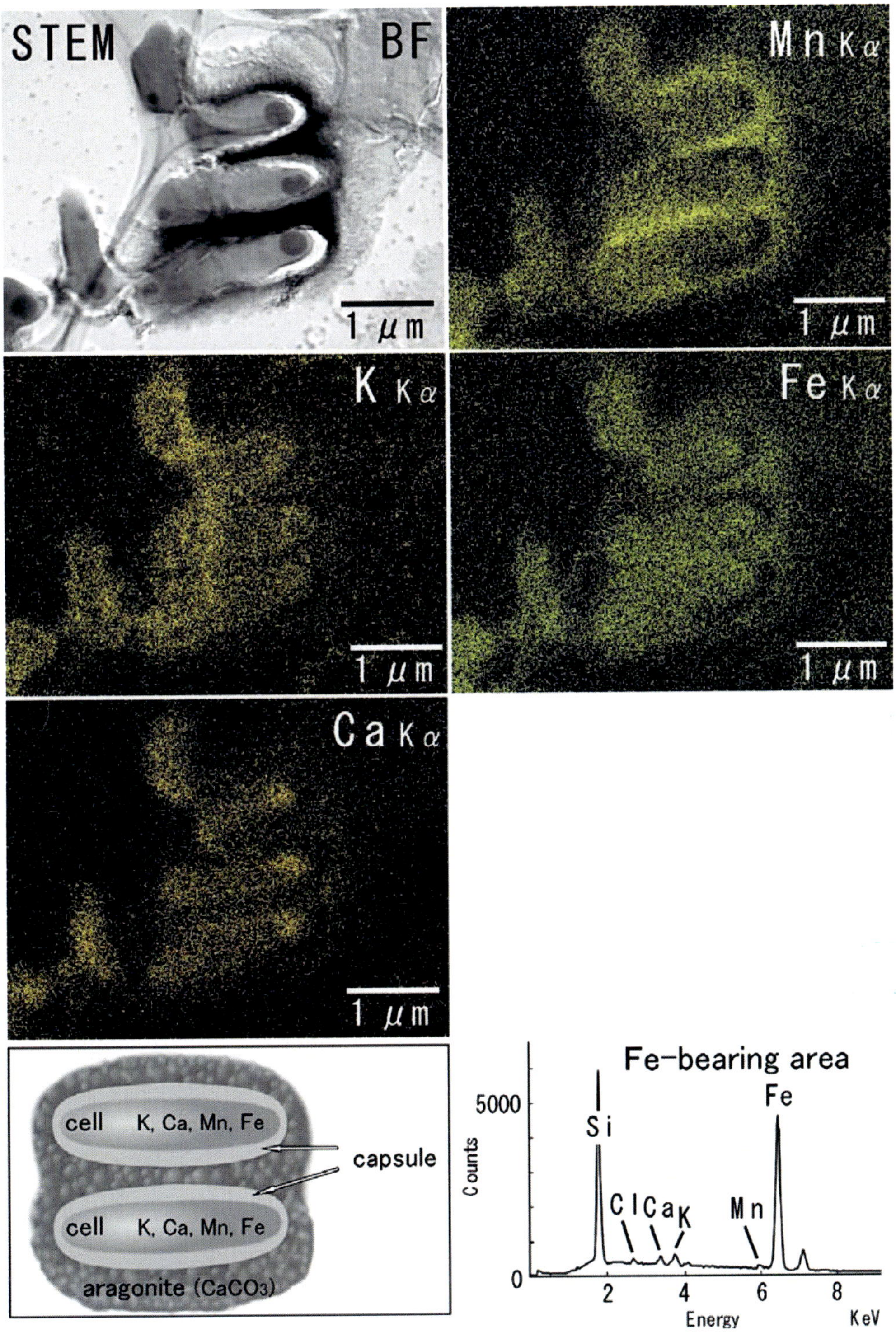

Figure 5. Zero loss images (BF) and the STEM-EDX elemental content maps of the pisolith. The bacterial area contains an abundance of, Ca, Mn, and Fe with light elements of C, N, and O, suggesting a bio-mineralization of pisolith. The Si and Fe elements are main components of Fe-rich bearing area in the pisolith.

The terrestrial hot spring pisoliths examined in this study could be a counterpart of the blueberries. Unfortunately, the internal structure of the Mars spherules could not be examined to see whether the spheres are solid, hollow, or layered. The morphology, size, and chemistry of hot spring pisoliths resemble the Martian spherules in many ways, although alternative hypotheses for their origin exist. A leading hypothesis holds that the beads were once-molten rock solidified in mid-air after an impact or volcanic eruption. Another hypothesis holds that the spherules are concretions that precipitated slowly around a central core (NASA, 2004). Mossbauer spectrometer data show the presence of an iron-bearing mineral jarosite, $KFe_3^{3+}(OH)_6(SO_4)_2$, which contains water in the form of hydroxyl as part of its structure. Soil X-ray spectra at Meridiani indicate the presence of abundant Si and Fe. These data suggest the presence of water and the availability of aqueous solution/precipitation processes on the surface or in the shallow sub-surface of Mars. This makes it likely that the spherules formed as the accretion of precipitated minerals in an environment with abundant water, suggesting that the spherules seen weathering out of the Martian sediment are concretions formed in place. The most plausible explanation would be water-soaked debris briefly gushing down a gully.

There appears to be up to 40% salt in the outcrop of the Arima hot spring area (Hyogo health environmental science center, 2002). This high concentration of salts, which would make the sample evaporated on Earth, also suggests the presence of sufficient water in the past. The Arima Hot Spring waters have very high salt contents, and an environment that produced microbially mediated pisoliths that are very similar to the Martian spherules in external morphology. Therefore, the complex interaction between microorganisms and variation in the hot springs environment may provide a model for the formation of the Martian spherules.

We would like to emphasize a probability that pisoliths identical to those reported in this paper will be found in hot spring on Mars. Hydrothermal activities have been present also on Mars, which are produced by two heat sources, volcanism and impact cratering (e.g., Brackenridge et al., 1985, Squyres et al., 1987). Hydrothermal deposits are a prime exobiological research target on Mars, since hot spring water has ideal conditions for microbiological activity, and early life on Mars might has evolved in hydrothermal settings. If life has been present and/or remains alive in hot springs, microorganisms should have built pisoliths as a part of the deposits. The formation of pisoliths can fossilize and preserve microbial life forms, life styles, and their chemical traces. It is fascinating to suppose that future explorations on Mars will discover pisoliths and retrieve information of the life on Mars.

4. Conclusions

Bio-mineralization of aragonite in pisoliths at Arima Hot Springs under high salt conditions can produce concentric laminae patterned structures. SEM and TEM electron microscope images show that microbes produced concentric laminar layers of aragonite alternating with Si- and Fe-rich layers. The filamentous microbes form the concentric framework for mineral laminae with a web-like network of microbial strands, suggesting the biological templating mechanisms. The complex interaction between microorganisms and biomineralization in the hot springs environment may provide a model for the formation of the Martian spherules.

ACKNOWLEDGMENTS

The authors thank Prof. Dr. David Dettman, and Loretta Li for reading and correcting the manuscript. We are also grateful for the cooperation and assistance of all students of the Tazaki laboratory. This study was funded by a grant from the Japanese Ministry of Education, Culture, Science, and Technology to Dr. Kazue Tazaki.

REFERENCES

Bell, J. (2006) Red planet's watery past. *Scientific American*, 50, December, 40-47.

Brackenridge, G., Newsom, H., and Baker, V. (1985) Ancient hot springs on Mars: origins and paleoenvironmental significance of small Martian valleys, *Geology*, 13, 859-862.

Capone, D.G. (2006) Ramping up the heat on nitrogenase. *Science*, 314, 1691-1693.

Chan, C.S., Stasio, G. De., and Welch, S.A. et al., (2004). Microbial polysaccharides template assembly of nanocrystal fibers. *Science*, 303, 1656-1658.

Fortin, D. (2004). What biogenic minerals tell us. *Science*, 303, 1618-1619.

Hyogo health environmental science center (2002). *Analytical data of Kinsen hot spring water.*

Kerr, R.A. (2004). Third success reveals a geologic delight. *Science*, 303, 602-603.

Kerr, R.A. (2006). Mars Orbiter's Swan song: The red planet is A-changin'. *Science*, 314. 1528-1529.

Lowe, D.R., and Braunstein, D. (2003). Microstructure of high-temperature (>73 °C) siliceous sinter deposited around hot springs and geysers, Yellowstone National Park: the role of biological and abiological processes in sedimetation. *Canadian Jour. of Earth Sciences*, 40, 1611-1642.

Malin, M., Edgett, K., Posiolova, L., McColley, S., and Noe Dobra, E. (2006) Present-day impact cratering rate and contemporary gully activity on Mars, *Science*, 314,1573-1577.

Renaut, R.W., Jones, B., and Rosen, M.R. (1996). Primary silica oncoids from Orakeikorako hot springs, North Island, New Zealand. *Palaios*, 11, 446-458.

Squyres, S., Wilhelms, D., and Moosman, A. (1987) Large-scale volcano-ground ice interactions on Mars, *Icarus*, 70, 385-408.

Tazaki, K. (1999). Architecture of biomats reveals history of geo-, aqua-, and bio-systems. *Episodes*, 22, 21-25.

Tazaki, K. (2003). $FeAs_2$ biomineralization on encrusted bacteria in hot springs: an ecological role of symbiotic bacteria. *Canadian Journal of Earth Sciences*, 40, 1725-1738.

Tazaki, K., Okuno, M., and Fyfe W.S. (2005) Proof for Bio-sedimentation in Carajas's BIF. *PEDOLOGIST* 49, 99-108.

Tazaki, K., Okuno, M., and Furumoto, M., and Watanabe, H. (2006) Biomineralization of pisoliths in hot springs. *Materials Science & Engineering* C, 26, 617-623.

In: Planet Mars Research Focus
Editor: L. A. Costas, pp. 117-164
ISBN: 978-1-60021-826-2

Chapter 5

THE SEARCH FOR LIFE ON MARS: RAMAN SPECTROSCOPY AND ANALYTICAL ASTROBIOLOGY

Howell G. M. Edwards* and Michael D. Hargreaves
Centre for Astrobiology and Extremophiles Research, Raman Spectroscopy Group, Division of Chemical and Forensic Sciences, School of Life Sciences, University of Bradford, UK

ABSTRACT

The analytical application of Raman spectroscopy to the identification of the survival strategies of terrestrial extremophiles through the characterisation of key biomolecular and biological signatures has provided novel information about the roles of protective chemicals synthesised by the organisms under environmentally stressed conditions. The non-destructive sampling requirement and the ability to discriminate between organic biomolecular spectral signatures and those from the inorganic host matrix in the geological record are essential requirements for the assessment of the survival strategies being employed against low wavelength ultraviolet radiation insolation, desiccation and extremes of temperatures found in terrestrial habitats termed Mars analogue sites (Table 1).

The recent announcement of the inclusion of a miniaturised Raman spectrometer as part of a combined Raman /LIBS instrument for the molecular and elemental analytical characterisation of specimens in the Martian surface and subsurface as a component of the European Space Agency ExoMars life-detection suite has focussed attention on the Raman spectra of key biomarkers of life signatures. To search for indicators of extinct or extant life signatures on Mars serves to highlight the importance of Raman spectroscopic techniques in this astrobiological mission. As a forerunner to the ExoMars mission it is imperative that the Raman spectroscopic data from extremophilic colonies from as wide a possible range of terrestrial sites of relevance to Mars is undertaken. This chapter will address the type of information that is required and which is accessible from these

* Tel: 00 44 (0)1274 233787; Fax: 00 44 (0)1274 235350; *E-Mail: h.g.m.edwards@bradford.ac.uk

terrestrial studies to adequately prepare the way for the interpretation of analytical scientific data that will emerge from the Mars exploration programme, commencing with the launch of ExoMars in 2013.

Table 1. Antartic Mars analogue cold desert sites.

State of liquid water	Antarctic site	Location (latitude and longitude)	Extra-terrestrial analogue	Mars habitat* Gya)	Putative conditions during existence Of Martian habitat	Reference
Aquatic cyanobacterial communities (saline and brackish)	Mars Oasis	71° 52'S, 68° 15'E	Early Mars	4.2-3.8	P > 5 atm T > 0°C CO2, volcanoes	Horneck (1995) Carr (1996)
Flowing surface water	Canada Stream	77° 37S, 163° 03E	Early Mars		Meteorites	McKnight *et al* (1993)
High UVB exposure	Mars Oasis (Ozone Hole)	71° 52'S, 68° 15'E	Earth orbit (ISS)			
Ice-covered lakes	Microbial (cyanobacteria) mats in Dry Valley lakes (Lake Hoare) Photosynthesis at low light. UV-tolerance	77° 38S, 161° 51E	Early Mars· Palaeolakes Mars: Gusev Crater (Map: USGS MC-23 SE I-2119)	3.8-3.1	T < 0°C (-20° C) Peak T > 0°C Carbonate Dissolved gases	Wharton *et al.* (1995) Grin and Cabrol (1997) Wharton (1994)
Liquid water in porous rocks Or aquifers	Endolithic microbes, inside Beacon sandstone at Battleship Promontory (cyanobacteria)	76° 55S, 160° 55E	Porous Martian lithic niches	3.1-1.5	Peak T < 0°C P c.100 mb	Cabrol and Grin (1995) Wynn-Williams et al. (1999)
No liquid water on surface.	Victoria Valley sand dunes (negligible microbes) Horseshoe Mountain (no endolithic microbes)	77° 20S, 161° 45E 77° 34S, 159° 57E		1.5 - present	P ~ 7.4 mb Low CO2 Vacuum	Cameron (1972) Friedmann *et al.* (1994)
	Palaeolake beds (L. Fryxell)	77° 37'S, 163° 09'E	Martian palaeolakes			Doran *et al.* (1998) Wharton *et al* (1995)
Permafrost.	Taylor Valley	77° 44S, 162° 10E				Vorobyova *et al.* (1997) Broady (1993)
Geothermal water?	Mount Melbourne (thermophilic microbes)	74° 20S, 144 40E				Farmer (1996) Shock (1997)

*nafter McKay et al. (1992)

INTRODUCTION

Raman spectroscopy is an analytical technique which provides information about the molecular structures and chemical environments of organic and inorganic molecules and molecular ions [1]. The generation of Raman spectra requires the illumination of target specimens by laser radiation, which can be focussed or non-focussed, and for which the "footprint" can range from about one micron to several hundred microns in cross-sectional diameter depending upon the optical design of the sample illuminator. Raman spectra are observed as two sets of wavelength shifted bands from the Rayleigh line, which occurs at the frequency of the incident laser radiation, the long wavelength set of Raman bands comprise the Stokes Raman spectrum and the low wavelength set that of the anti-Stokes Raman spectrum, which is relatively weaker in intensity compared with the Stokes Raman spectrum [2]. Raman spectra are generally measured as wavenumbers (cm^{-1}) which are shifted from the laser excitation wavenumber, namely that of the Rayleigh line. Since Raman spectra are significantly weaker in intensity compared with the Rayleigh line intensity, it is important that the irradiance of the incident laser being used for the observation of the Raman spectra is high, consistent with the requirement that sample damage does not occur. Because the Raman effect has an inverse dependence upon the fourth power of the laser wavelength, much Raman spectroscopy has been carried out in the visible region of the electromagnetic spectrum using blue and green wavelengths. However, the advent of non-visible wavelength diode lasers operating in the ultraviolet and the near-infrared regions has provided some novel sources of excitation of Raman spectra; currently, there are reports of Raman spectra being excited at wavelengths of 240 nm in the deep ultraviolet and at 1330 nm in the near infrared region and with all other instrumental factors being equal, power for power, the Raman spectrum excited in the ultraviolet will be some 950 times more intense than that recorded in the near infrared region over this wavelength range.

Despite the advantage of greater intrinsic Raman scattered intensity at low excitation wavelengths, there are several good reasons for the adoption of Raman laser excitation at longer wavelengths, as will emerge from the discussion that follows, not the least of which is the sensitivity of the materials under study to the high laser energies in the ultraviolet and the onset of fluorescence associated with the presence of low energy molecular electronic states that are more likely to be probed using exciting radiation at low wavelengths. Fluorescence emission is several orders of magnitude greater than the corresponding Raman scattering generated by the same laser wavelength and relatively intense, broad-band fluorescence backgrounds can generally swamp the much weaker Raman bands from the same sample. An example of this is provided in figure 1; here, the Raman spectra of a *Caloplaca saxicola* epilithic lichen from Crater Cirque, Northern Victoria Land, Antarctica, are shown. The details of the sample location and habitat are provided in Table 2.

Table 2. Sample location and habitat

	Acarospora chlorophana	*Caloplaca saxicola*	*Nostoc commune*	Cryptoendolithic lichen	*Chroococcidiopsis sp.*
Location	Football Saddle, Northern Victoria Land	Crater Cirque, Northern Victoria Land	Mars Oasis, Alexander Island	East Beacon, McMurdo Dry Valleys	Balham Valley / Barwick Valley
Latitude	72° 31'		71° 53' S	77° 50' S	77° 20' S
Longitude	169° 46'		68° 14' W	160° 53' E	161° 05' E
Altitude (m asl)	c. 800	c. 800	c.50	2200	
Distance to coast (km)	c. 10	c. 10			
Aspect, facing	S	N		N	N
Substratum, exposure and climatic factors	Pitted substratum comprising alkalibasalt, trachyandesite and phonolite. Sheltered micro-habitats on southern slope of saddle	McMurdo volcanics comprising alkalibasalt, trachyandesite and phonolite. Sheltered north-east facing slope	Aquatic and loosely settled on silt. Sheltered to south and west, within zone of enhanced UVB levels during ozone depletion	Sedimentary orthoquartzite sandstone. Exposed ridge, iron – stained outcrop	Sedimentary orthoquartzite sandstone. Sheltered valley, little iron staining
Moisture (range)	Not known. Rock surface	Not known. Rock surface	Fringes of melt pond	0.13-1.12 % dw	~1%
Habitat	Crustose	Epilithic, crustose	Aquatic and littoral	Endolithic	Endolithic

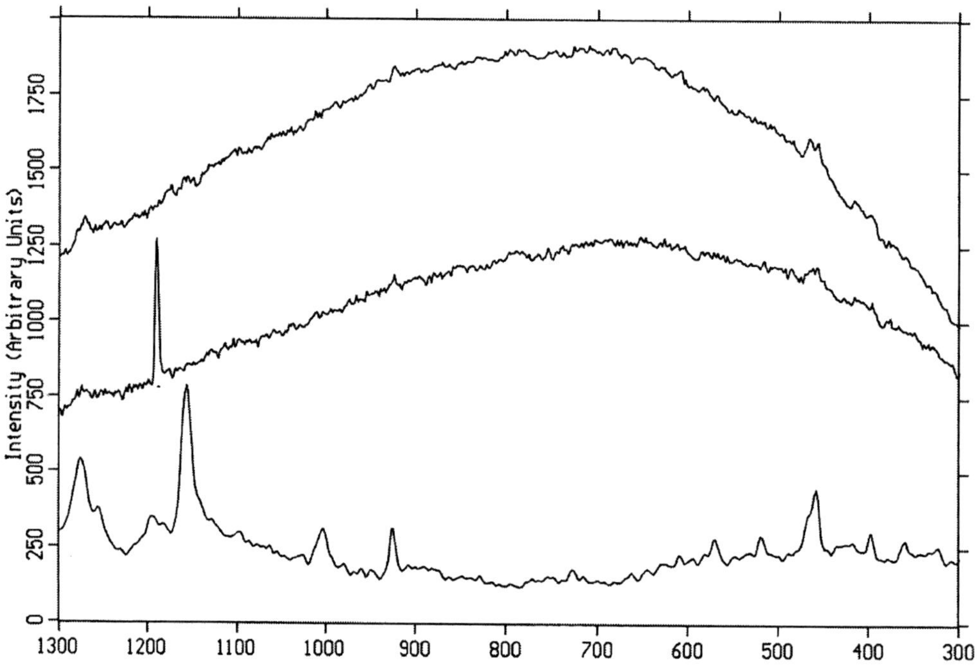

Figure 1. Raman spectra of *Caloplaca saxicola,* an Antarctic extremophilic epilithic lichen, recorded with different laser excitation wavelengths; from top, 633 nm, 785 nm and 1064 nm. The swamping of the weaker Raman bands by the fluorescence emission background is noted for the first two wavelengths of excitation

The upper and middle spectra in Figure 1 have been excited using 633 and 785 nm radiation, respectively and in both cases the broad-band fluorescence emission dominates the spectrum over the wavenumber range 300-1300 cm $^{-1}$. In contrast, the spectrum excited using 1064 nm radiation in the near-infrared region shows Raman bands which can be used as biomolecular diagnostics of the protective biochemicals that are being produced by the extremophile in response to the stressed environmental conditions prevailing in the Antarctic site (Table 3). Hence, the selection of the optimum instrumental requirements and experimental conditions for the generation of Raman spectra is not a trivial task and is compounded even further by the additional constraints that are now being exercised by the proposed adoption of miniaturised Raman spectrometers for incorporation into robotic instrumentation suites for planetary landers and rovers [3]. The purpose of this chapter, therefore, is to consider the major factors which affect the observation of Raman spectra and the application of Raman spectroscopy to the characterisation of molecular entities which have relevance to the remote exploration for signs of relict or extant life in extraterrestrial scenarios.

Table 3. Raman band with assignments for protective bio-elements associated with quartz in *Acarospora chlorophana*

Excitation wavelength (nm)			Approximate assignment
1064	852 mini	830	
3069 w			ν(C=CH) aromatic
3036 w			ν(CH) aromatic
2918 s, br			ν(CH_3) asymmetric
1672 s		1672 w	ν(C=O) conjugated, free; parietin
1631 w, sh			ν(C=O) conjugated, H-bonded; parietin
1612 ms			ν(C=C) aromatic; parietin
1553 s		1554 mw	ν(C=C) aromatic; parietin
1520 vs			ν(C=C) β-carotene
1477 w, sh			Ring stretch coupled with (OH); parietin
1450 m			Ring stretch; parietin
1380 ms			Ring stretch, in plane / ν(C-O) phenyl
1370 ms			Ring stretch, in plane / ν(C-O) phenyl
1322 w			Ring stretch, in plane; parietin
1276 vs		1277 ms	Ring stretch, in plane; parietin
1254 ms, sh		1256 m	ν(C-O) aromatic ether; parietin
1198 mw			Ring stretch, C-C chelate; parietin
1180 w			Ring stretch, C-C chelate; parietin
	1161 mw, br		α-Quartz
1156 s			ν(CC) β-carotene
1137 w, sh			δ(C-CH), in plane; parietin
1105 w			δ(C-CH), in plane; parietin
	1080 vw		α-Quartz
	1062 vw		α-Quartz
1004 mw			ν(CC) ring breathing; ar. / β-carotene
926 s		927 ms	δ(C-H) out of plane; parietin
	794 mw, br		α-Quartz
	693 mw		α-Quartz
631 mw		633 m	Skeletal deformation; parietin
611 mw		611 m	Skeletal deformation; parietin
572 m		572 ms	Skeletal breathing; parietin
519 m			?; parietin
465 w, sh	470 vs*	467 m, sh	?; parietin / *α-Quartz
458 s		459 ms	δ(C=O), in plane; parietin
416 mw		423 m	?; parietin
397 m	405 mw*	400 m	Skeletal deformation; parietin / *α-Quartz
	367 w		α-Quartz
241 w		240 m, br	?; parietin

Signal strength: vs = very strong; s= strong; mw = medium; w = weak; vw = very weak; br = broad band; sharp

FACTORS AFFECTING RAMAN SPECTRAL INTENSITY

Raman spectra are generally weak and special techniques are required for their generation and observation but the information which can be derived from Raman spectra amply repays the special attention given to their study. Raman spectra will routinely cover the wavenumber shift region from about 50 to 3500 cm $^{-1}$, which will contain all the important vibrational spectroscopic information relating to organic biomolecules and their inorganic counterparts. An example of the Raman spectrum obtained from the extremophilic endolithic cyanobacterial colonisation of Beacon sandstone (Table 1) by *Chroococcidiopsis* sp. shown in Figure 2 demonstrates this aspect very clearly. Generally, the CH stretching modes of organic aliphatic and aromatic hydrocarbon derivatives will occur [4] in the wavenumber region 2800-3150 cm^{-1}, amide stretching bands of proteins near 1650 cm-^{1}, the C=C modes of unsaturated lipid chains and carotenoids between 1500-1640 cm^{-1}, quadrant ring stretching bands of aromatic polycyclic hydrocarbons near 1600 cm^{-1}, and various characteristic modes belonging to C-C, C-O, C-N, C-P and C-S moieties at lower wavenumbers down to 500 cm^{-1}, including the uniquely observable S-S modes of cysteine residues in keratotic molecules in the region 460-520 cm $^{-1}$. At the same time the vibrational modes of molecular inorganic materials and ions such as the Fe-O in haematite, P-O in hydroxyapatite, S-O in sulfate, N-O in nitrate, Si-O in quartz and silicates such as pyroxene and olivine, C-O in carbonates, Ti-O in rutile and anatase and Fe-S in pyrites can all be characterised and identified [5] in the wavenumber range 100-1100 cm^{-1}. In the particular example of *Chroococcidiopsis* sp. shown here in Figure 2, the data given in Table 4 illustrate the information that is contained in the Raman spectrum and the vibrational spectroscopic assignments that can be made of the Raman bands to the biological and geological features seen in this spectrum.

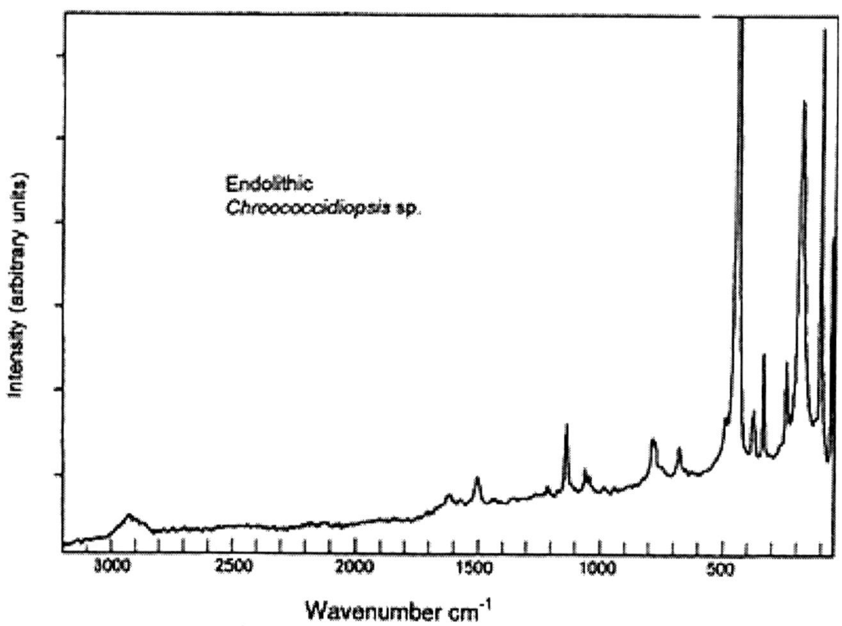

Figure 2. FT-Raman spectrum with 1064 nm excitation of *Chroococcidiopsis* cyanobacterial cryptoendolithic colony in sandstone from East Beacon, Antarctica. The full wavenumber range covers the organic Bands and those of the host geological matrix, the strongest feature being that of quartz at 463 cm $^{-1}$.

Table 4. Raman spectra band assignments for cryptoendolithic colonies in Beacon sandstone

Excitation wavelength (nm)				Approximate assignment
1064	852 port	830 substratum	514.5	
2935 m, br				ν(CH) aliphatic
1633 w, br				
1585 vw				
1521 w, br			1518 s, br	ν(C=C) β-carotene
1450 vw, br				
1230 vw				α-quartz
1159 m		1160 mw		α-quartz / ν(C-C) β-carotene
			1155 ms, br	ν(C-C) β-carotene
1081 mw, d				α-quartz
1064 mw, d		1066 w, br		α-quartz
1005 w			1006 mw	ν(C-C) aromatic
805 mw, d				α-quartz
795 mw, d		795 mw, br		substratum
695 mw		695 w		α-quartz
679 w, sh				
464 vs	467 vs	465 vs		α-quartz
400 mw, d	400 ms, br	399 mw, d		α-quartz
393 mw, d		394 mw, d		α-quartz
354 m		355 mw		α-quartz
		320 w		
263 m		264 m		α-quartz
205 ms				α-quartz

Signal strength: vs = very strong; s= strong; mw = medium; w = weak; vw = very weak; br = broad band; sh = sharp

The occurrence of both the organic and inorganic signatures in the same Raman spectrum gives a valuable and powerful analytical advantage over many other analytical techniques since now not only the presence of organics in a geological environment can be assessed but the interaction between the organics and the inorganic matrix can be examined without the necessity of extraction or separation of the phases being undertaken. Several examples of case studies will be given below to illustrate this application of Raman spectroscopy to the detection of organic molecules of relevance to life in terrestrial geological environments, which clearly have a fundamental application as a test-bed for future planned Raman analytical experiments on Mars. Other useful factors which can advocate the adoption of Raman spectroscopy as an analytical technique for potential life detection experiments in stressed terrestrial environments relate to the low Raman scattering cross-section for water and hydroxyl groups which immediately has importance for the detection of organics near deep-sea hydrothermal vents, the use of a Raman probe for molecular detection in the subsurface of glaciated lakes, the detection of life signatures in permafrost, the analysis of lacustrine sediments and for the detection of micrometeorites in Arctic and Antarctic ice-cores. In none of these examples would there be a need for specimen desiccation to record the Raman spectra, which would realistically be extremely difficult to achieve for several of the scenarios mentioned here.

There follows a consideration of the instrumental features which could dictate the selection of a particular Raman spectroscopic system, then we shall consider the use of several Raman systems in case studies to illustrate the operation of these principles in practice for the detection of biomaterials and biomolecules in real-world samples.

Wavelength of Excitation

Because of the dependence of the Raman scattering upon the fourth power of the wavenumber of the laser excitation (and inverse proportionality to the fourth power of the wavelength) the recording of Raman spectra hitherto at low wavelengths in the visible region has been strongly favoured, especially taking into account the sensitivity of radiation detectors in the visible and low wavelength regions of the electromagnetic spectrum. However, the generation of fluorescence emission, which is several orders of magnitude greater than the Raman scattering, from the population of excited vibronic states lying close in energy to the ground electronic state by high energy radiation at low wavelengths can compromise the observation of Raman spectra, as can be seen in Figure 1, for example. The onset of fluorescence emission is minimised by excitation at longer wavelengths, despite the inherently weaker Raman effect at these wavelengths, as shown in Figure 1 with the excitation effected at 1064 nm, and the manufacture of diode lasers which operate in the near infrared has created a new opportunity for the observation of quality Raman spectra which are essentially fluorescence – free from long wavelength excitation sources. This is especially useful for the recording of Raman spectra from organic biomolecules which exhibit fluorescence with short wavelength excitation, for example, porphyrins and polycyclic aromatic hydrocarbons. Another method which has been suggested for the combating of fluorescence emission in Raman spectra excited at low wavelengths is the probing of the anti-Stokes Raman spectral region, which occurs at higher energies (lower wavelength shift) than the laser line unlike the fluorescence which is a broad-band emission shifted towards longer wavelengths. However, since the anti-Stokes Raman bands are normally much weaker in intensity than their Stokes counterparts at ambient temperatures because of their dependence on the Maxwell-Boltzmann population of vibrational energy levels [2], the anti-Stokes Raman band intensities rapidly diminish with wavenumber shift from the exciting line, unless the thermal population of the excited molecular vibrational levels is increased by heating the system. Clearly, such an application, which does find practical favour in the spectroscopic determination of species temperatures in heated gases and plasmas, is not a realistic proposition for thermally cool planetary surfaces where most of the molecules will be situated in lower energy levels in the ground electronic state.

Nevertheless, the selection of a suitable laser wavelength for Raman excitation is not trivial and involves a careful consideration of the scattered Raman intensity and data acquisition over a range of selected materials; in this respect, some recent experiments have been described which inform the adoption of Raman instrumentation that will be suitable for a robotic lander on planetary surfaces.

An illustration of the effect that different wavelengths of excitation have upon the recorded Raman spectra is provided in Figures 3 and 4, which have been obtained from cyanobacterial colonisation of Antarctic rock substrates (Table 2), namely, Beacon sandstone endolith and *Chroococcidiopsis* sp., respectively. Figure 3 clearly demonstrates the inablility of the visible wavelengths of excitation to confirm the presence of geological or biological components of the specimen, whereas the Raman spectrum excited at 1064 nm provides diagnostic information about both. Figure 4 shows a similar effect for the 514.5 nm and 1064 nm excitation wavelengths, but the presence of three weak Raman features can now be seen superimposed upon the more intense background fluorescence emission at 514.5 nm; these are three characteristic Raman bands for beta –carotene, observed at 1515, 1157 and 1007 cm^{-1}, which are excited preferentially by the incident green radiation through the process of resonance Raman scattering [2], in which the intensity of certain vibrational modes involved in an electronic absorption near the excitation wavelength is enhanced by several orders of magnitude over that of normal Raman scattering.

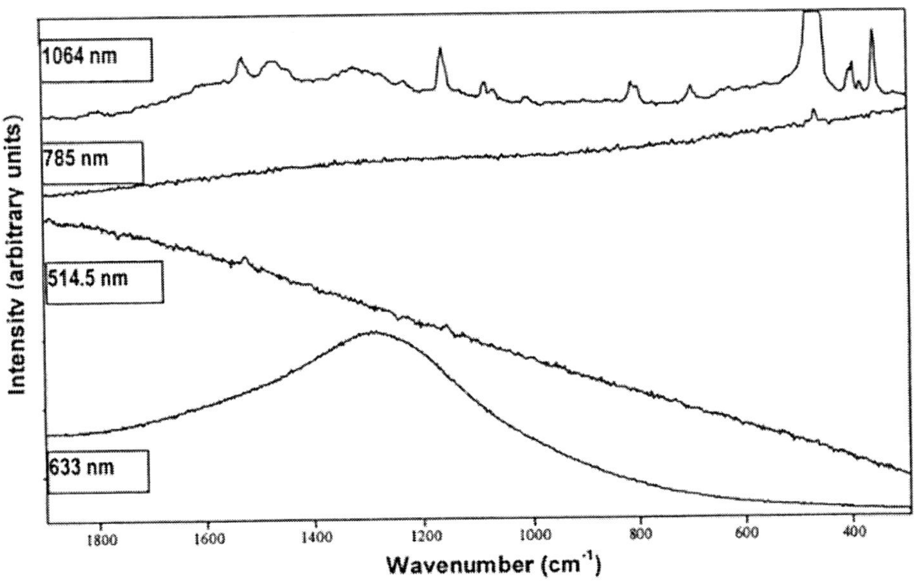

Figure 3. Raman spectral stack-plot of endolithic green algal layer in Beacon sandstone from Mars Oasis, Antarctica, recorded using laser excitation wavelengths of 514.5 nm, 633 nm, 785 nm and 1064 nm. Fluorescence emission prevents the Raman spectra of the biochemicals and of the geological matrix being shown, except for the near infrared excitation at 1064 nm.

It is also appropriate to consider the possibility of laser damage to the specimen under investigation; a good example of this is provided in Figure 5, where the excitation of a specimen of lead (II) lead (IV) tetroxide, Pb_3O_4, is shown with *a* green (514.5 nm) and *b* red (633 nm)excitation wavelengths. The spectra are very different in appearance; the genuine Raman spectrum is that of *b,* whereas that of *a* which has been excited with the green wavelength radiation is actually that of massicot, PbO, caused by the photochemical degradation of the Pb_3O_4 specimen by the high energy laser excitation in the green. The effects upon biological species of high energy laser radiation can be even more profound and the D and G bands of amorphous carbon at 1320 and 1590 cm^{-1} are seen as evidence that extensive photolytic damage has been caused to an organic biochemical through laser irradiation at high energies. This effect is minimised using low energy radiation for the excitation of Raman spectra and must be a critical point of consideration for biological studies when evaluating laser wavelength options.

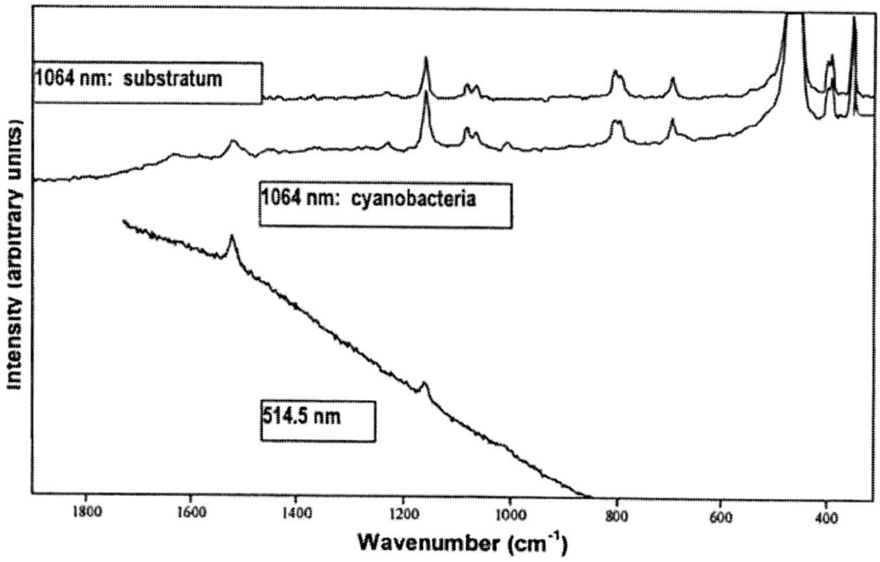

Figure 4. Raman spectral stack-plot of *Chroococcidiopsis* from Barwick Valley, Antarctic Dry Valleys, where the excitation at 514.5 nm now shows the presence of three weak Raman bands from beta-carotene at 1515, 1157 and 1007 cm^{-1} superimposed on the fluorescence emission background, compared with the essentially fluorescence –free Raman spectra obtained with 1064 nm excitation.

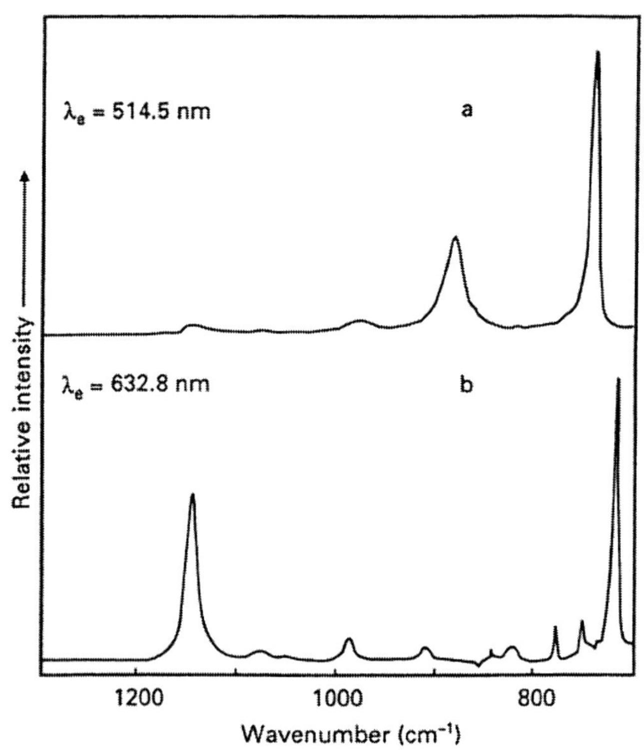

Figure 5. The effect of wavelength of laser excitation upon the observed Raman spectrum of lead(II) lead (IV) tetroxide, Pb_3O_4, excited with *a* 514.5 nm and *b*633 nm radiation. The genuine Raman spectrum is, whereas that of *a* is that of massicot, PbO, formed by degradation of the specimen.

Comparison of Dispersive-Raman and FT-Raman Spectroscopies

The point has already been made in this article that for Raman spectroscopic studies of biological materials the adoption of a system with a near-infrared excitation wavelength, for example using a Nd^{3+}/YAG laser operating at 1064 nm, obviates the problems associated with fluorescence emission that bedevils Raman spectra recorded in the blue and green regions of the electyromagnetic spectrum. A disadvantage of long wavelength excitation of Raman spectra is the fourth power dependence of the scattered Raman intensity on the wavenumber of the exciting radiation, i.e. Raman spectra excited in the ultraviolet will be significantly stronger than those excited in the near infrared. Despite this disadvantage, however, the advent of FT-Raman spectrometers operating in the near infrared at 1064 nm, which combined long wavelength laser excitation with the spectroscopic precision of interferometric measurements and Fourier transforms, provided a renaissance for Raman spectroscopic studies of sensitive biological materials which were not accessible to the technique hitherto. The development of longer wavelength diode lasers and charge-coupled detector (CCD) systems have since given additional performance attributes to conventional dispersive Raman spectroscopy, which when coupled with microscope capabilities have revolutionised Raman spectroscopy through a rapidity of data acquisition from specimen footprints of one micron or so. This has led to two-dimensional and three –dimensional Raman mapping experiments for the location of materials on heterogeneous surfaces and in transparent specimens. Whereas a detailed comparison between conventional dispersive CCD and FT-Raman techniques is often complicated by the correlation of instrumental factors with specimen requirements, the major features are as follows:

The operation of a Raman spectrometer at 1064 nm confers a reduced Raman scattering intensity over equipment that offers excitation in the visible or in the red /near infrared regions, e.g. the 633 or 758 nm wavelengths which are commonly used in CCD dispersive instruments. There is no doubt, however, that even with 633 or 785 nm excitation, some fluorescence or degradation problems in Raman spectroscopy can still occur. The major drawback to the use of CCD techniques further into the infrared is the development of stable diode lasers and effective detectors with usable Raman wavenumber shift ranges.

Despite the slower speed of data accumulation using FT-Raman instrumentation, where it is quite common to acquire data from 2000 scans, each taking one second –giving a total spectral acquisition time of about one hour, compared with a ten-spectral accumulation acquisition from a CCD-dispersive instrument, each requiring some ten seconds to achieve – giving a total spectral acquisition time of only two minutes, other experimental factors can dictate the use of the FT-Raman instrument.

In a scanning system as used by conventional CCD-spectrometers, only a specific preset wavenumber range is scanned at one time, unlike the Fellget advantage of the interferometer which means that all spectral data elements are acquired simultaneously. Hence, in the FT-Raman system, all Raman spectral data from 3500-100 cm^{-1} shift are acquired in one spectral scan, which may only be taken with say 200-2000 cm^{-1} for the CCD instrument. This means that careful consideration needs to be given in the latter case as to which parts of the Raman spectrum are to be accessed, since possible key vibrational indicators which lie outside the preset wavenumber range will not be recorded, whereas this does not apply to the FT-instrument that records the whole wavenumber range, and hence acquires all the vibrational information, simultaneously.

The Jacquinot throughput of the interferometric system is significantly greater than the CCD slit; in the former; the J stop is typically about 10 mm diameter in an interferometer compared with a slit of 100 microns in a dispersive system –in both cases, however, the effective achievable spectral resolutions are similar, being about 2 – 4 cm^{-1}.

The FT-system constantly provides an internal calibration of the data against a helium-neon laser source whereas the Raman wavenumber signals for the visible CCD-system needs external monitoring of calibration on a regular basis for spectral data correction—this is of importance for the automatic data recognition of spectral features that will be a prerequisite for remote robotic instrumentation on planetary surfaces.

A major advantage generally of Raman spectroscopy is its low sensitivity towards water and hydroxyl groups; the Raman spectrum of water is weak and much use has been made of this for the study of solutes in aqueous solutions and materials embedded in ice, for biomolecular and biopolymeric studies and for the examination of archaeological artefacts from waterlogged sites which cannot be desiccated. However, the moving of the laser excitation wavelength into the near infrared can severely restrict the Raman data that can be acquired in the presence of water because of absorption by the overtone and combination bands of water in the near infrared. Even at 1064 nm the effect of water-based absorptions upon the higher Raman wavenumber shifts can be marked and for this reason the substitution of heavy water in the form of deuterium oxide is sometimes advocated for the improvement of long-wavelength excited Raman spectra. Clearly, this is not possible for ex-laboratory and field applications but the possible resultant dimunition in quality of the Raman spectra in these circumstances should be borne in mind.

Hence, it may be concluded that for the application of Raman spectroscopic techniques to space missions the selection of a suitable wavelength of laser excitation is critically important, depending upon whether one is seeking information about the geology or the biology of specimens at a planetary surface. Also, the miniaturisation of Raman spectrometers to a size and power compatibility commensurate with a rugged and robust construction for adoption on remote landers and rovers is now advanced and several possibilities exist in a prototype stage of development. All of these involve diode lasers and simplified detectors which are tuned to operate over specified, often narrow, wavenumber ranges.

Concentration of Molecular Species

Raman band intensities are dependent upon several factors which include the irradiance of the laser ($W\ m^{-2}$) and a linear dependence on the concentration of the molecular scattering species [2]. The former factor necessitates the use of a reasonably high-powered, focussed laser beam incident upon the specimen consistent with the survival of the molecular integrity (and this can be troublesome for sensitive biomolecules which can decompose under laser irradiation) whereas the latter factor can be extremely useful for quantitative and semi-quantitative analysis of the Raman spectra. Unlike infrared optical absorption spectra, the Beer-Lambert Law relating absorbance and species concentration does not apply to Raman scattering and there can be no virtue therefore in increasing the path length of the specimen for Raman scattering to provide enhanced band intensities. The result is that Raman spectra are relatively easy to quantitate and for example, to a very good approximation, ten times the amount of material in the scattering volume will give ten times the intensity of the observed

Raman band; however, this linearity of response means that there is a threshold for the observation of Raman spectra below which the observation of Raman bands is impractical. This sensitivity will be different for every molecular system and will depend upon molecular Raman scattering coefficients, which effectively describe the response of molecules to laser Raman scattering. Normally, a minor species concentration in a molecular system which represents about 0.01%-0.1% (about 100-1000 ppm) of the major component will be at the lower limit of detection in the Raman spectrum. However, as will be demonstrated below, there are several ways in which the detection of minor components in a system can be enhanced, for example using surface-enhanced Raman spectroscopy (SERS), resonance Raman spectroscopy (RRS) and a combination of these, surface – enhanced resonance Raman spectroscopy (SERRS) [6]. It has been claimed that single –molecule detection has now been achieved using the SERRS technique by coating the surface of the specimen with a specially prepared silver colloid which enhances the surface plasmon waves generating the SERS effect and judicious selection of a laser wavelength to invoke the resonance enhancement of the scattered Raman intensity [7].

Perhaps a more practical application for planetary exploration of an increased sensitivity towards the minor components in heterogeneous specimens is provided by the adoption of Raman microscopy, where the laser is focussed into a very small spectral "footprint" which may be of the order of only several microns diameter and individual particles can be examined preferentially from the surrounding matrix; the Raman scattered intensity from these particles or small aggregates, which only represent effectively nanogram or picogram quantities of material, is enhanced by the high irradiance used in the sample illumination. Hence, good quality Raman spectra can be achieved from apparently low laser powers at source of approximately 1 mW, which imaged into a specimen particle of one cubic micron gives an irradiance of more than 1 MW m^{-2}, resulting in a Raman spectrum being obtained from only one picogram of material at a "concentration" in its matrix of less than one part-per-billion. Of course, the major problem here is the location and identification of a suitable particle to undertake the spectroscopic study! An example of the Raman spectra which can be acquired from coarse grains (ca. 0.2 mm diameter) in a heterogeneous specimen of a lacustrine sediment from Lake Hoare, an Antarctic analogue of putative Martian palaeolakes, is provided in Figure 6. Here, the diversity of geological composition is demonstrated by the very different geological signatures that are being presented in the spectra and the major mineral component in each spectrum is indicated in the figure.

Molecular Effects

The concept of molecular scattering coefficients and the sensitivity of molecules towards excitation in Raman spectroscopy has already been raised and it is appropriate here to examine the practical consequences that these parameters have upon the ability to record the Raman spectra of different materials. The sensitivity of Raman spectroscopy for the observation of Raman bands depends upon the polarisability change in electronic charge density of the chemical bonds within a molecular system on excitation with laser irradiation.

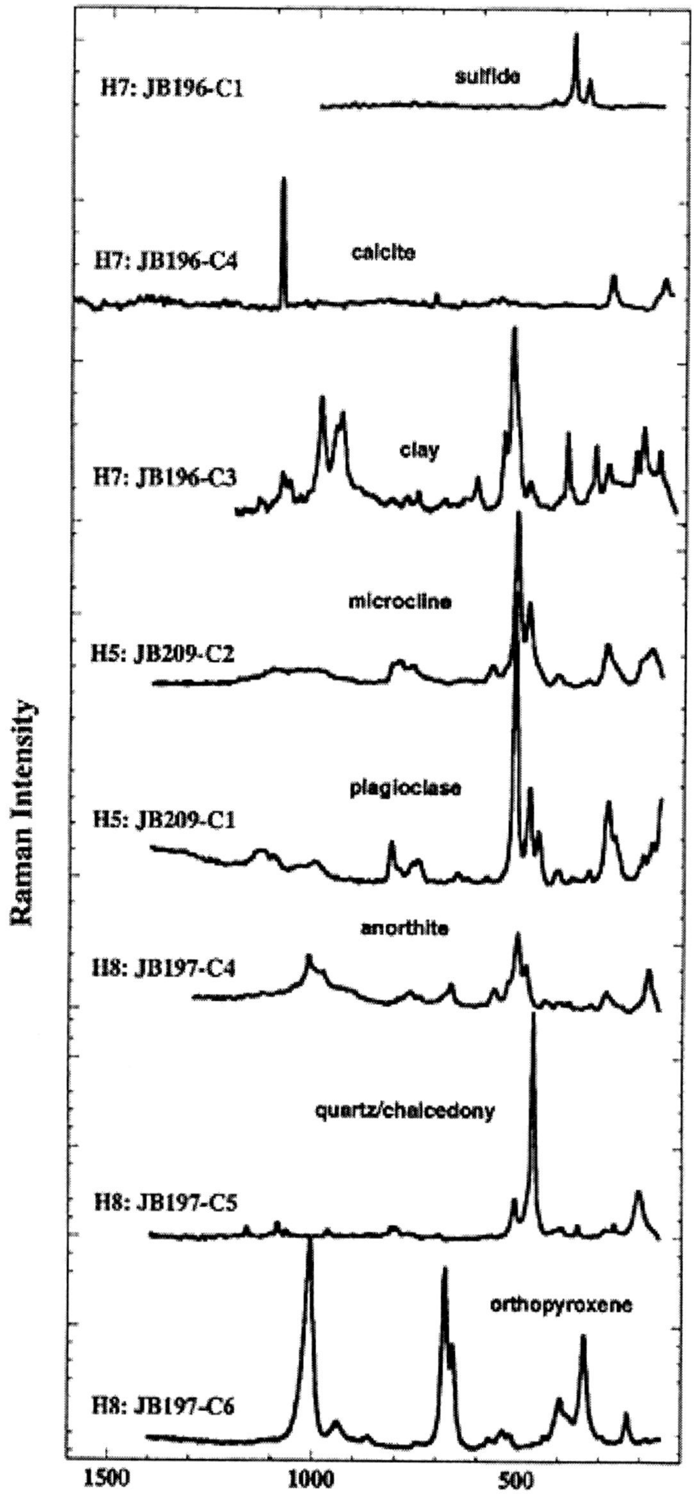

Figure 6. Raman spectra of lacustrine sediments from an anoxic region of Lake Hoare, Antarctica. The spectra are of individual grains found in three specimens, H5, H7 and H8, and the dominant mineral has been component identified in each case.

Generally, bonds between large, polarisable atoms can give rise to strong Raman spectra where the number of Raman bands observed is dictated by the molecular symmetry; the more symmetric a molecule, the fewer the Raman bands that are obtained in the spectrum. Most molecules, however, are of low symmetry or have no symmetry at all –hence, they have relatively rich Raman spectra in terms of the number of bands displayed. The wavenumber positions of the Raman bands are also dictated by the type of molecular bond that is being irradiated; Raman bands originating from bonds between heavy atoms will occur in a low wavenumber shift region of the spectrum, for example, mercury sulfide has a very strong and characteristic Raman band at 253 cm^{-1}, whereas a carbon-hydrogen bond in an aliphatic hydrocarbon will occur near 2950 cm^{-1}. This gives rise to the molecular specificity of Raman spectra, whereby it is possible to recognise the presence of molecules or chemical moieties from the wavenumber positions of their Raman bands [1].

In some cases, the Raman bands coincide in wavenumber with their infrared counterparts, but occur with significant intensity changes which arise from the different spectroscopic selection rules operating for absorption spectroscopy and scattering processes. An excellent example of this is the ketonic carbonyl, C=O, vibration which is found in proteins at 1665 cm^{-1} and is characteristic of an alpha-helical tertiary structure; although this band theoretically occurs at exactly the same wavenumber in both Raman and infrared spectroscopy, it is significantly stronger in the infrared because of the large dipole moment displayed during the bond vibration. In some cases, bands appear in the Raman spectrum, which do not appear in the infrared and *vice versa*; this occurs usually because of the operation of symmetry restrictions and is much used in the definition of molecular structures [8]. As a diagnostic device, however, the presence of key features in the Raman spectra of molecules is a very important process. Hence, the presence of a band in the Raman spectrum of keratotic proteins at about 500 cm^{-1} is a diagnostic verification of S-S bonding; furthermore, the change in wavenumber of this feature occurs in a verifiable way [9] with conformation about the C-S-S-C linkage in the protein and localised *cis, trans* and *gauche* conformational structures can be identified diagnostically between 480 and 520 cm^{-1}.

The low molecular scattering coefficient of water in Raman spectroscopy is useful for cases where desiccation of the specimen is not practicable, such as field applications in ice-laden environments or in contaminated groundwaters, such as Polar glaciated regions and volcanic geysers, respectively. The presence of water molecules in different chemical or physical sites in crystal structures is an important parameter and can be recognised through the environmental molecular effects on other neighbouring molecular ions, such as mineral hydrates, for example. Figure 7 shows the Raman spectra of two key biogeological materials that are seen to be possible signatures of extinct or extant life, whewellite (calcium oxalate monohydrate) and weddellite (calcium oxalate dihydrate), excited using 488 nm radiation . Although the water bands are not visible *per se* in these Raman spectra, the effect of the coordinated water in different sites in the crystal matrix upon the vibrational modea of the oxalate molecular ion is apparent in the wavenumbers of the stretching and bending bands of the $C_2O_4^{2-}$ ions. The recognition of these two materials in the geological record is significant fro life-detection experiments since these materials remain even after the life processes which have produced them have ceased to exist.

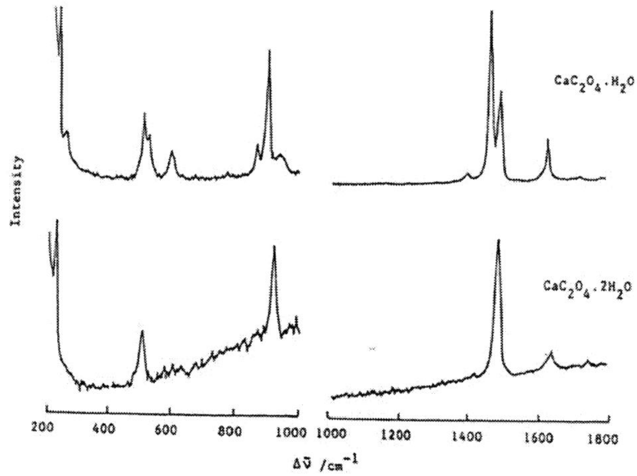

Figure 7. Laser Raman spectra of whewellite (upper) and weddellite (lower), excited using 488 nm radiation.

It should be noted that Raman spectroscopy depends upon the interactions of electromagnetic radiation with chemical bonds for the production of a spectrum. Hence, a Raman spectrum would not be expected from monatomic ions, such as Ca^{2+}, Na^{+} and Mg^{2+}. However, as has been noted in the case of the molecular electronic environments in the vicinity of water molecules, the electronic environmental effects of monatomic cations upon molecular anions can also be observed diagnostically in the Raman spectrum even when these cations themselves have no Raman spectrum. This is generally the basis of mineral diagnostic differentiation in Raman spectroscopy; in Figure 8 are shown the Raman spectra of calcite and aragonite, two polymorphs of $CaCO_3$. Here, the Raman C-O symmetric stretching band of the CO_3^{2-} ion which is common to these two minerals occurs at 1086 cm^{-1}. It is interesting that although calcite and aragonite have an identical stretching band wavenumber of 1086 cm $^{-1}$, which effectively means that this is not diagnostically useful for discriminating between these two polymorphs and can merely serve to identify calcium carbonate in a mineral matrix, the two polymorphs can be distinguished from the different wavenumbers exhibited by the other bands in their spectra, namely 712 /704 and 283 /206 cm^{-1}, respectively for calcite/aragonite. A similar situation exists for the two magnesium minerals dolomite, $CaMg(CO_3)_2$, and magnesite $MgCO_3$, where the CO_3^{2-} stretching band occurs at 1094 cm-1 for both minerals and it is now essential to use the other Raman-active bands for differentiation purposes, e.g. 713/738 cm^{-1} and 300,177,156/330,120 cm^{-1}, respectively for dolomite /magnesite. This highlights a very important point in diagnostic spectroscopy, namely that it is often not practicable to rely on just one defining band for characterisation purposes and that sometimes up to three features need to be considered. Another example of this from the area of geology and mineralogy, Figure 9, is provided by the two naturally occurring polymorphs of titanium (IV) oxide, all have characteristic Raman signatures for diagnostic bands in the spectrum, but their discrimination requires the selective observation of several features.

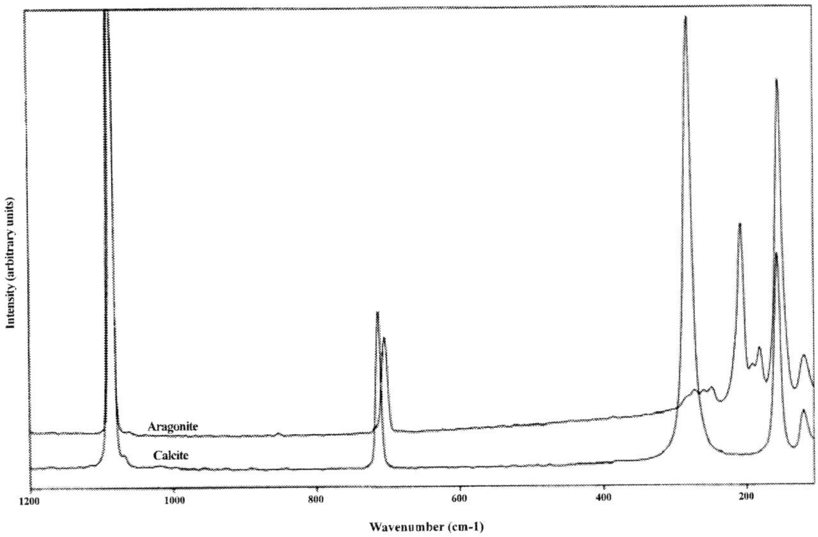

Figure 8. Laser Raman spectra of calcite and aragonite, 1064 nm excitation; 500 scans accumulated at 4 cm^{-1} spectral resolution; wavenumber range 100-1200 cm^{-1}.

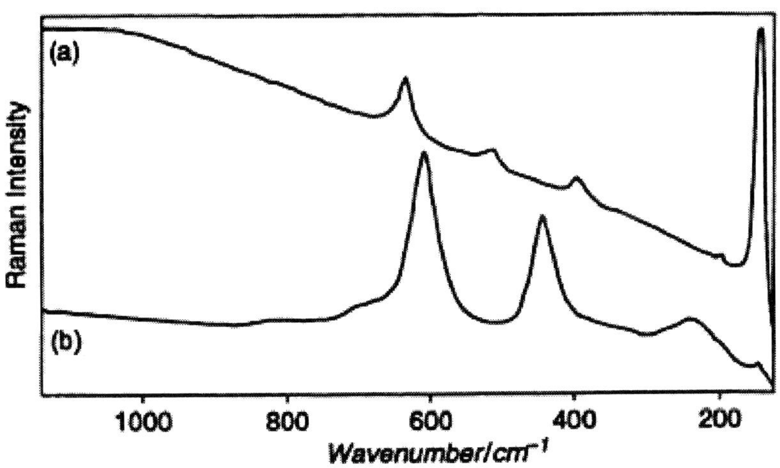

Figure 9. Laser Raman spectra of two polymorphs of titanium (IV) oxide minerals: the spectral stack plot illustrates the discrimination and specificity of the Raman data for materials which analyse as elementally identical.

The specific information provided by Raman spectroscopic analysis of geological samples directly relevant to mars can be illustrated here by the spectral stackplot of the Raman spectra recorded from the Nakhla SNC Martian meteorite shown in Figure 10. The rich diversity of chemical information that is provided by the micro-raman spectra of this specimen exemplifies the geological applications of Raman spectroscopy to Martian materials.

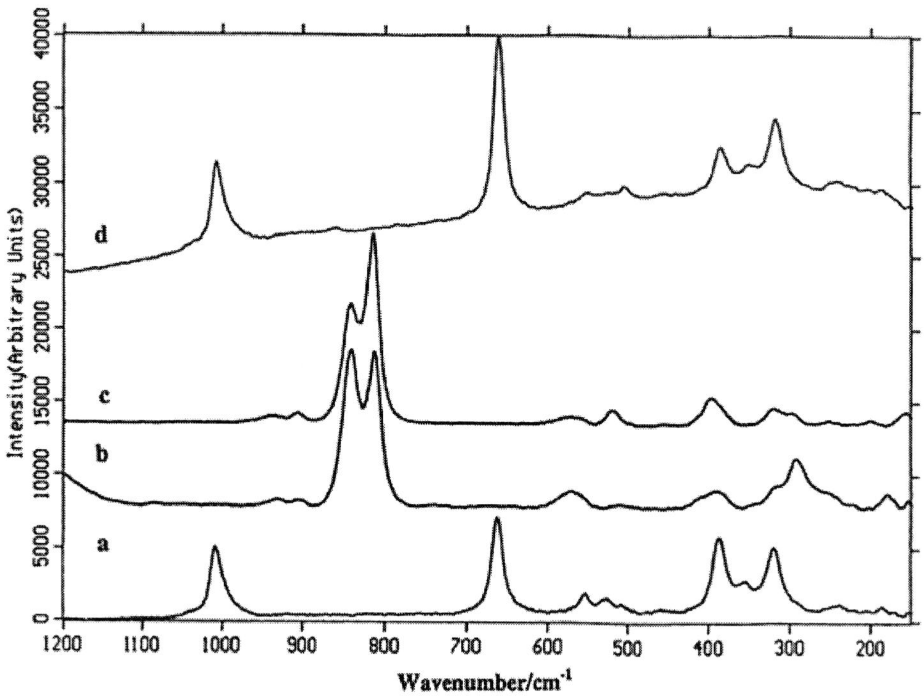

Figure 10. Raman microscope spectral stackplot of the *Nakhla* meteorite, an SNC Martian meteorite, illustrating the discrimination offered between different regions; in each case a spectral footprint of about 2 microns is taken in the specimen and the geological heterogeneity should be noted. *a* black particle, *b* white particle, *c* brown glassy particle, *d* clear glassy particle. Raman bands at 320, 350 386, 662 and 1009 cm^{-1} are characteristic of clinopyroxene, those at 815 and 841 cm^{-1} of olivine and those at 508 and 520 cm^{-1} of plagioclase. No bands arising from biomolecules can be identified in these spectra.

The construction and effective use of automatic database comparison for the recognition of Raman band signatures will therefore need to address this issue of multiple band occurrence. Although the change in vibrational band wavenumbers in the Raman spectrum can be used to provide some definitive diagnostic applications in many areas of chemical science, it does create a problem for the construction and operation of databases and especially for automated recognition of molecular entities from established databases – essentially, there is no single wavenumber position that characterises every molecule with respect to its Raman bands [8]. For example, a "free" C=C is found in the Raman spectrum at 1625 cm^{-1} for ethylene but substitution of chemical groups on the adjacent C-H bonds in the alkenyl molecule will change the wavenumber of this band, especially if the groups concerned have the ability to strongly influence the electronic environment of the isolated C=C bond through conjugation, in which the electronic density on the C=C bond is affected. A C=C-C=O moiety, therefore, will exhibit a C=C Raman band at 1605 cm^{-1} and a C=O band at 1695 cm^{-1}, both shifted down from their respective wavenumber positions in their "free" molecules. Although this is expected theoretically from molecular dynamics, it would be quite difficult to design a suitable database that would immediately recognise these bands as arising from individual C=C and C=O vibrations. Taking this discussion further to the important case of carotenoids, biomolecules based upon the chemical structures of beta-carotene and its congeners with extended chain conjugation of C=C units with each other through a C=C-C=C type coordination, the C=C vibration now appears in the Raman spectrum between 1510-1534 cm $^{-1}$ depending upon the length of the chain and end-substituents [10]; this should be compared with the case of a free, isolated C=C in ethylene at 1625 cm $^{-1}$. Clearly, some sophistication in the automatic database for recognition of this spectral signature and its attribution as a C=C mode diagnostically would be required --- and carotenoids are identified as very significant biomolecules, having a major role in prebiotic chemical evolution discussions of the origin of life in the Solar System [11].

Particle Size

The use of microscope objective lenses of high power and numerical aperture to record the Raman spectra of heterogeneous materials has produced some significant applications for Raman spectroscopy in the field of structural analysis and molecular species identification and have compensated for the rather poor sensitivity directed at minor components in these specimens. As illustrated above, a microscope objective lens of 100 X magnification can produce a focussed laser beam of waist diameter about 1 micron; in a confocal microscopic mode, the acquisition of spectral information theoretically in addition from a one micron segment through a sample has contributed to significant applications in depth profiling because of the real foreshortening of the dimensions of the laser focal cylinder.

However, the adoption of Raman microscopy for specimen analysis can create several problems which need to be recognised:

The high power densities at the specimen, where irradiances of MW m^{-2} can easily be realised; whilst contributing effectively to the scattered Raman scattered intensity this can also result in specimen damage or alteration – a good example of this is provided by the change of goethite (FeOOH) to haematite (Fe_2O_3) at 633 nm with increased laser irradiance [12]. This could be very troublesome the presence of a goethite/haematite mixture on a rock surface could be indicative of endolithic colonisation in the substrata.

Particle size is a rather less well understood aspect of Raman spectra and finely divided materials. Wang *et al.* [13] observed that if the mean particle size in a specimen was greater than that of the laser excitation wavelength then the Raman band intensity decreased with the particle size. However, Hendra [14] found that the Raman intensity increased with the average particle size and with the wavelength, but if the average particle size was > 3 lambda then no significant change occurred in the Raman intensity with increase in the particle size. At particle sizes of less than 3 lambda the intensity of Raman scattering decrease with the particle size was not demonstrated. It was concluded that for a laser wavelength of 785 nm in the near infrared the limiting particle size below which the Raman intensity decreased with decreasing particle size is 2.4 microns and that as this laser wavelength is approached the Raman intensity increased for a particle size below 0.8 microns. It is clear that the influence of the particle size on the resultant Raman spectrum could be highly significant for the examination of the fine debris or detritus from drilling operations carried out using a planetary lander and that this now needs to be evaluated under terrestrial test conditions.

Laser Focal Dimensions

The Raman spectra obtained from a specimen are collected not from a point source but rather from a focal cylinder which may be defined as follows [15]:

Length of waist = *w* = 0.64 lambda f/D
Depth of focus = *l* = lambda f/D

where the numerical aperture of the objective lens and its magnification, the focal length and the wavelength of the laser radiation define the *l* and *w* parameters of the focal cylinder, Figure 11. It should be realised that some scattered radiation will be originating from out-of-focus planes above and below this focal cylinder, normally dependent upon a $1/e^2$ integration factor.

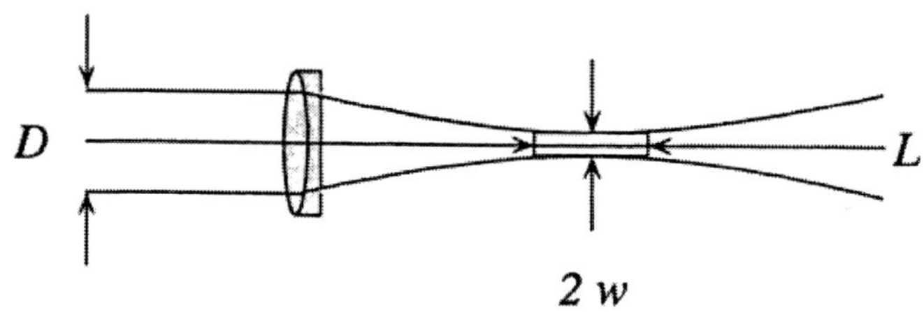

Beam waist at focus $2w = (1.27\lambda)(f/D)$

Depth of focus $L = (2.54\lambda)(f/D)$

Figure 11. Laser focal cylinder with key dimensions.

Tabaksblat *et al* [15] found that for a 40 microns particulate sample some 20 % of the total radiation collected through a 300 microns pinhole for an incident laser wavelength of 514.5 nm came from scatter planes that were > 5 microns from the focal cylinder waist plane. This measurement has given rise to much controversial argument about the actual dimensions of the focal cylinder in Raman microscope spectra which have crucial relevance particularly for the concept of *confocality*. It is now generally accepted that the definition of microscopic sample location and dimensions in confocal microscopy is not straightforward. Wijnaendts van Resandt *et al.* [16] found that for 300 mm lens objectives of 50 X and 100 X magnifications the actual lengths of the focal cylinders were 6 and 3 microns, respectively, exactly twice the values predicted theoretically. There followed much discussion [17, 18, 19] about the factors that were responsible for this inability to realise practically the theoretical optical definition of confocality. Everall [19] attributed the discrepancy between the theoretical and practically achievable confocal dimensions in Raman microspectroscopy to the effect of refraction at the surface of the specimen. He defined the true depth of focus as:

$$\text{Depth of focus} = \text{Delta}\,\{\,(NA^2\,(n^2-1)/(1-NA^2)+n\,)^{1/2} - n\,\}$$

where Delta is the distance below the sample interface of the lens focus, n is the refractive index of the specimen and NA the numerical aperture of the lens. The effect of refraction at the air-specimen interface causes the focal point to be greater that previously assumed and results in a poorer axial resolution (depth) than predicted theoretically. Another factor that emerged from this treatment was that the depth resolution also was predicted to change with the focal distance and depth within the specimen itself. From measurements on a silicon wafer, Everall[19] showed that a calculated 2 micron depth of focus actually gave a true depth of focus (focal plane) of 2.2 Delta, which evaluates as 11 microns for 514.5 nm laser radiation.

In a subsequent article Everall [20] extended and modified his original calculations which then were confirmed with the findings of Baldwin and Batchelder [21] that the depth of focus calculations were under estimated for larger specimen sections but were realistic approaches for thin films. The real situation is even more complex when the distortion produced by spherical aberration mismatch in the refractive index of the specimen resulting in a focal smearing occurs; this is very dependent upon the numerical lens aperture and is pronounced for large numerical apertures [22, 23]. The end result is that one should not be too definitive about the precision of the depth of focus in a confocal Raman microscopy experiment and that the Raman spectra recorded from subsurface layers in transparent materials are subject to "spillover" from adjacent layers which are not formally considered to be significant theoretically butcertainly can be critical experimentally.

SPECTRAL RESOLUTION

The effect of the instrumental spectral resolution upon the appearance of the resultant Raman spectra with regard to the capability of discrimination between the biological and geological band signatures is a critical one. Although the spectral resolution, measured in cm $^{-1}$, is operator adjustable in laboratory instrumentation it will normally be a fixed parameter consequent upon the design features in the case of miniaturised instruments. The effect of changing the spectral resolution for 1064 nm laser excitation upon the Raman spectrum of olivine illustrates this point. The characteristic doublet at 820 and 860 cm $^{-1}$ using a spectral resolution of 4 cm^{-1} provides a key diagnostic indicator for this mineral. At a spectral resolution of 32 cm^{-1}. However, this doublet collapses to a broader singlet band with a peak wavenumber of 844 cm^{-1}. The result of the recording of the Raman spectrum of olivine at this lower resolution of 32 cm^{-1} is that a database search mechanism would probably fail to assign the specimen to olivine. Figure 14 shows a similar situation applies for biomolecular analysis, namely an epilithic lichen, *Acarospora* sp. form Vegetation Island, Antarctica, excited using 1064 nm radiation with a spectral resolution of 4 cm $^{-1}$ and also acquired with a spectral resolution of 32 cm $^{-1}$. An even more serious situation can now be envisaged than was the case for the geological specimen of olivine considered earlier, in that many more key biological signatures that are essential for the identification of the complex biomolecular composition of the organism are now obscured at the lower resolution; the key features due to rhizocarpic acid at 1664, 1595 and 1001 cm^{-1} are seen at higher resolution, but with poorer resolution data is observed at 1663 and 1599 cm^{-1} only. The sharp band at 1001 cm^{-1}, a key feature for monosubstituted aromatic ring compounds, is now totally absent at lower resolution. Similarly, the quadruplet in the neighbourhood of 1300 cm^{-1} has now collapsed to two broader bands at 1346 and 1287 cm $^{-1}$, from 1347 and 1280 cm^{-1} in the spectrum recorded at higher spectral resolution. The bands at 1159 and 1520 cm $^{-1}$ from a carotenoid component are also absent at lower resolution, as are features at 944 and 787 cm $^{-1}$.

Clearly, the operating resolution under which the Raman spectra are being recorded will be vital for the effective database discrimination capability that will be an essential component of miniaturised spectrometer design. Instrumental evaluation of proposed miniaturised spectrometers must be an integral part of their incorporation into planetary landers. As the instrument size is diminished physically, each pixel in the diode array detector covers a larger wavenumber range and a choice facing instrument constructors is the compromise between wavenumber shift range versus the spectral resolution -- the former will decrease the spectral coverage for materials of interest whereas the latter can distort the observed spectra which may result in a mis-attribution of the data or even missing features, the presence of which may be vital particularly for the biological assignment.

Figure 12: Mars Oasis, Antarctica; a terrestrial niche for the survival of endolithic cyanobacteria in a "*limits of life*" situation; a translucent Beacon sandstone outcrop containing endolithic microbial communities on top of a dolerite sill. Dr David Wynn-Williams, Head of Antarctic Astrobiology at the British Antarctic Survey, Cambridge, is prospecting for endoliths. Reproduced from the book Astrobiology: The Quest for the conditions of Life, Eds: Gerda Horneck, Christa Baumstark-Khan, 2003, Springer. With kind permission of Springer Science and Business Media.

Adaptation to Mars

Rovers have revitalised the possibility of finding the signs of life on Mars –extinct or extant. We now have evidence of images of a planetary surface [24, 25] that clearly has suffered water erosion, subsurface aquifers, the possibility of subglacial polar lakes (similar to Ellsworth and Vostok in Antarctica) and a tantalising glimpse of geological niche scenarios that match terrestrial models, for example Mars Oasis, Antarctica [26] (Figure 12), where biological adaptation and survival to these harsh Martian environments may have occurred. The new phase for Martian exploration now firmly rests on the molecules that have been synthesised by extremophiles for their protection against low wavelength radiation and desiccation and the associated auxiliary molecules that are required to harness the limited energetics and nutrients that are available in their protected niche environments [27, 28, 29]; finally, in a "limits of life" scenario, where life has become extinct, the remote analytical observation of residual molecular signatures in the geological record may perhaps indicate that the evolutionary experiment on Mars commenced but was not successful in its survival (Figure 13).

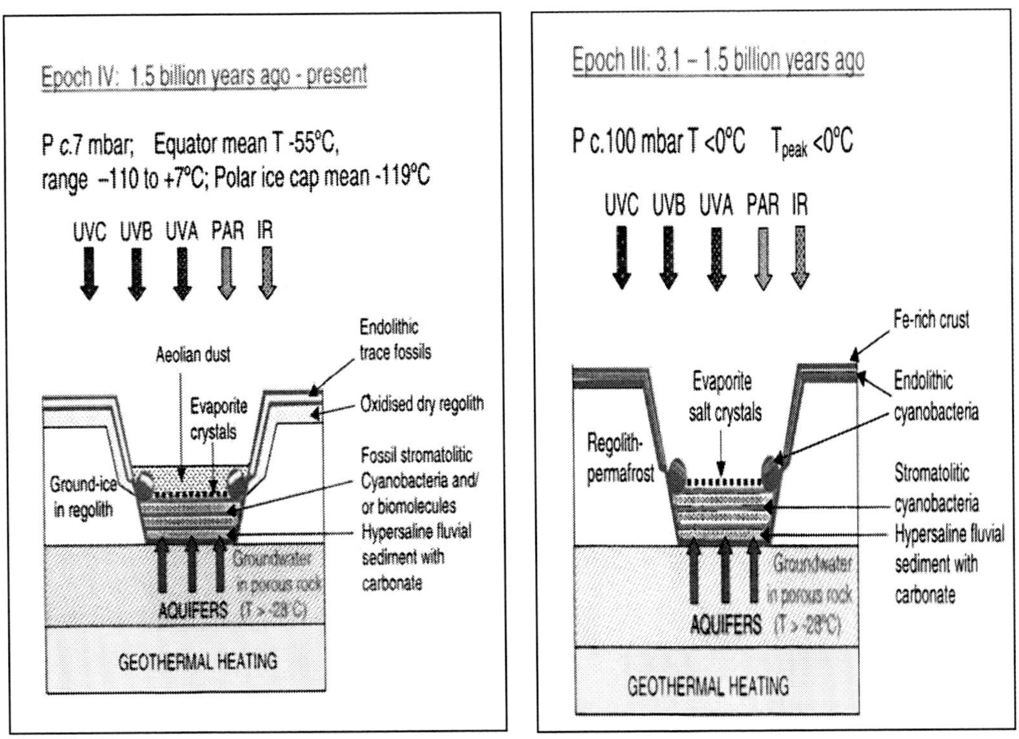

Figure13: Mars Epochs III and IV; highlighting the geological survival possibility of endoliths on the planet. Reproduced from ICARUS, 144, David Wynn-Williams, Howell G. M. Edwards, Proximal Analysis of Regolith Habitats, p486-503, 2000, with permission from Elsevier.

EXTREMOPHILES

Extremophiles are organisms that can survive where humans cannot; their survival in hostile terrestrial environments depends critically upon their ability to synthesise sophisticated suites of protectant biomolecules (Table 5) and the adaptation of geological and hydrological niches for colonisation to combat low wavelengths of insolar radiation, desiccation, and extremes of temperature, pH, salinity, pressure and heavy metal toxins [29,30]. The search for life on Mars is linked closely to the discovery of novel strategies adopted by extremophiles in terrestrial scenarios where previously the existence of life was believed impossible. A selection of extremophiles and their terrestrial habitats will be considered here, which has provided astrobiologists with the term *"terrestrial Mars analogues"* on which the experimental remote analytical projects envisaged for robotic Mars explorations are now being evaluated. The driving force behind the detection of life signatures on Mars is really two-fold, in that the search for a biological evolutionary occurrence other than our own will undoubtedly enhance the definition of our own early prebiotic chemistry but it will also engage with the efforts of space exploration in the placement of a human being on the surface of Mars, which will require a fuller knowledge of the planetary composition and its potential biological dangers than we have currently [31, 32].

In this respect, the study of extremophile survival strategies which operate in some of the most inhospitable places on Earth will further our understanding of our own precarious existence in a changing natural world.

Extremophiles that have been studied in our laboratories and which address several scenarios that could be expected on Mars, i.e. *terrestrial Mars analogues*, include the following:

- Life in rocks: Antarctic endolithic microbes;
- Shocked rocks: pore colonisation;
- Subglacial lakes: polar colonisation.
- Meteorite impact craters: halotrophic cyanobacteria;
- Lacustrine sediments: microbial colonies;
- Volcanic lava deposits: vacuole colonies;
- Stromatolites: cyanobacterial residues;

Chasmolithic Community

A chasmolithic community living in a fracture inside a marble rock was collected during the AMASE expedition (Arctic Mars Analogue Svalbard Expedition) in August 2004 to the Vest Spitzbergen Island in the Svalbard Archipelago, sited 80°N, inside the Arctic Circle. Although the temperatures are not as cold as those in Antarctica (at Spitzbergen Island, the winter temperature reaches minus 30-35 °C), the site and the samples are considered extremophiles because of the low temperatures and the absence of sunlight for almost six months each year.

Table 5. Raman spectral signature bands and UV-visible absorption maxima for functional microbial pigments in extreme terrestrial Antarctic habitats

Physiological Function	Pigment	Pigment Type	Raman vibrational bands (wavenumber cm^{-1})					Absorption peaks (nm) UVC <280	UVB 280-320	UVA 320-400	Visible >400	
Ultraviolet Screening pigments	Usnic acid	**Cortical acid**	2930		1607	1322	1289	>220	290	325	<400	
	Pulvinic dilactone	**Pulvinic acid derivative**		1672		1405		>246	290	367		
	Paretin	**Anthraquinone**		1675		1099	551	>257	288		431	422
	Calycin	**Pulvinic acid derivative**			1611		1379	269				
	Atranorin	***para*-depside**	2942	1666	1303	1294	1266	<274			>400	
	Gyrophoric acid	**Tri-depside**		1661		1290		275	304			
	Fumarprotocetraric acid	**Depsidone**		1642	1630	1290	1280	273	315			
	Emodin	**Quinone**		1659					291		440	
	MAA *(Nostoc)* 7437	**Mycosporine Amino Acid**	2920		1400		820		(>310)	330		
	Scytonemin	**8-ring dimmer**		1590	1549	1323	1172	252	300	370	>400	
Anti-oxidants	*β*-carotene	**Carotenoid**			1524	1155		<246	283	384	429	
	Rhizocarpic acid	**Isoprenoid**		1665	1620	1596		>200				
Photo-synthesis	Porphyrin	**Tetrapyrrole ring**			1453							
	Chlorophyll, (Cyano bacteria)	**Tetrapyrrole ring**			1360	1320					680	700
	$BChl_a$ *(Rhodopseudomonas)*	**Tetrapyrrole ring**	na								850	870
	Chlorobium Chl	**Tetrapyrrole ring**	na								650	660
Accessory Light-harvesting	Phycocyanin	**Phycobilin**		1638	1369						560	620
	Phycoerythrin	**Phycobilin**	na								544	

The sample was found in a glacier moraine in Bockfjorden, at an altitude of 30 metres above sea level, and shows an epilithic community on the crust which gives a greyish tonality to the white stone substrate, and a chasmolith inside a crack where several coloured strata, namely pinkish, black, green and brown areas, are visible.

The Raman spectrum of the substratum presents bands at 1097 cm^{-1}, which is seen as a shoulder on the stronger signature at 1086 cm^{-1}, 713 cm^{-1} with a shoulder at 725 cm^{-1} and 157 cm^{-1}. This provides an excellent example of the use of Raman spectroscopy to distinguish between isomorphic minerals. The bands at 1097, 725, 299 and 177 cm^{-1} are characteristic of dolomite, a calcium magnesium carbonate ($CaMg(CO_3)_2$), whereas calcite, the calcium carbonate isomorph, shows signatures at 1086, 713, 281 and 157 cm^{-1}, very close in wavenumber, but readily distinguishable from the spectral signatures of the dolomite. Figure 14 gives a Raman spectral stackplot of magnesite, $MgCO_3$, with Raman bands at 1094, 738, 326, 242 and 119 cm^{-1}, dolomite, $CaMg(CO_3)_2$, with bands at 1098, 725, 299 and 177 cm^{-1}, aragonite, $CaCO_3$, with bands at 1086, 704, 208 and 154 cm^{-1}, and calcite, $CaCO_3$, with bands at 1086, 713, 283 and 156 cm^{-1}.

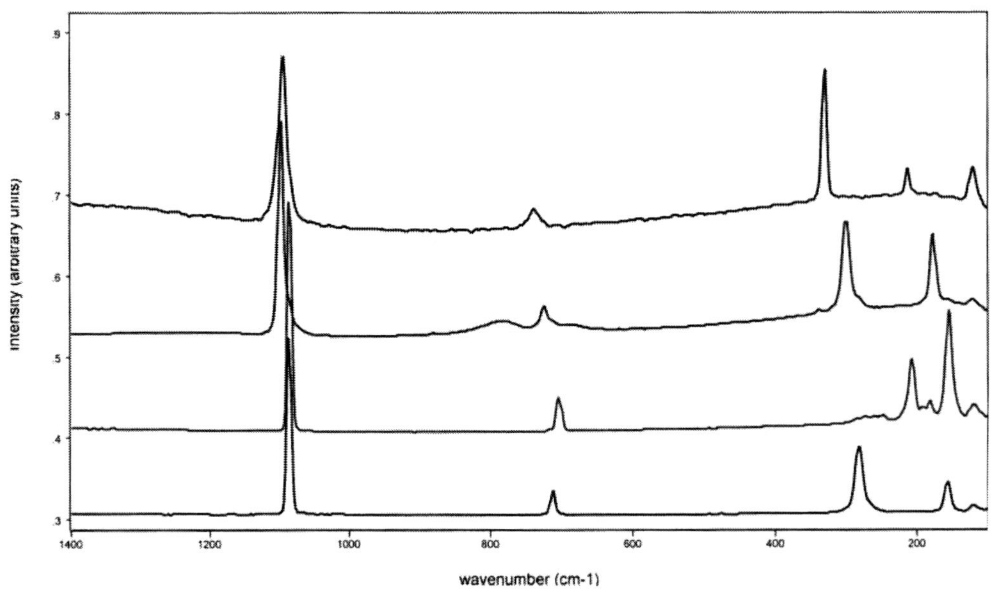

Figure 14: Raman spectra of, from the top: magnesite, dolomite, aragonite and calcite-indicating the spectral discrimination between these carbonates.

On the surface crust, the grey tonality is the result of cyanobacterial colonisation, which gives the characteristic Raman spectral bands of scytonemin. This UV-protective pigment is not found in the chasmolithic organisms inside the crack and has probably been produced as a strategic response to the increased radiation levels experienced at the rock surface.

Figure 15 shows a section of chasmolithic feldspar from Lake Hoare, Antarctica, in which two zones of bacterial colonisation can be clearly seen, from which the Raman spectrum of calcite can be clearly identified closely associated with the colonisation zones signatures of quartz and feldspar can also be seen in the spectrum.

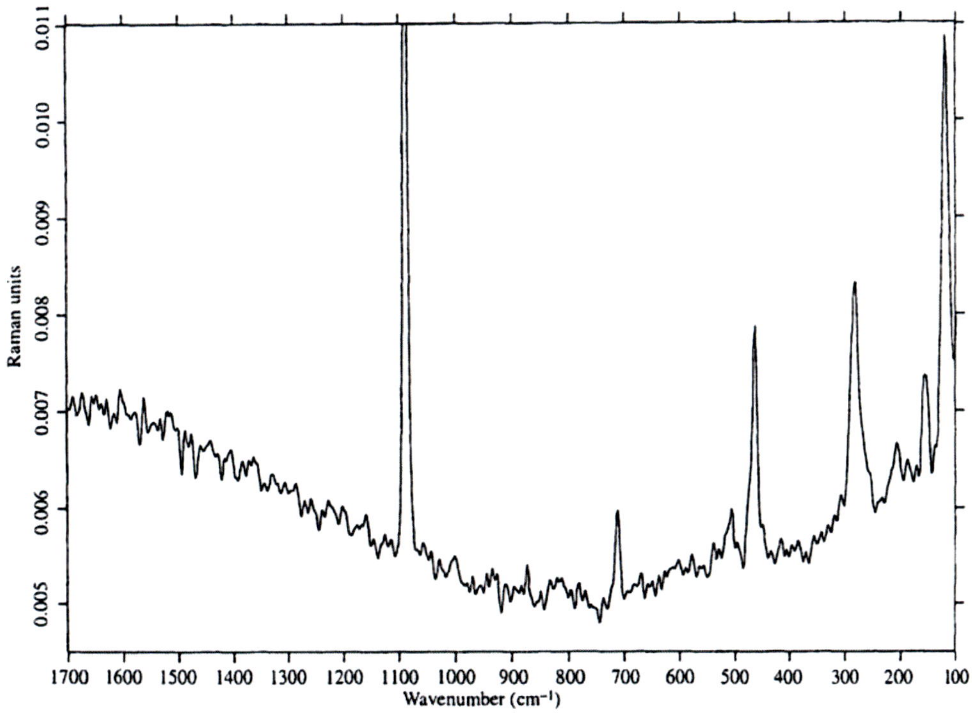

Figure 15: Chasmolith in feldspar, Lake Hoare, Antarctica: with the Raman spectrum of calcite, quartz and feldspar from the cyanobacterial colonisation zone.

In the chasmolithic pink coloured zone area, the Raman spectra show bands of chlorophyll and a carotenoid, but the fluorescence emission is very strong here, which reduces the quality of the Raman signatures. The green layer gives a strong fluorescence emission with 785 nm laser excitation and no spectral signatures are visible, but selection of

514 nm laser excitation provides bands from a carotene which are clearly recognizable at 1524, 1157 and 1000 cm^{-1} (Figure 16); these signatures are consistent with an assignment to astaxanthin. The same applies to the black zone in the chasmolith, but now the carotene gives bands at 1516, 1154 and 1004 cm^{-1} and so it can be identified as beta-carotene. Although no spectral signatures could be recorded for this system with the near-infrared laser at 1064 nm, three different micro-organisms are differentiated with 514 nm excitation because of the different carotenes produced, two discrete organisms occurring in the green band the third in the black band.

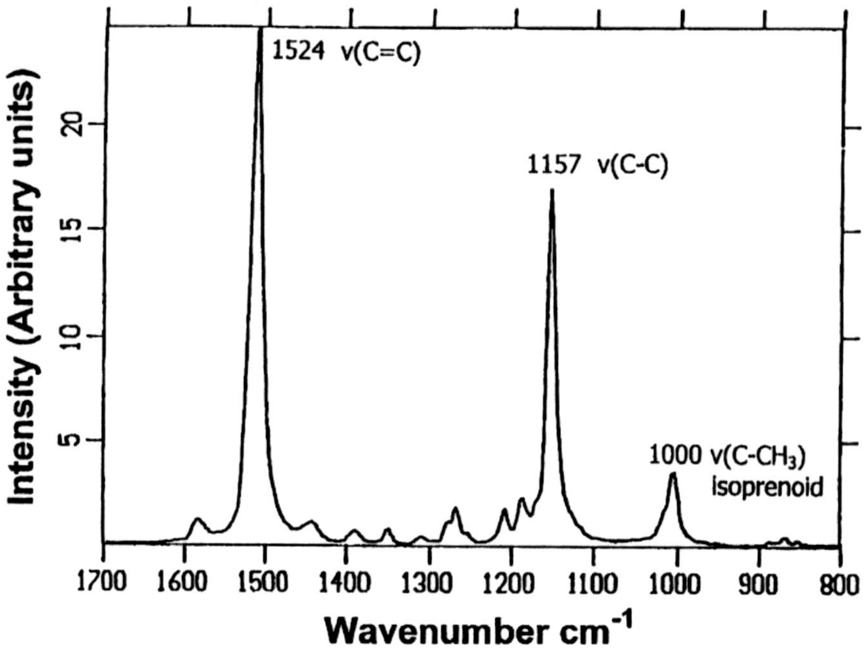

Figure 16: FT-Raman spectrum of a carotene found in an Antarctic Beacon sandstone endolith, Battleship Promontory, Alatna Valley.

Endolithic Community

This specimen was collected at Mars Oasis in Antarctica, the coldest desert on Earth, during the summer season of 2002. The sandstone shows a normal orange-red colouration but a depletion of the red mineral (haematite) is clearly visible in the vicinity of the endolithic organism (Figure 17).

Quartz is the main component of the rock, with signatures at 128, 206, 355, 542, 696, 795, 807, 1064, 1081, 1161 and 1227 cm^{-1}; the red colour is recognizable as haematite because of its Raman bands at 223, 291, 404, 495 and 609 cm^{-1}. This is abundant below the organism stratum but has been removed from the immediate vicinity, probably to permit photosynthetic-active radiation to reach endnolithic community. The specificity of Raman spectroscopy permits one to distinguish the very weak signature of haematite in the crustal spectrum (Figure 18); haematite has been proved to be a UV-screening compound and it is interesting that the organisms can maintain this mineral on the crust for the purpose of protecting against radiation that reaches the terrestrial surface (exacerbated by the depletion of Antarctic ozone). This particular crust colouration, modified by the underlying biological colony, was used by the prospecting geologist as a visual clue to the presence of endolithic communities inside the rock.

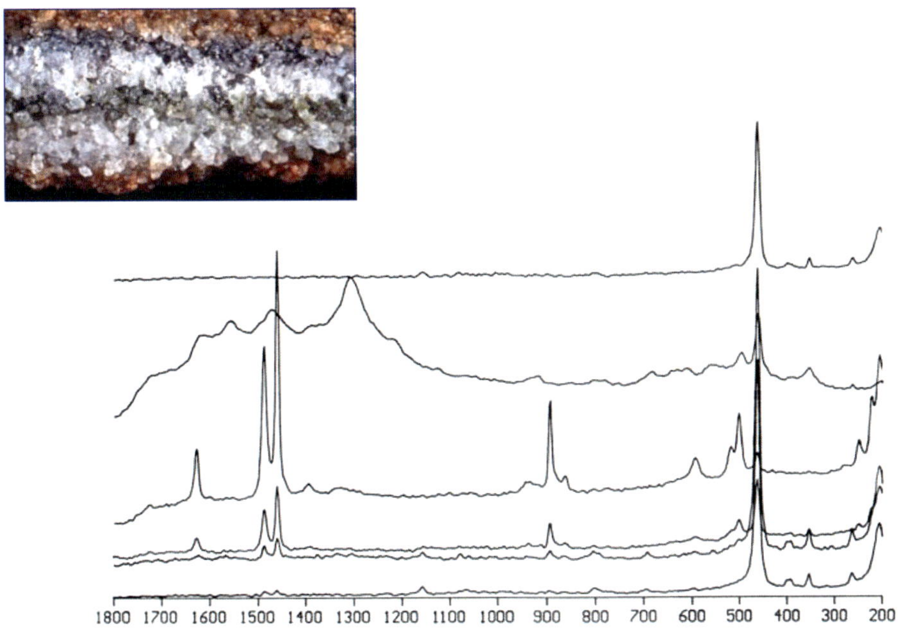

Figure 17: Endolithic colonisation of Beacon Sandstone, Mars Oasis, Antarctica, showing zonation. The lower interface of the algal zone dominated by cyanobacteria occurs 8 mm below the upper surface of the rock.

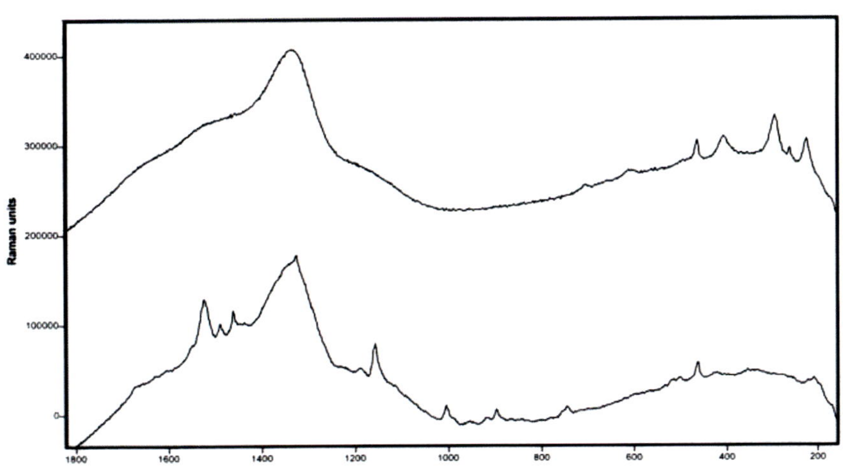

Figure 18. Continued on next page.

Figure 18: Raman spectrum of haematite crust on sandstone endolith. The Raman spectral stackplot shows the spectra obtained from a vertical transect through the surface crust, the biological hyaline layer, algal layer, depletion zone and basal rock, from the top.

The stackplot of spectra collected from the endolithic specimen shows bands of calcium oxalate monohydrate at 1490, 1463, 896, 502 and 206 cm^{-1}in the depleted pigment area. Two different calcium oxalates are identified by Raman spectroscopy because of their signatures in extremophile exemplars; whereas in the calcium oxalate monohydrate (whewellite) Raman spectrum, apart from some weaker signatures, two bands are shown at 1490 and 1463 cm^{-1}, for calcium oxalate dihydrate (weddellite) only one characteristic strong signature, centred at 1475 cm^{-1}, is visible. Even when they appear in admixture whewellite and weddellite are clearly distinguishable using the Raman spectroscopic analytical technique. In addition, beta-carotene with bands at 1516, 1154 and 1004 cm^{-1} and chlorophyll, with a strong and broad signature at 1327 cm^{-1} and weaker signals at 915, 744 and 517 cm^{-1}, have also been identified, as well as quartz.

Volcanic Rock Colonisation

A primary colonization in fumarole minerals from a recently active and still warm (two week-old) lava flow was collected by the author during a field trip to the Kilauea volcano, Hawaii, in 2004 and was analyzed using Raman spectroscopy about seven days after collection, figure 19. Besides identifying the mineral components, a carotenoid was recognized because of the Raman signatures at 1530 and 1143 cm^{-1}, and a chlorophyll-like compound could be related to the signatures at 1449, 1342, 952, 834, 748, 680, 595, 438, 257 and 173 cm^{-1}. This result indicates that the organisms needed only a very short time indeed to colonize an inhospitable niche such as a fumarole in a still very warm and active lava flow.

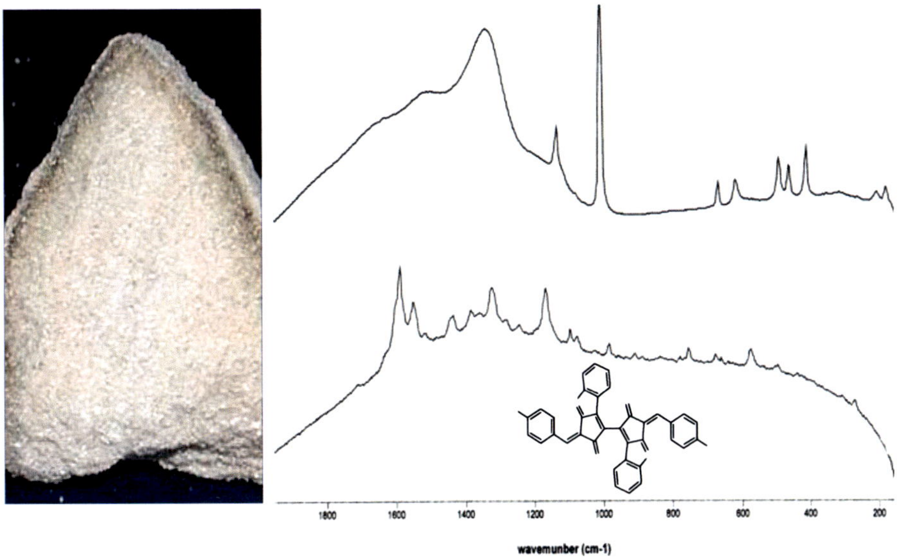

Figure 19. Antarctic endolith in gypsum, with Raman spectra of, upper spectrum, the surface (gypsum) and lower spectrum, the underlying cyanobacterial zone (scytonemin). The chemical structure of scytonemin is also given here.

A second sample of lava (Figure 20) was collected during the AMASE expedition of 2004 in Scott Keltiefiellet, in the Vest Spitzbergen island of Svalbard [40] at 80°N at about 1,000 m altitude. Organisms were found inside a vacuole of about 2 cm diameter in a black lava basalt matrix; the vacuole possessed a simple orifice of less than 1 mm diameter through which the organisms, sun, water and gases could penetrate. Technically, endoliths are organisms living inside the pores of a sedimentary rock, but this new kind of "endolithic" community affords a novel opportunity to look for life inside volcanic rocks that would not normally have been considered to be a geological niche for endolithic colonisation.

Three different carotenes were observed in the vacuole when 514 nm laser excitation was used, but there were no chlorophyll signals. With 785 nm excitation only two carotenes were detected but chlorophyll signals appeared alongside the carotenes and characteristic bands from c-phycocyanin, an accessory light-harvesting pigment, were also detected.

Halotrophs

The presence of salts in SNC (Shergotty-Nakhla-Chassigny) Martian meterorites [32-35] and the acknowledgement of the presence of sulphates in the Martian regolith [35-38], together with the recognition of the existence of lacustrine water sediments at or below the surface of Mars [39-44] makes terrestrial halotrophic organisms potential Martian analogues.

Apart from the Raman spectroscopic identification of two different types of bacteria in a gypsum crystal (*Nostoc* and *Gloeocapsa*), it was also possible to detect organic signatures from bacterial colonies sited several millimetres below the surface in a transparent crystal of selenite from a 26 Mya meteoritic impact crater [38] at Devon Island in the Canadian High Arctic (Figure 21).

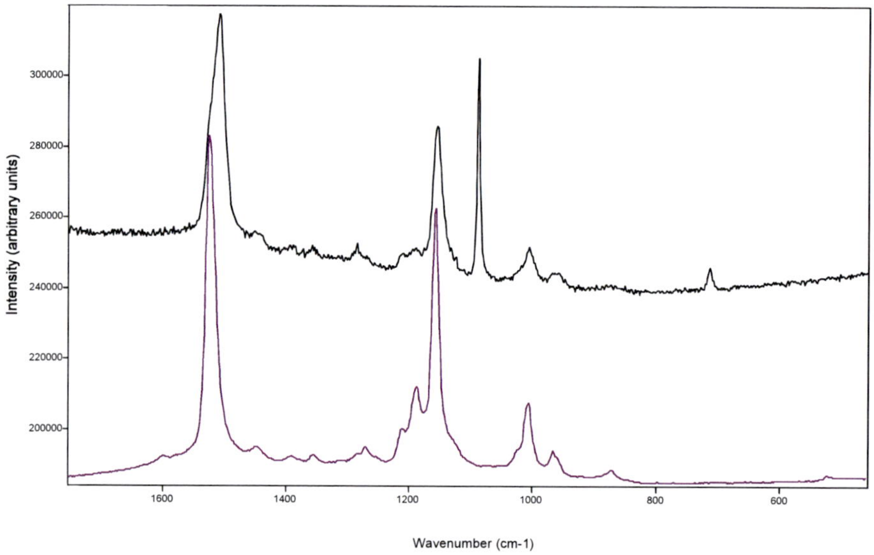

Figure 20. Novel endolith in a vacuole in a volcanic basalt lava matrix, Svalbard, Spitsbergen, Norwegian Arctic, showing two different carotenoids in admixture. Accessory photopigments are also identified in this system (not shown here).

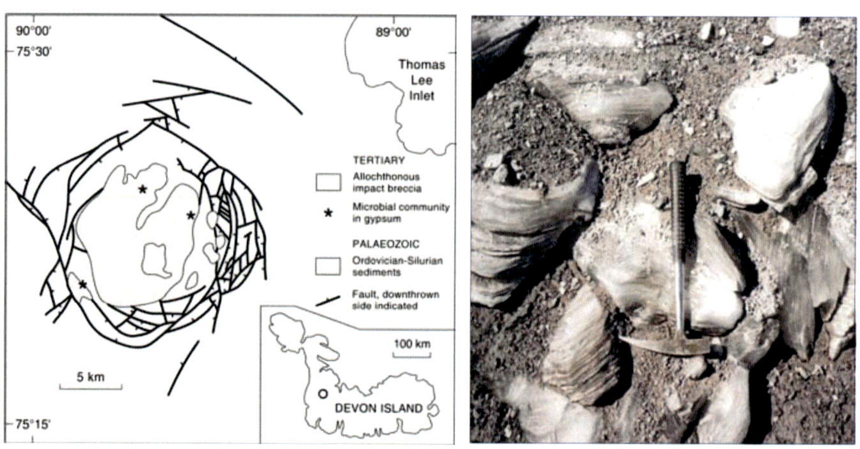

Figure 21: Selenite var. gypsum) deposits, Haughton meteorite impact crater, Devon Island, Canadian High Arctic, 26 Mya. Also shown are details of selenite crystals within surface melt breccia deposits.

The Raman spectrum of the selenite crystal surface shows the strongest gypsum band at 1006 cm^{-1} and other minor gypsum signatures; Raman spectra taken from a vertical transect through the crystal from the surface, using a confocal laser beam to obtain spatial information from within the crystal, provides evidence of additional features in which the Raman bands, can be identified with the cyanobacterial signatures. Confocal laser probing of the dark-coloured inclusion within the crystal reveals that it is a halotrophic cyanobacterial colony containing the radiation protectant scytonemin (Figure 22).

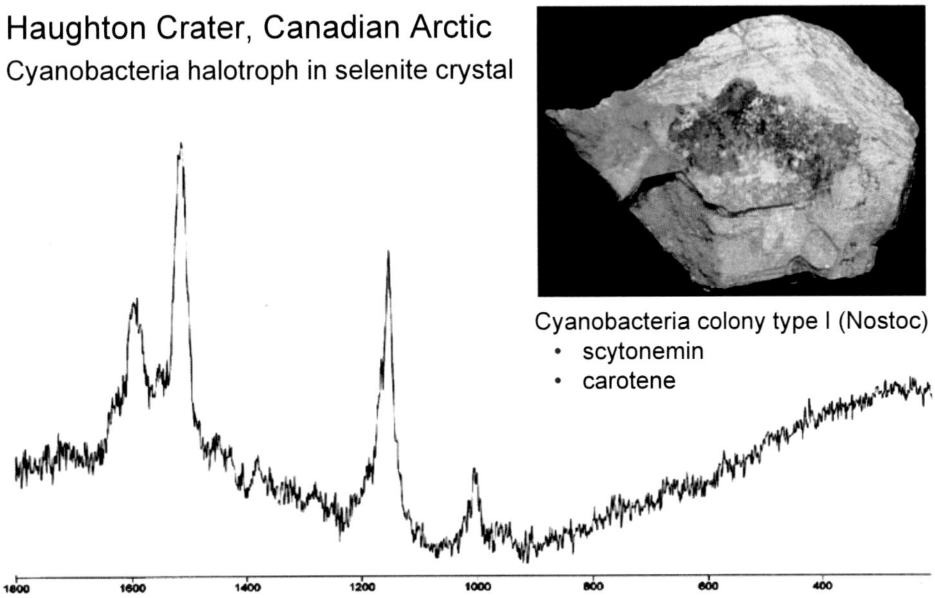

Figure 22: Raman spectra of halotrophic cyanobacterial colony inside selenite crystal, located approximately 5-8 mm below the upper surface. The Raman spectrum of the cyanobacterial inclusion obtained using confocal laser imaging shows the presence of scytonemin and a carotenoid.

A parallel study of an Antarctic gypsum endolith shows evidence of the same radiation protectant, scytonemin, in the lower spectrum with that of the surface gypsum for comparison in the upper spectrum; the different strategies adopted for the chemical protection of the colonies against low wavelength radiation for the two Antarctic endolithic systems, involving in one case the presence of haematite at the surface and in the other gypsum, is noteworthy.

Stromatolites

Shade adaptation is required by bottom-dwelling cyanobacteria in ice-covered lakes, whilst enabling them to avoid UV-stress. During seasonal changes in the flow of melt-water, the thick cyanobacterial mats at the bottom of ice-covered Antarctic Dry Valley lakes becomes buried in silt. However, they permeate through the mineral layer to form a new mat each season, resulting in stratified stromatolites. The translucency of the 3 m thick ice cover on Lake Vanda permits penetration of solar radiation through 70 m of water column to the bottom mats where the receipt of photosynthetically active radiation (PAR) is remarkably up to ~ 120 μmol m^{-2} s^{-1}. Even the PAR level of ~ 15 μmol m^{-2} s^{-1} measured at the bottom of

Lakes Hoare and Fryxell, which have thicker ice covers (about 4.5 m), is more than enough to sustain active photosynthesis by oxygenic mats. This has resulted in a wide diversity of cyanobacterial mat communities in many Antarctic Dry Valley lakes. Analogous habitats on Mars, such as Gusev and Noachis Craters, may provide suitable sources of fossil microbes and biomolecules such as porphyrins and cyanobacterial scytonemin. Modern stromatolites are considered to be analogues of *Conophyton*, which is a columnar stromatolite abundant in Precambrian rock when UV stress would have been much greater.

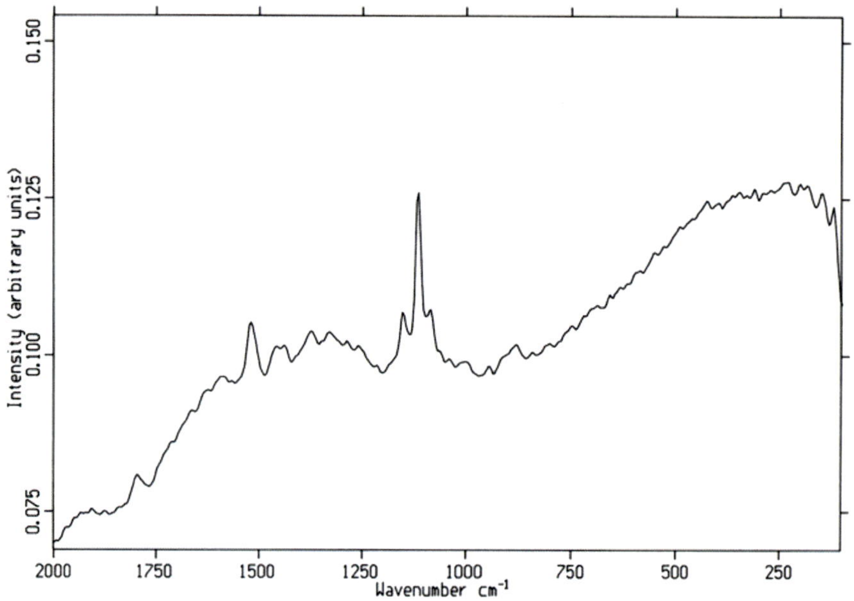

Figure 23a. Salda Golu Lake, Turkey: hydromagnesite colonised stromatolites in saltern. The FT-Raman spectrum of the cyanobacterial colonised zone in a stromatolite is shown.

Figure 23b. Salda Golu Lake, Turkey: hydromagnesite colonised stromatolites in saltern. The FT-Raman spectrum of the cyanobacterial colonised zone in a stromatolite is shown.

We have studied columnar endolithic colonisation of carbonates from Salda Golu Lake in Turkey, where stromatolitic outcrops of magnesite and hydromagnesite have formed (Figure 23a/23b). Here, the identification of the biological component in the Raman spectrum of the inorganic carbonate matrix is a good terrestrial model for the survey of potential stromatolite sites on Mars – such as that identified at Juventus Chasmae in Sabaea Terra (Figure 24) near the Schiaparelli Crater; in the figure, the light-coloured area in the meteorite impact crater is believed to be a magnesium carbonate deposit..

Figure 24. Juventus Chasmae, potential stromatolite region, on Mars. Reproduced from the Journal of the Geological Society, 156, Michael J. Russell *et al.*, Search for signs of ancient life on Mars: expectations from hydromagnesite microbialites, Salda Lake, Turkey, 1999, with permission from the Geological Society.

With climatic changes on a geological scale, these lakes may dry up and their cyanobacteria become desiccated fossils. This process may have helped to preserve cyanobacteria-like organisms in chert (flint) from the Australian Apex formation in Queensland for 2.5 Gya. Their pigments, especially early porphyrins which are recalcitrant components of chlorophylls, are recognizable in oil-bearing shales. A primitive photosynthetic system capable of delivering an electric charge across a membrane bi-layer could have consisted of a porphyrin (pigment), a quinone (electron donor) and a carotene (electron acceptor) to provide a chain of conjugated bonds. Such molecules are valuable biomarkers on Earth and may fulfil a similar function for their recognition by remote life-detection systems on Mars. A recently recorded microRaman spectrum of the Trendall Pilbara stromatolite has identified a carotene and a porphyrin together in a fissure inside the rock (Figure 25). This stromatolite is one of the oldest rocks found on Earth, with an age of 3800 M years, dating from a very early period in the geological evolutionary history of our planet. However, whilst it would be facile to assume that the bioorganic signatures that have been detected here are also of this age, this example does demonstrate very nicely the viability of Raman spectroscopy for the detection of organic signatures in ancient rocks. In another sampling region of the same specimen, there is clear spectroscopic evidence for the presence of beta-carotene, scytonemin, dolomite and goethite (Figure 26).

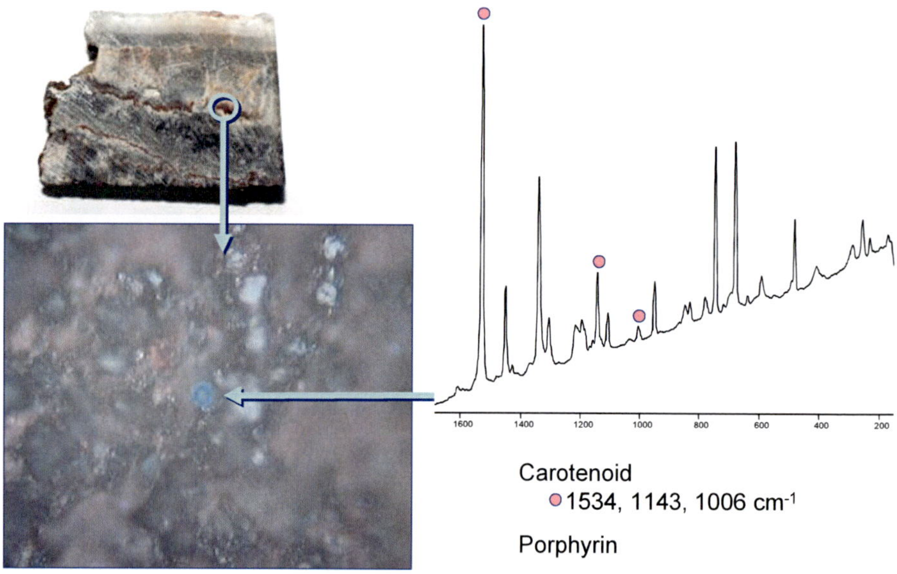

Figure 25. Trendall Pilbara stromatolite, Western Australia, 3800 Mya; Raman spectroscopic evidence for protective biochemicals, carotene and porphyrin, produced in a protected niche.

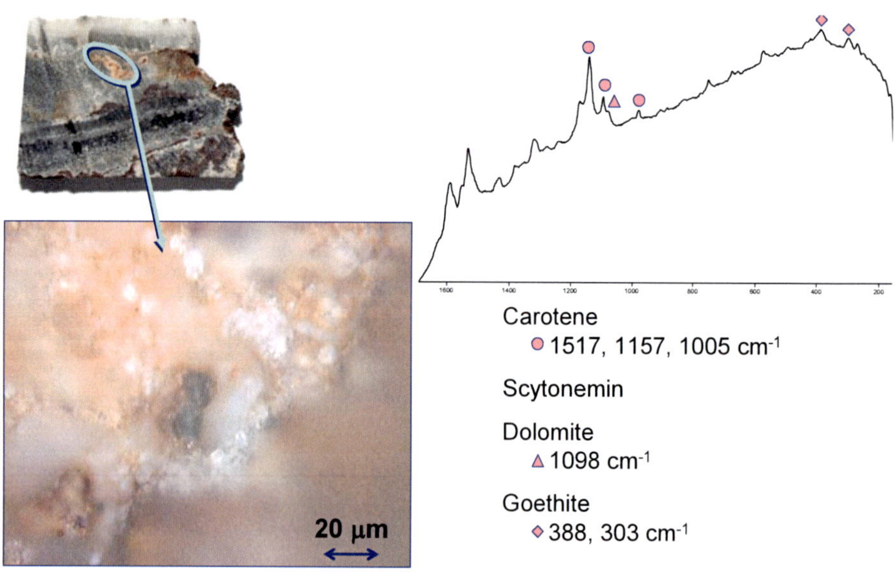

Figure 26. Specimen of Trendall Pilbara stromatolite shown in Figure 15, but different niche; different chemical suite identified –scytonemin, carotene, dolomite and goethite.

Figure 27. *Sabkha* surface crust with gypsum and halite crystals, Rhub-al-Khalil, Arabian desert; cyanobacterial colonisation at interface with subsurface dolomitized calcite. The Raman spectrum shows the presence of scytonemin and carotene.in the biological zone.

Salt Pan Evaporites

The dried terrestrial salt pans with their extremophilic halotrophic cyanobacterial colonisation are good models for the study of Martian sites such as Gusev Crater. Here, an example from the Rhub-al-Khalil in the Arabian Desert shows a cyanobacterial colony in a dolomitic zone, several cm below the surface of a large gypsum crystal embedded in a halite matrix (Figure 27). The Raman spectra of the biological component clearly demonstrate the presence of photoprotective pigments such as scytonemin and carotenoids.

Remote Sensing Probes

Difficulties experienced in the enclosure of a specimen in the sample illuminator of a laboratory – based Raman spectrometer has led to the evolution of remote sensing probes which bring the laser beam onto the specimen and collect the scattered Raman radiation. Typically, these probes are made of fused silica and are comprised of a single optical fibre which conducts the laser onto the specimen through an objective lens surrounded concentrically by six similar fibres which collect the scattered radiation and image this into the spectrometer- this is described as a 6/1 probe.

This arrangement for remote sensing is seen to be the method of choice for the remote sensing Raman analytical experiments envisaged for extraterrestrial exploration. A problem with remote sensing using optical fibres is the superposition of the background spectrum of the silica fibre upon the Raman spectrum; an example of a remote Raman spectrum recorded with the aid of a 6/1 probe of 1 metre length is shown in Figure 28. Here, the broad bands from the Si-O stretching of silica are seen below 1000 cm^{-1}; these could obscure the mineralogical features of a specimen, but spectral subtraction of the probe background can provide a means of removing this intereference, albeit with the production of increased spectral noise.

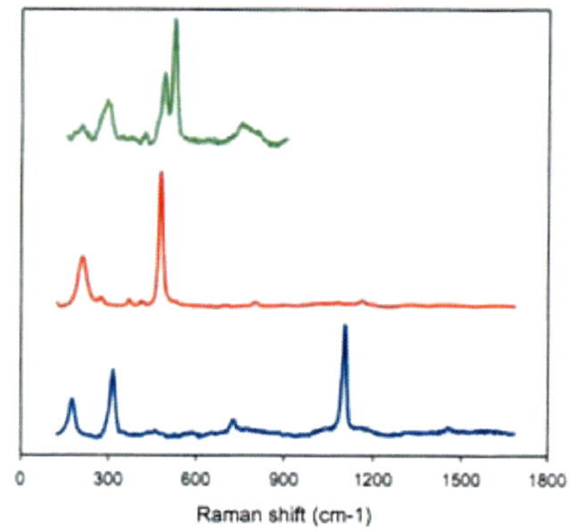

Figure 28. Prototype advanced Raman field instrument constructed for NASA adoption by JPL/Washington University St Louis; the laser excitation is at 532 nm and the spectra from the top are of feldspar, quartz and calcite.

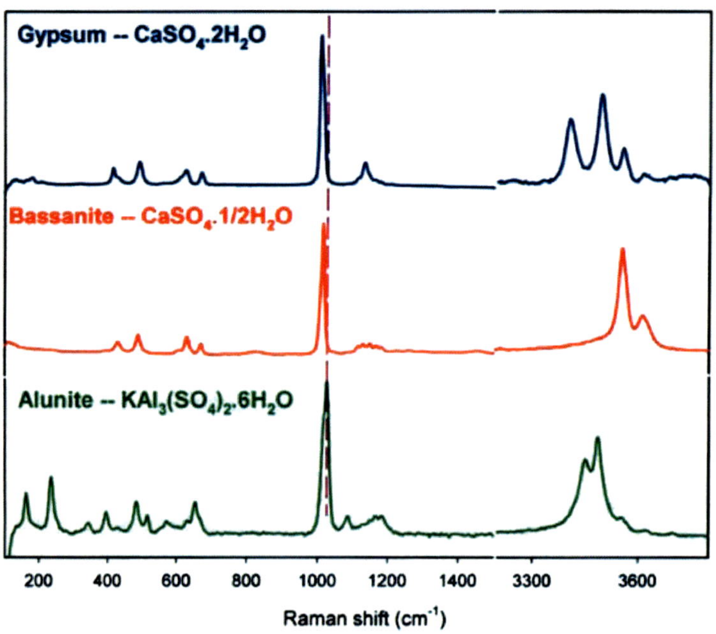

Figure 29. Raman spectra of sulfates recorded using the advanced prototype instrument shown in Figure 28; top, gypsum. middle, bassanite, lower, alunite.

MINIATURISED RAMAN INSTRUMENTATION

Several miniaturised prototype Raman spectrometers have been constructed and are being developed and evaluated by groups for possible space mission adoption by NASA and ESA. The ExoMars/PASTEUR and AURORA ptojects are currently targeting the inclusion of novel miniaturised Raman spectroscopic instrumentation in mineralogical and life-detection suites on robotic planetary landers and rovers for the exploration of Mars.

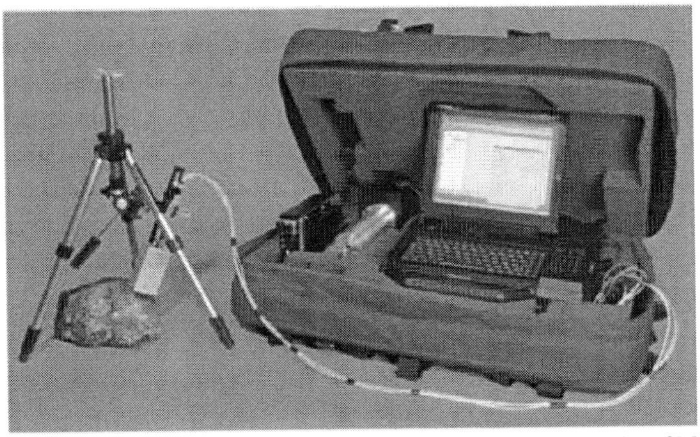

Figure 30. Prototype Raman field instrument constructed for NASA by the University of Montana; operating with laser excitation of 852 nm.

Figure 28 shows a prototype advanced field instrument for mobile remote Raman spectroscopy from JPL/University of Washington St. Louis; this compact unit delivers green radiation at 532 nm (frequency-doubled Nd^{3+}/YAG laser radiation) onto a specimen through the probe head of a sensing device and the results from three geological samples are shown. The upper spectrum is that of feldspar, the middle spectrum is that of quartz and the lower spectrum is calcite. The quality of the data produced by this miniature spectrometer can be gauged from the high level of discrimination evidenced in Figure 30 from the three mineral sulfates, gypsum, bassanite and alunite (results reproduced courtesy of Dr. Alian Wang [49], Washington St. Louis University). Another miniature Raman instrument that has ben constructed as a joint venture between NASA and the University of Montana [50] (Professor David Dickensheets) is shown in Figures 30 and 31. Here, 852 nm radiation in the near infrared region is used to excite the Raman spectra of organic molecules; Figure 33 shows the Raman spectrum of *Acarospora chlorophana,* in which several key features are indicated over the wavenumber range 500-1500 cm^{-1} with an operational spectral resolution of 12 cm^{-1}. This arrangement is seen not to compromise the retrieval of quality spectral data in the lower wavenumber region in comparison with the spectra that were excited at 1064 nm with the laboratory-based FT-Raman system on the same sample. However, the decreased response and sensitivity of the miniaturised instrument in the higher wavenumber region is noticeable. A comparison of the bands recorded using this miniature Raman spectrometer operating at 852 nm and laboratory instruments using other lasing wavelengths of excitation is given in Table 4; this confirms that the identification of the protective biochemicals produced by the extremophile lichen from our standards in the database can be accomplished successfully using 852 nm radiation.

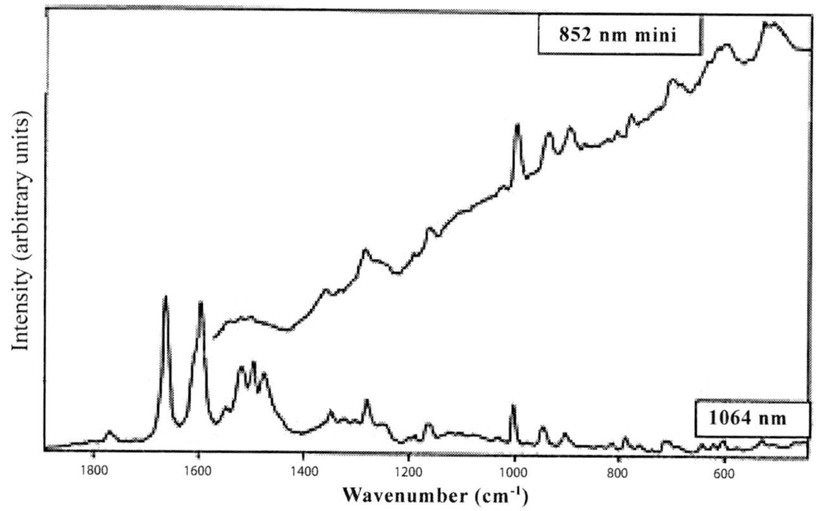

Figure 31. Raman spectra of *Acarospora chlorophana* recorded using the miniature Raman instrument shown in Figure 30; despite the limited operational range of this instrument compared with the laboratory-based FT instrument using 1064 nm laser excitation, the characteristic features of the biomolecules are clearly observed.

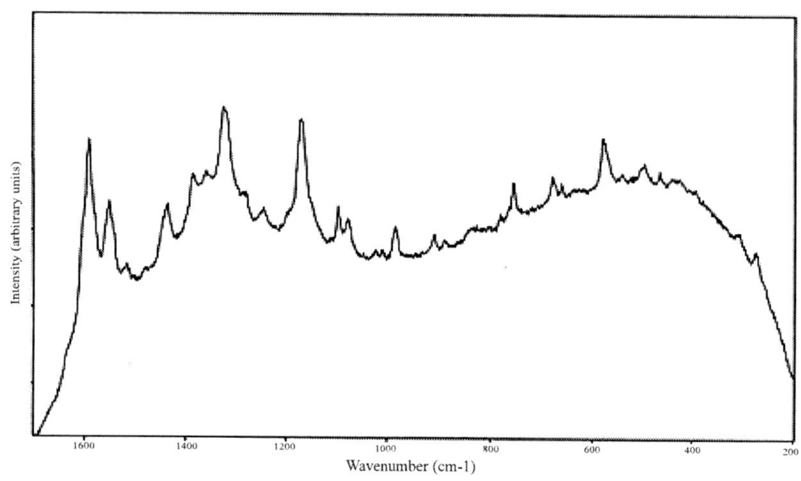

Figure 32. FT-Raman spectrum of scytonemin, the UV-radiation protectant biomolecule synthesised by several extremophilic species in hot and cold deserts

The importance of the evaluation of miniaturised Raman prototypes using terrestrial Mars analogues, particularly of extremophilic biological organisms, is stressed so that the advantages and limitations of parameters such as laser excitation wavelength, spectral wavenumber range , spectral resolution, scan characteristics and the speed of data acquisition can be properly quantified in the design of future experimental spectrometers for space mission adoption.

Table 6

Excitation Wavelength /nm					Assignment
1064	*852/ mini*	*785*	*633*	*514.5*	
711 w	-	713 w	713 w	-	
702 mw	701 m	703 w	703 w	-	CCH aromatic def
687 mw	687 w	687 w	687 w	-	CCH aromatic def
640 w	636 w	639 w	641 w	-	CCH aromatic def
619 w	616 w	619 w	618 w	-	CCO ring bend
599 w	600 mw	601 w	603 w	-	CCO ring bend
527 mw	529 m	529 w	529 w	-	CCO bend
508 mw	510 m	-	-	-	Calcium oxalate , bend
464 w	464 vw	-	-	-	Si-O , quartz , substrate
449 vw	-	447 w	450 w	-	
406 w	406 mw	-	-	-	Haematite , substrate
206 w	-	-	-	-	Si-O , quartz , substrate

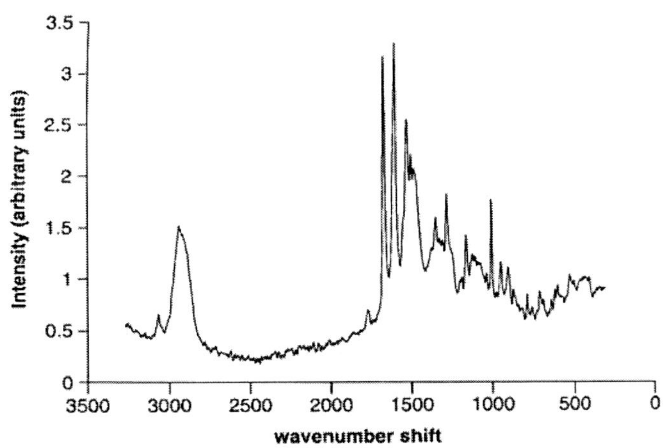

Figure 33. Raman spectrum of *Acarospora chlorophana*, an epilithic extremophile from Antarctica, recorded with the miniature Raman spectrometer shown in Figure 30 using 1064 nm laser excitation.

A detailed knowledge of the different chemical strategies adopted by terrestrial extremophiles, which are considered Martian analogues, is a prerequisite for planetary exploration. Geo – and biomarkers must be studied together to provide a better understanding of the adaptation abilities of the organisms. Geo- and bio-strategies are complementary and the survival of the organisms depends critically upon their being able to modify their microniche environments.

Table 7. Raman bands of the most common and geo-and biomarkers in extremophile examplars, and their chemical formulae

Calcite	$CaCO_3$	**1086**	**712**	**282**	156								
Aragonite	$CaCO_3$	**1086**	**704**	**208**	154								
Dolomite	$CaMg(CO_3)_2$	**1098**	**725**	**300**	177								
Magnesite	$MgCO_3$	**1094**	**738**	**330**	**213**	119							
Hydromagnesite	$Mg_5(CO_3)_4(OH)_2\ 4H_2O$	**1119**	**728**	326	**232**	202	184	147					
Gypsum	$CaSO_{4.}\ 2H_2O$	**1133**	**1007**		669	628	**492**	**413**					
Anhydrite	$CaSO_4$	**1015**	674	628	500	416							
Quartz	SiO_2	1081	1064	808	796	696	500	542	**463**	354	263	**206**	128
Haematite	Fe_2O_3	**610**	500	**411**	**293**	245	**226**						
Limonite	$FeO(OH)nH_2O$	693	**555**	481	**393**	**299**	**203**						
Apatite	$Ca_5(PO_4)_3(F,CI,OH)$	1034	**963**	**586**	**428**								
Weddellite	$Ca(C_2O_4)2H_2O$	1630 **1475** 1411		**910**	869	597	**506**	**188**					
Whewellite	$Ca(C_2O_4)H_2O$	1629	**1490**	**1463**	1396	942	**896**	865	596	521 **504** 223 207 **185 141**			
Chlorophyll	$C_{55}H_{72}O_5N_4Mg$	1438	1387	**1326**	1287	1067	1048	**988**	**916**	**744**	**517**	351	
c-phycocyanin	$C_{36}H_{38}O_6N_4$	1655	**1638**	1582	1463	**1369**	1338	**1272**	1241	1109	1054	**815 665** 499	
Beta-carotene	$C_{40}H_{56}$	**1515**	**1155**	**1006**									
Rhizocarpic acid	$C_{26}H_{23}O_6$	**1665**	1610	**1595**	**1518**	**1496**	1477	1347	1303	1002	944	902 768 448	
Scytonemin	$C_{36}H_{20}N_2O_4$	1605	**1590**	**1549**	1444	**1323**	1283	1245	**1172**	1163	984 752 675 574 270		
Calycin	$C_{18}H_{10}O_5$	1653	**1635**	**1611**	**1595**	**1380**	1344	1240	1155	1034	**960**	878 498 484	
Paretin	$C_{16}H_{12}O_5$	**1671**	1631	1613	**1153**	1387	1370	**1277**	1255	**926**	571	519 467 **458**	
Usnic acid	$C_{18}H_{16}O_7$	**1694**	1627	**1607**	**1322**	**1289**	1192	1119	**992**	959	846	602	540
Emodin	$C_{15}H_{10}O_5$	**1659**	1607	1577	1557	**1298**	**1281**	942	**565**	**467**			
Atranorin	$C_{19}H_{18}O_8$	**1666**	**1658**	1632	**1303**	**1294**	1266	**588**					
Pulvinic dilactone	$C_{18}H_{10}O_4$	**1672**	**1603**	1455	**1405**	1311	**981**	**504**					
Gyrophoric acid	$C_{24}H_{20}O_{10}$	**1662**	1628	1612	1334	1304	**1291**	**1235**	**1138**	**561**			

Corroborative bands appear in bold

Conclusion

Raman spectroscopy is a useful analytical technique for planetary exploration because it does not require sample manipulation, and macroscopic and microscopic analyses are possible when identifying components in the specimen. It is sensitive to organic and inorganic compounds and is not destructive to samples that are strictly limited in quantity or accessibility and so they can be used afterwards for other, perhaps more destructive, analyses. In this respect the situation of a Raman spectrometer on a planetary rover or on a fixed lander, to which specimens are brought by a rover, are both acceptable space mission scenarios.

Laser excitation wavelength is an important parameter to consider, not only because fluorescence emission can swamp the weaker Raman scattering, but also because of the different sensitivities shown towards compound identification by different excitation wavelengths. In our assessment 785 nm laser excitation gives a good balance between organic and inorganic compounds, but 514 nm excitation induces a resonance Raman effect for carotenes whose identification is reasonably straightforward, even in small concentrations, although the higher fluorescence emission from some systems can still mask several weaker Raman bands. The observation of more than two spectral biomarker bands is necessary normally for the unambiguous identification of the components of a specimen, particularly when there is a complex mixture of organics and inorganics present.

Spectral resolution is another critical parameter that can affect observed band wavenumbers due to the overlapping of close bands, which can cause erroneous identifications of the material when comparing with standard wavenumber tables and spectral databases. The technical requirements for the construction of a miniaturized Raman spectrometer suitable for extraterrestrial planetary exploration could force a rather low spectral resolution, and so new databases adapted to low resolution spectral data would then be necessary. Nevertheless, good quality spectra are now being obtained from miniaturised Raman spectrometers such as that shown in Figure 30, which weighs one kilo and operates with diode laser excitation at 1064 nm. An example of the Raman spectrum of an extremophile obtained from this instrument in our laboratories is shown in Figure 32.

The ability to identify organic molecules and minerals present in admixture in a specimen at the same time is an advantage of the Raman analytical technique in astrobiology. Extremophiles are closely related to rock substrata, and biological and geological spectral signatures are shown together in the Raman spectrum; these are distinguishable and can be unambiguously assigned, even if some appear in a relatively low proportion. A list of some of the most important biogeological Raman spectral markers is provided in Table 7; this can be regarded as a prototype database for the spectral recognition of indicators from remote life detection instrumentation on planetary landers and rovers.

The Future

In 2006, the European Space Agency announced that significant funding had been allocated for the forthcoming ExoMars mission on Project AURORA – the most ambitious scientific analytical programme yet devised to detect the signatures of extinct or extant life on Mars (Figure 32). A miniaturised scientific instrumental payload on a planetary rover will use a sophisticated arsenal of analytical techniques, some of which have never before been used in space missions. The scientific instrumentation suite on ExoMars (Pasteur) has to fulfil three specific functions, namely, the characterisation of the geological context of the selected landing site, the detection of life signals and organic molecules in the samples from that site and the evaluation of the hazards to the planned human exploration of Mars. An integral part of this instrumentation will be a novel combined miniaturised Raman/LIBS spectrometer which will be designated the specific task of searching for biogeological marker signatures in the Martian surface and subsurface regolith (Figure 33). The Raman spectroscopic data presented in this paper from terrestrial Mars analogues and models will form an important part of the analytical scientific preparation for Mars exploration involving the search for life in the ExoMars mission.

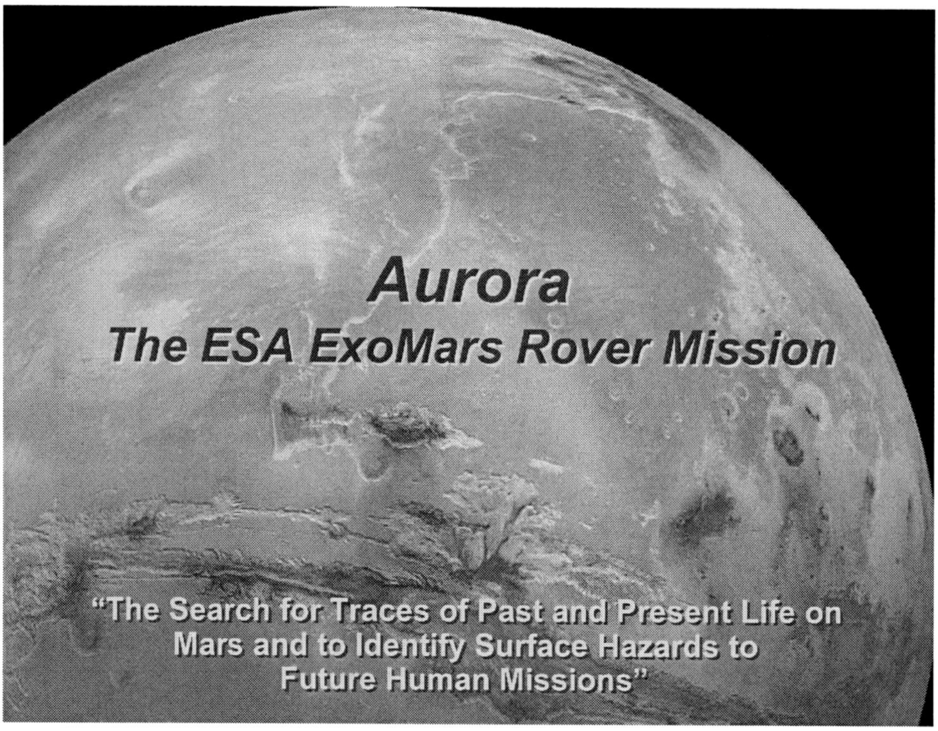

Figure 34. AURORA: the European Space Agency /NASA programme, incuding the ESA ExoMars mission.

The ExoMars mission and others that follow it, including the manned lunar landings, will culminate in a knowledge base that is essential for the planned human landings and exploration of Mars scheduled for 2033 – the first time that human beings will have set foot on another planet in our Solar System. A symbolism of this leap forward in analytical scientific astrobiology is depicted in Figure 36.

The authors record their appreciation of funding over many years for the basic analytical Raman spectroscopic studies of terrestrial extremophiles and Mars models from the European Space Agency and UK organisations (EPSRC, NERC and PPARC) which have enabled the preliminary work to be carried out to establish the viability of the Raman technique for incorporation into extraterrestrial life-detection instrumentation and for the opportunity to take novel analytical chemical science to the "last frontier".

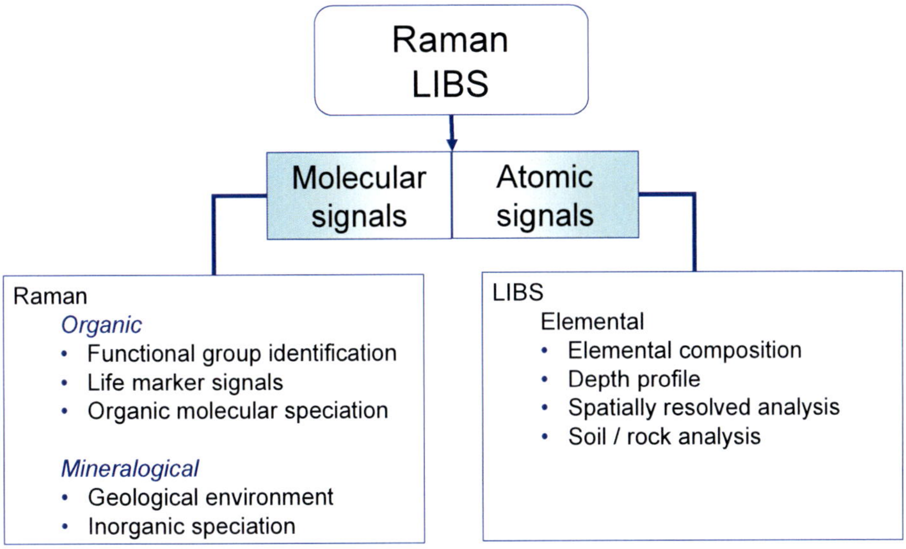

Figure 35. The principle of the novel miniaturised Raman/LIBS combined spectrometer for the ExoMars mission to Mars.

Figure 36. Symbolic depiction of the Mars mission phases, 2007-2033, involving a revisitation of manned lunar exploration before embarking upon the manned missions to Mars.

In this they are especially grateful for scientific collaboration with the late Dr David Wynn-Williams, Head of Antarctic Astrobiology at the British Antarctic Survey, Cambridge, and for the ongoing support of Professors Charles Cockell, John Parnell, Monica Grady, Fernando Rull and Drs Cynan Ellis- Evans, Roger Worland, Andre Brack, Frances Westall, Susana Jorge, Derek Pullan, Beda Hoffmann, Jan Jehlicka and Liane Benning—a range of expertise encompassing geology, microbiology, geochemistry, meteoritics, earth sciences and biomolecular sciences that is so critical for the effective study of *astrobiology,* the quest for life in the Solar System and beyond.

REFERENCES

[1] I.R. Lewis and H.G.M. Edwards (Eds.), *Handbook of Raman Spectroscopy: From Research Laboratory to the Process Line*, M. Dekker, New York, 2000.
[2] D.A. Long, *Raman Spectroscopy*, John Wiley and Sons, Chichester, UK, 2002.
[3] ftp://ftp.estec.esa.nl/pub/aurpra/Rover/PPP-IDD.pdf; ESA Progress Letter (2004): Pasteur Instrument Payload for the ExoMars Rover Mission. ESA Publications Division, Noordwijk, The Netherlands; *http://esamultimedia.esa.int/docs/Aurora/Pasteur_ Newsletter_4.pdf.*
[4] D. Lin-Vien, N.B. Colthup, W.G. Fateley and J.G. Grasselli, *The Handbook of Infrared and Raman Characteristic Frequencies of Organic Molecules*, Academic press, London, (1991).
[5] M. Bouchard and D.C. Smith, *Spectrochimica Acta, Part A,* 59, 2247-2266 (2003).
[6] G. Dent and W.E. Smith (Eds.), *Modern Raman Spectroscopy*, John Wiley and Sons, Chichester, UK (2005).
[7] pp128-129 in ref.6.
[8] H.G.M. Edwards, "Spectra –Structure Correlations in Raman Spectroscopy", in *Handbook of Vibrational Spectroscopy*, Volume 3, eds. J.M. Chalmers and P.R. Griffiths, John Wiley and Sons, Chichester, UK (2000).
[9] H.G.M. Edwards and E.A. Carter, "Raman Spectroscopy of Biological Molecules", in *Infrared and Raman Spectroscopy of biological Materials*, eds. H.-U. Gremlich and B. Yan, M.Dekker, New York (2001).
[10] M. Veronelli, G. Zerbi and R. Stradi, *J. Raman Spectroscopy,* 26, 683-692 (1995).
[11] A. Brack, B. Fitton, F. Raulin and A Wilson (Eds.), Exobiology in the Solar System and the Search for Life on Mars *ESA Special Publication* SP-1231, ESA Publications Division, Noordwijk, The Netherlands (1999).
[12] D.L.A. de Faria, S.V. Silva and M.T. de Oliveira, *J. Raman Spectroscopy*, 28, 873-878 (1997).
[13] H. Wang, C.K. Mann and T.J. Vickers, *Applied Spectroscopy*, 56, 1538-1544 (2002).
[14] P.J. Hendra, in *Handbook of Vibrational Spectroscopy*, Eds. J.M. Chalmers and P.R. Griffiths, Volume 2, John Wiley and Sons, Chichester, UK (2000).
[15] R. Tabaksblat, R.J. Meier and B.J. Kip, *Applied Spectroscopy,* 46, 60-68 (1992).
[16] R.W. Wijnaendts van Resandt, H.J.B. Marsman, R. Kaplan, J. Davoust, E.K.H. Steilzer and R.Sticker, *J. Microsc*., 38, 29-34 (1985).
[17] T. Wilson (Ed.), *Confocal Microscopy*, Academic Press, London (1994).

[18] S. Hadjoost and J.A. Yarwood, *Applied Spectroscopy,* 50, 558-564 (1996).
[19] N.J. Everall, *Applied Spectroscopy,* 54, 773-782 (2000).
[20] N.J. Everall, *Applied Spectroscopy,* 54, 1515-1520 (2000).
[21] K.J. Baldwin and D.J. Batchelder, *Applied Spectroscopy,* 55, 517-524 (2001).
[22] S. Michielsen, *J. Applied Polymer Science*, 81, 1662-1669 (2001).
[23] A.M. McDonald, A.S. Vaughan and P. Wyeth, *Applied Spectroscopy,* 57, 1475-1481 (2003).
[24] M.C.Malin and K.S.Edgett, *Science*, 288, 2330 (2000).
[25] M.H.Carr, "*Water on Mars*", Oxford University Press, New York (1996).
[26] D.D. Wynn-Williams and H.G.M.Edwards, Environmental UV Radiation:Biological Strategies for Protection and Avoidance, in *"Astrobiology: The Quest for Life in the Solar System"*, eds. G.Horneck and C. Baumstarck-Khan, Springer-Verlag, Berlin, pp. 244-260 (2000).
[27] C.S.Cockell, *Planetary and Space Science*s, 48, 203 (2000).
[28] R.Caricchidi, *Astrobiology*, 2, 281 (2002).
[29] C.S.Cockell and J.Knowland, *Biol.Revs*., 74, 311 (1999).
[30] D.D.Wynn-Williams and H.G.M.Edwards *Icarus*, 144, 486 (2000).
[31] A.Brack, B.Fitton, F.Raulin and A.Wilson, (eds.), "Exobiology in the Solar System and the Search for Life on Mars", *ESA Special Publication* (SP-1231), ESA Publications Division, Noordwijk, The Netherlands (1999).
[32] S.E.Jorge Villar and H.G.M.Edwards, *Analytical &Bioanalytical Chemistry*, 384, 100-113 (2006).
[33] I.R.Lewis and H.G.M.Edwards (eds.),"Handbook of Raman Spectroscopy: From the Process Line to the Laboratory", Marcel Dekker, New York, (2000). S.E.Jorge Villar, H.G.M.Edwards and L.G. Benning, *Astrobiology*, submitted for publication (2006).
[34] M.D.Lane and P.R.Christensen, *Icarus*, 135, 528 (1998).
[35] D.C.Catling, *J.Geophys. Res-Planets*, 104, 16453-16469 (1999).
[36] J.C.Bridges, D.C.Catling, J.M.Saxton, T.D.Swindle, I.C.Lyon and M.M.Grady, *Space Sci. Revs*., 96, 365-392 (2001).
[37] M.E.E.Madden, R.J.Bodnar and J.D.Rimstidt, *Nature*, 431, 821-823 (2004).
[38] H.G.M.Edwards, S.E. Jorge Villar, J.Parnell, C.S.Cockell and P.Lee, *Analyst*, 130, 917-923 (2005).
[39] D.T. Vaniman, D.L.Bish, S.J.Chipera, C.I. Fialips, J.W. Carey and W.C. Feldman, *Nature*, 431, 663-665 2004).
[40] R.E.Arvidson, F.Poulet, J.P.Bibring, M.Wolff, A.Gendrin, R.V.Morris, J.J.Freeman, Y.Langevin, N. Mangold and G.Bellucci, *Science*, 307, 1587-1591 (2005).
[41] Langevin, F.Poulet, J.P. Bibring and B.Gondet, *Science*, 307, 1584-1586 (2005).
[42] J.B.Murray, J.P. Muller, G.Neukum, S.C.Werner, S.van Gasselt, E.Hauber, W.J. Markiewicz, J.W.Head, B.W.Fong, D.Page, K.L.Mitchell and G.Portyankina, *Nature*, 434, 352-356 (2005).
[43] N.A.Cabrol, E.A. Grin and W.H.Pollard, *Icarus*, 145, 91-207 (2000).
[44] J.M.Moore and D.E.Willhelms, *Icarus*, 154, 258-276 (2001).
[45] D.A.Paige, *Science*, 307, 1575-1576 (2005).
[46] H.Hiesinger and J.W.Head, *Planetary and Space Science*s, 50, 939-981 (2002).

[47] S.W.Ruff, P.R.Christensen, R.N.Clark, H.H.Kieffer, M.C.Malin, J.L.Bandfield, B.M.Jakosky, M.D.Lane, M.T.Mellon and M.A.Presley, *J.Geophys.Res-Planets*, 106, 23921-23927 (2001).

[48] H.G.M.Edwards, S.E. Jorge Villar, J.Parnell, C.S.Cockell and P.Lee, *Analyst*, 130, 917-923 (2005).

[49] A. Wang, L. Haskin and E. Cortez, *Applied Spectroscopy,* 52, 477-487 (1998).

[50] D.L. Dickensheets, D.D. Wynn-Williams, H.G.M. Edwards, C. Crowder and E.M. Newton, *J. Raman Spectroscopy*, 31, 633-677 (2000).

In: Planet Mars Research Focus
Editor: L. A. Costas, pp. 165-185

ISBN: 978-1-60021-826-2

Chapter 6

MANNED MISSIONS TO MARS: THE LIKELY IMPACT OF MICROBIAL INFECTION

Peter W. Taylor and Helena Rosado
School of Pharmacy, University of London, 29-39 Brunswick Square, London, UK

ABSTRACT

Manned missions to Mars will result in a diminution of immune status among the crew and cause profound changes to the composition of the bacterial flora in the intestinal tract, nasal passages and upper airways, leading to an increased risk of infection. Experiments conducted during short-term flight and aboard orbiting space stations suggest that growth in microgravity leads to increases in bacterial antibiotic resistance and to cell wall changes. Growth under low-shear modelled microgravity (LSMMG) has convincingly shown that a reduced gravitational field acts as an environmental signal for expression of enhanced bacterial virulence in Gram-negative pathogens such as *Salmonella enterica*. Recent microbiological surveillance of the International Space Station has identified members of the bacterial genus *Staphylococcus* as major contaminants of working and living quarters. We therefore examined the effect of LSMMG on parameters of antibiotic susceptibility and virulence in clinical isolates of *Staphylococcus aureus*. This bacterial species is likely to be the most problematical staphylococcal opportunist encountered in transit. Three strains were grown under LSMMG in a High Aspect Ratio Vessel and compared with cells grown under normal gravity (NG) in the same vessel. There were no significant differences in the antibiotic susceptibility, growth rate or morphology of staphylococci grown under LSMMG compared to NG. LSMMG-induced reductions in the synthesis of the pigment staphyloxanthin were noted. Strains secreted less protein under LSMMG and reductions in hemolysin secretion were found. We noted reduced expression of hemolysins under LSMMG conditions; these virulence determinants are key components of the staphylococcal armamentarium, enabling the pathogen to cause damage to host cells and tissues. Thus, in contrast to published data on Gram-negative pathogens, modelled microgravity reduces the expression of key virulence determinants of *S. aureus*. The relevance of these findings to the risk of infection is discussed in relation to other factors that will impinge on the success of Martian expeditions.

INTRODUCTION

The desire to explore far beyond the boundaries of our native planet, with its concomitant economic and societal aims, is coalescing into strategies to explore and perhaps later colonize Mars. Our nearest planetary neighbour, unlike most if not all other extraterrestrial bodies in our solar system, possesses an abundance of the raw materials required to support life. The announcement in January 2004 by President George W. Bush of the *Vision for Space Exploration* set the National Aeronautics and Space Administration (NASA) on course to replace the Space Shuttle with the Crew Exploration Vehicle Orion in 2010 and undertake further manned lunar missions by 2015, with the goal of living and working on the Moon for increasingly extended periods of time. These activities will serve as a foundation for manned missions to Mars and to worlds beyond. In similar fashion, the European Space Agency (ESA) has committed to a program of Martian exploration with the intention of launching a human mission to Mars in 2030-2033. NASA's strategy to assess, understand, mitigate and manage the risks associated with long-term exposure of astronauts to the space environment is guided by the Bioastronautics Critical Path Roadmap (BCPR). In this context, the International Space Station (ISS) constitutes a key platform for the validation of countermeasures for the circumvention of risks associated with travel beyond Earth orbit [1].

Current and foreseeable limitations of propulsion design mean that manned missions to the Red Planet are likely to be launched from the lunar surface or the ISS and, with outbound and inbound flight durations of 5-6 months each and a projected surface stay of up to 570 days, astronauts will be away from Earth for around 1,000 days. During this time they will be exposed to microgravity approaching zero in deep space and 0.38 g on the Martian surface and will be at severe risk from well-documented microgravity-induced decreases in bone mineral density, with high probability of bone fracture. In addition, psychological problems due to isolation, noise, confinement and a restricted working environment, muscular atrophy, decreased cardiovascular performance, renal stone formation due to elevated urine calcium resulting from bone resorption, and prolonged exposure to ionising and non-ionising forms of radiation will impact on the well being of the crew [2]. These risk factors are well recognized and form key elements of the BCPR. *In vitro* studies have demonstrated that conditions associated with space flight have the capacity to significantly reduce the efficacy of cellular immune function and it is anticipated that, in addition to increasing the risk of allergic and autoimmune disease, decreased immune function could lead to an increased risk of microbial infection or cancers.

Although there is no documented evidence that an onboard infection has resulted in the abortion of a space flight, microbial infections have caused significant problems during orbital missions. The Institute of Medicine, in a wide-ranging examination of the potential medical risks likely to be encountered during exploration missions into deep space [1], drew attention to a report of a severe dental infection suffered by Yuriy Romanenko during a 96-day flight of Salyut 6: the Soviet cosmonaut experienced debilitating toothache for over two weeks that was only remedied after scheduled touchdown. The Soviet authorities had no contingency plan in place to deal with dental emergencies and Romanenko's ordeal was the subject of a televised interview in his own country and of published accounts in the Western dental literature [3]. Analysis of medical events among cosmonauts aboard the Russian space station Mir from March 1995 to June 1998 indicated that a significant number of episodes of infection occurred, including conjunctivitis and acute respiratory and dental infections [1]. Although data from U.S. missions is sketchy, 26 instances of infection were reported for American astronauts during the Space Shuttle program STS-1 to STS-89 over the period April 1989 to January 1998 [1]. Such incidences are reminiscent of the patterns of illness encountered in other extreme environments by submariners during extended submarine missions [4] and by members of Antarctic expeditions [5].

There is a growing body of evidence suggesting that the risk of infection during extended flight may be compounded as a result of the influence of microgravity and the deep space environment on the ecology and behavior of microorganisms, particularly bacteria. In addition to its negative impact on the immune response, space flight leads to profound changes in the composition of the bacterial microflora of the intestinal and nasal tracts, to changes in bacterial growth kinetics, and to reduction of the susceptibility of bacteria to commonly used antibiotics. There are indications that some bacteria become more virulent when cultured under conditions of modelled microgravity. This contribution will evaluate the potential impact of these effects on extended missions into deep space and describe some recent studies in our laboratory on the biology of *Staphylococcus aureus*, an opportunistic pathogen of major importance in various terrestrial environments and a major bacterial contaminant of the ISS.

Space Flight and Immune Status

It has been frequently recorded that individuals, such as polar explorers, who have lived for many months in extreme and isolated environments develop infections immediately after resumption of human contact, suggesting some degree of immune deficit. Although there is good evidence that space flight also impacts on the integrity of the immune response, there is little to suggest flight-induced immune deficits acquired during short- to medium-duration missions result in serious illness. For example, occupation of Skylab for periods of 30 to 84 days resulted in an increased incidence of relatively minor dental problems such as gingival inflammation and dental calculus indicative of some decline in immune function, but more serious manifestations of immune dysfunction were not apparent [6]. Similarly, around one half of the astronauts who flew Apollo missions reported minor bacterial or viral infections within a week of their return that had no lasting impact on their well being [7]. However, it is clear that the likelihood of the appearance of serious immune-related disorders would be much higher during a 1,000-day Martian mission.

Sonnenfeld and Shearer [8] have reviewed current evidence showing the impact of space flight or analog environments on human and animal immune function. Proliferation of human leukocytes in response to mitogenic stimulation, reductions in the synthesis of interferons α and β, inhibition of natural killer cell activity, depression of delayed-type hypersensitivity reactions and alteration of leukocyte sub-populations in the marrow and spleen are just some of the factors that, if maintained for the duration of a long mission, would be likely to eventually compromise astronaut health. The cellular components of the immune system play a central role in the control of bacterial and viral pathogens, limiting their ability to colonize, invade and spread within the body. The reactivation of latent herpes virus during space flight is a clear manifestation of changes in the immune status of astronauts and an indication that extended flight may challenge the immune defenses of even the fittest. There are a range of opportunistic pathogens, including bacteria, fungi and viruses that depend on reduced immune function to cause serious infectious disease, and some of these will inevitably accompany the crew into deep space.

Fellow Travellers - Microbes on Board

In line with COSPAR recommendations regarding the preservation of pristine environments in our solar system, spacecraft are manufactured in ultra-clean facilities comparable to those used for the manufacture of medicines. In spite of these precautions, low-level contamination of spacecraft often occurs. For example, the surfaces of Mars Odyssey spacecraft assembled at the Kennedy Space Center Spacecraft Assembly and Encapsulation Facility exhibited very low levels of contamination by a range of robust microbial species, often of the genus *Bacillus* [9, 10].

Microbes originating from the commensular flora of the crew will inevitably colonize the spacecraft, and onboard supplies of air, food and water will provide further sources for contamination of the fabric of the vessel. Pierson [11] reported that potable water generated by the fuel cells aboard Space Shuttle flights was commonly contaminated with very low levels of *Burkholderia cepacia* and other bacteria. Water brought from ground sources and stored aboard Mir for potable use tended to display higher bacterial counts [11]. Similarly, aboard the ISS, potable water reclaimed from humidity condensate yielded very low levels of bacteria, although water transported to the platform by the Space Shuttle contained somewhat higher levels [12]. On extended duration missions to Mars, microbes are likely to be used in solid waste remediation and may serve as a source of food [11]. The attendant microbial contamination associated with the onboard cultivation of plants, as a food source or as a component of a bioregenerative life-support system [13], as well as the transport of agricultural materials for cultivation in Martian greenhouses [14], will also contribute to the complexity of the microbial population within the closed environment of the vessel. On a 1,000 day mission to Mars, experience suggests that the extent and complexity of microbial contamination will increase with time away: the cloudy humidity condensate collected from behind a service panel during the NASA 6 mission to Mir in January 1998 contained a wide variety of bacteria, including a range of Gram-negative bacteria that are only infrequently associated with contamination aboard short duration missions [15]. Analysis of one sample suggested the presence of a member of the genus L*egionella*, bacteria that can cause lethal infections, although the isolate was refractory to subculture. The consortium of organisms that had accumulated over the twelve years since the launch of Mir also included fungi of medical importance, protozoa, dust mites and spirochetes. Some bacteria were embedded in a glycocalyx to form biofilms of interconnected bacterial cells that are known to be more resistant to disinfectants and antibiotics than less developed bacterial communities. Although the threat to crew health of the development of such diverse microbial populations is unclear, these findings indicate that microbial monitoring and vessel disinfection should be significant factors in the design, engineering and operation of vessels commissioned for missions to Mars.

The continuous human occupation of the ISS since November 2000 makes the station an excellent test bed for the prediction of likely microbiological problems on future Mars-bound exploration missions. The approach to microbiological risk on the ISS is one of prevention rather than reliance on in-flight solutions, and highly efficient air filtration systems, microbiological monitoring, and features to minimize the accumulation of moisture have been incorporated into platform design [11]. The microbiological status of the air, water and surfaces inside the ISS has been regularly monitored over the past six years [16] and some 500 samples collected at different locations. The number of viable microbes in potable water never exceeded 100 per ml and the concentration of airborne bacteria and fungi were 710 and 44 per m^3 respectively. Bacterial contamination on surfaces, presumably resulting from deposition of airborne cells, fluctuated between 25 and 43,000 per 100 cm^3, depending on sampling location. By far the most predominant bacteria were members of the genus *Staphylococcus*; they were isolated from 84% of samples. *S. aureus* and other opportunistic pathogenic species were frequently encountered. Bacteria with the capacity to degrade material components of the station such as electronic insulation cables and damage metal surfaces were also isolated. The microbiota of the environment within the ISS appears to be dominated by bacterial species associated with the skin and mucous membranes, in a fashion not dissimilar to that of a medical care unit. In total, over 70 species of microorganisms have been found in the environment of the orbital complex, about half being bacteria and half fungi, demonstrating the appearance of remarkable biodiversity over a relatively short period of time.

The resident intestinal, oral and nasal flora of the crew will make by far the largest contribution to the microbial reservoir on board an extended mission. Around 10^{14} microorganisms, the overwhelming majority bacteria, are found in the intestinal tract and constitute 50-60% of the fecal mass [17]. The population is comprised of about 150 bacterial species, although 30-40 species account for 99% of the total; most are anaerobes and the colon contains more bacterial mass than other regions of the intestinal tract. There is complex and poorly understood interplay between the intestinal microflora, the immune system and various external factors that include diet, stress and behavior. Even prior to launch, significant reductions in the numbers of beneficial lactobacilli in the gut has been observed and attributed to pre-launch stress [18]. As early as two weeks into the confinement imposed by space flight there is a significant reduction in the number of bacterial species that can be isolated from the intestinal tract as well as an interchange of intestinal bacteria between crew members [19, 20]. A simplification of the intestinal bacterial flora was also noted in Apollo and Skylab crews, favoring the emergence of robust Gram-negative aerobic species such as *Klebsiella pneumoniae*, *Pseudomonas aeruginosa* and various Enterobacteriaceae [7]. These observations parallel terrestrial experiments in which volunteers were kept in isolation and fed only sterilized, dehydrated foods: these conditions resulted in sharp decreases in the numbers of bifidobacteria, lactobacilli and other beneficial bacteria. The ingestion of bacteria present in food appears to assist in the maintenance of a healthy and diverse intestinal flora; withdrawal of an exogenous source of gut-colonizing bacteria leads to undesirable changes that affect immunocompetence and nutritional status, increasing the risk of opportunistic infections, poor wound healing and metastatic disease. For these reasons, it has been proposed

[20, 21] that probiotic preparations, containing viable *Lactobacillus casei* and other beneficial bifidobacteria that protect the gut from colonization by potentially harmful species, might prevent adverse changes in gut flora composition.

In light of the nature of the bacterial contamination of the interior surfaces of the ISS, the microbiota of the skin and oral and nasal passages warrant attention. Carriage of staphylococci in the nares and upper airways is ubiquitous in the general population. Confinement is known to result in quantitative and qualitative changes to the microbial flora of the skin, mouth and pharyngeal mucosa, with increases in the numbers of opportunistic pathogens, such as *S. aureus*, and a reduction in the number of non-pathogenic bacteria. Similar changes have been noted with regard to the composition of the nasal flora in Soviet cosmonauts [19]. Less extensive changes in the upper respiratory tract microflora were reported in a study of Apollo 13 through 16 missions of 140-150 h duration [22], although a marked decrease in the isolation frequency of streptococci, corynebacteria and Gram-negative rods and a concomitant increase in the frequency of *S. aureus* were noted. The data support the notion that significant transfer of *S. aureus* occurred between members of the crew on these missions, a process that may have been accelerated in a microgravity environment by the generation of larger respiratory droplets, with a longer settle time, compared to those found on Earth. Qualitative differences between the Soviet and American studies are most likely due to differences in atmospheric composition, environmental control systems and crew hygiene regimens [22] although a key finding in both studies was the propensity of the troublesome opportunistic pathogen *S. aureus* to emerge, spread and colonize most if not all crew members. This organism may cause serious, even life-threatening infections in individuals with even moderately compromised immune systems [23] and the infections they cause are often difficult to treat in terrestrial healthcare institutions, let alone within the confines of a space vessel.

EFFECT OF SPACE FLIGHT ON THE BACTERIAL PHENOTYPE

The environment within the confines of a spacecraft travelling beyond Earth's orbit is unique and one that fellow-travelling microbes would not have previously encountered. Manned flight brings further complexity and there is convincing evidence that the combination of zero or microgravity and cosmic and solar radiation has significant effects on the properties of bacteria and fungi [24, 25]. Together with flight-induced changes to immune function, alterations in the phenotype of normally harmless or low-grade opportunistic pathogens could result in the emergence of episodes of infection.

Mishra and Pierson [24] have reviewed early Soviet and American studies on the effects of short orbital flights on the viability and growth of a range of microorganisms. Although much of this data is conflicting, there are indications that metabolic activity (reflected in a shorter lag phase before the onset of exponential cell division) and rates of growth increase as a result of exposure to flight conditions in comparison to terrestrial controls. Although clear evidence is lacking, these effects are usually attributed to the influence of low gravitational fields on the external fluid environment surrounding cells in suspension culture [25]. However, *E. coli* and *B. subtilis* cultured on solid medium during flight grew at the same rate and to the same extent as terrestrial controls, suggesting that fluid dynamics and extracellular transport phenomena, rather than cellular dynamics, contributed to the microgravity-induced differences in growth kinetics seen in earlier studies that used liquid culture [26]. More recent studies on Space Shuttle flights (STS-50, -54, -57, -60, -61 and –63) demonstrated that a range of bacteria grown under batch culture conditions yielded higher final cell concentrations and total biomass, with increased growth rate and a shorter lag phase in comparison to terrestrial laboratory cultures [27-29]. These studies were undertaken using *Escherichia coli* and *Bacillus subtilis* grown in minimal salts medium.

Treatment of on-board infections could be compromised by reversible or irreversible increases in antibiotic resistance and there is some evidence (not reported in the open literature) that bacteria cultured from astronauts in flight were more resistant than isolates obtained from the same individuals either pre- or post-flight [30]. For these reasons, experiments were carried out in July 1982 aboard Salyut 7 as part of the Cytos 2 programme: *S. aureus* and *E. coli* isolates from the commensal flora of the French astronaut Jean-Loup Chrétien were examined for their antibiotic susceptibility during orbital flight and compared to ground controls [30, 31]. The minimal inhibitory concentrations (MICs) for colistin and kanamycin determined onboard against the *E. coli* isolate were recorded as >16 $\mu g.ml^{-1}$ compared to control values of 4 $\mu g.ml^{-1}$ for both antibiotics. The *S. aureus* isolate was tested against oxacillin, chloramphenicol and erythromycin: control values of 0.16, 4 and 0.5 $\mu g.ml^{-1}$ were found that increased around twofold when MICs were undertaken aboard the orbital station. It is generally accepted that a twofold difference in MIC values is not significant given the technical limitations of the bioassay [32], notwithstanding the difficulty of making accurate in-flight determinations and the fact that readout was based on pH-induced color change rather than the recommended turbidity endpoint. After recovery of the samples, the ultrastructure of the *S. aureus* isolate was examined by electron microscopy after fixation and sectioning. Images from this experiment (Figure 1) show that terrestrial controls have an appearance typical of *S. aureus*, with clearly differentiated peptidoglycan cell walls and septum formation in the plane of cell division. In-flight cultures have a somewhat different morphology and this has been interpreted as showing very large increases in the thickness of the staphylococcal cell wall [31, 33]. A thickened cell wall is associated with increases in vancomycin susceptibility [34, 35] and changes in staphylococcal susceptibility to β-lactam antibiotics due to growth in the presence of polyphenols [36]; enzymes involved in the synthesis of cell wall peptidoglycan are, directly or indirectly, the targets for these agents [23, 36]. However, the in-flight image does not resemble staphylococcal cells, even accounting for the fact that the bacteria may have undergone extensive phenotypic modification: the layers external to the cytoplasmic membrane appear much less dense than those associated with staphylococci with thickened cell walls [37] and the cell surface appear to be undergoing "blebbing" reminiscent of that associated with Gram-negative bacterial cells [e.g. 38]. These doubts raise the possibility that Figure 1b may represent an image of a Gram-negative contaminant and not that of *S. aureus* cells. In light of numerous technical advances in this area over the last quarter-century, it would be of value to repeat these experiments aboard the ISS. Flight-induced changes were phenotypic, as antibiotic susceptibility patterns reverted to pre-flight values when in-flight cultures were sub-cultured in a terrestrial environment.

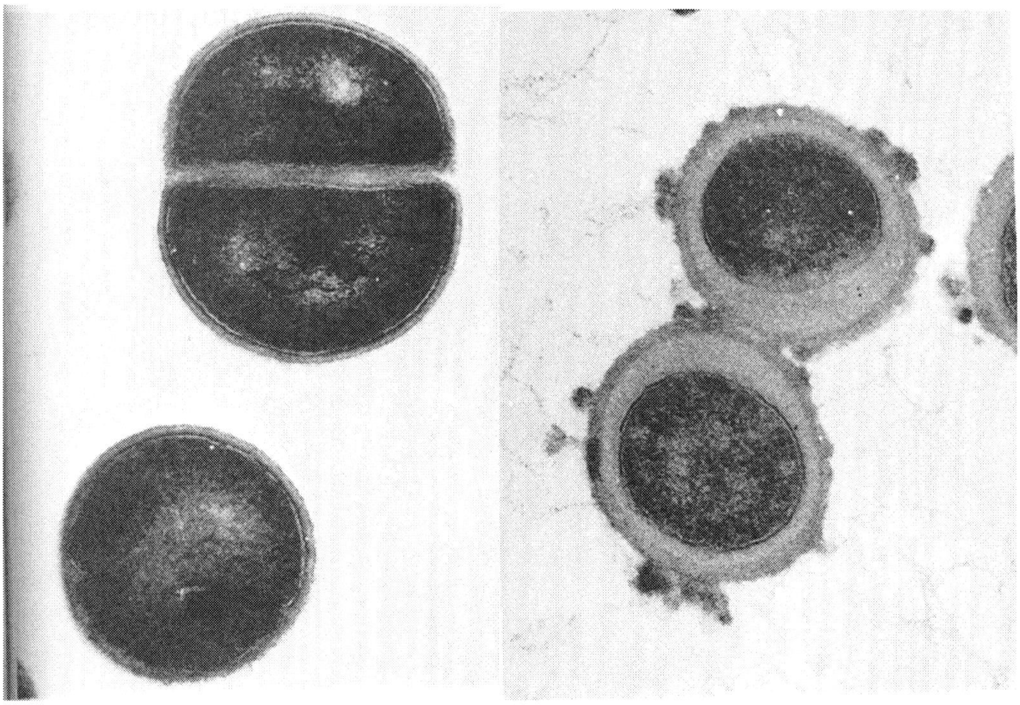

a b

Figure 1. Ultrathin sections of (a) terrestrial and (b) in-flight-cultivated *S.aureus*. Experiment undertaken aboard Salyut 7 by J-L. Chrétien during the Cytos 2 program in 1982. Reprinted from Tixador et al. [29] with permission from Elsevier.

Similar experiments were conducted aboard Space Shuttles Challenger [33] and Discovery [39]. In contrast to the Franco-Soviet studies, these included on board controls that took into account additional effects such as vibration and acceleration in-flight compared to ground controls. *E. coli* strain ATCC 25922 was examined for susceptibility to colistin in-flight under microgravity conditions and in centrifugal simulated 1 g and compared to static rack (1 g) and centrifugal (1.4 g) conditions on Earth. In flight MIC determinations (microgravity and 1 g) yielded MICs of 2 $\mu g.ml^{-1}$; these are double the values obtained with the Earth controls [33]. This data has been viewed as indicative of a microgravity-induced increase in antibiotic resistance [33, 40]; however, caution must be exercised in interpreting such small differences in MIC. Determination of the effect of flight board STS-42 on the growth rate of ATCC 25922 in the presence and absence of sub-inhibitory concentrations of dihydrostreptomycin yielded inconclusive results [39]: differences between in-flight and ground controls were small. Although the MICs appeared identical at 8 $\mu g.ml^{-1}$, there were differences in the growth curves at 6 $\mu g.ml^{-1}$ that suggested subtle changes in the interaction between the antibiotic and the ribosomal target within the bacterial cell; this difference could be related to the chemical stability of the antibiotic in space (see below). Kacena and Todd [41], in demonstrating that *E. coli* grown on solid agar medium aboard STS-69 and -73 were as susceptible to gentamicin as ground controls, highlight the caution that must be taken in the interpretation of in-flight antibiotic sensitivity tests, particularly when suspension culture techniques are used. Clearly, more work is needed to clarify the anomalies presented by in-flight data in this important area. A systematic approach would shed light on the impact of microenvironments within bacterial suspension cultures on MIC values, and clarify whether the nature of the antibiotic used and its molecular target within the bacterial cell are important factors that determine the outcome of interactions of antibiotics and bacteria in space.

Bacteria and Modelled Microgravity

The factors that may impact on bacterial physiology, leading to changes in gene expression and behaviour, can only be examined simultaneously during actual space flight. These include zero and microgravity, acceleration, vibration, radiation, electromagnetism and additional environmental stresses associated with the closed environment of the space vehicle. However, opportunities to examine these parameters in space are limited by the constraints of in-flight experimentation. These include the need for the development of specialized equipment and restrictions imposed on power, weight and volume; intense competition for the crew's time dictates that experiments are simple to perform with little or no crew involvement [24]. It is, therefore, essential that data from infrequent in-flight experiments are supplemented by studies employing simulated flight parameters.

The development of clinostats, in particular the High Aspect Rotating Vessel (HARV), has stimulated a considerable amount of research into the effect of modelled microgravity on the physiology of a wide range of cells, including microorganisms. The HARV was developed at the NASA Johnson Space Center [42] and consists of a hollow cassette completely filled with growth medium that slowly rotates on an axis parallel to the ground such that the bacteria are continually suspended in the medium, falling through a sustained low-shear (<1 $dyn.cm^2$) environment that simulates true microgravity. The excellent review by Nickerson and colleagues provides a detailed account of this technology [40]. Using the HARV, the Nickerson group have provided clear evidence that microgravity acts as an environmental signal for expression of enhanced bacterial virulence in Gram-negative bacteria [43]. They determined that *Salmonella enterica* serovar Typhimurium, a human and animal pathogen frequently causing gastrointestinal infections, displayed enhanced virulence for mice when grown under low-shear modelled microgravity (LSMMG) compared to bacteria grown under normal gravity (NG) [44]. LSMMG-grown cells had a decreased LD_{50}, produced shortened time to death of infected test animals, and more readily colonized the liver and spleen. The processes involved in LSMMG-induced gene regulation appear to be novel and may be microgravity-specific [45]: the authors hypothesize that, in response to LSMMG-induced stimuli, the organism is "fine tuning" the expression of virulence mechanisms in a novel fashion. These groundbreaking studies have cast new light on novel bacterial phenotypes that may arise during extended space travel and highlight the urgent need for such work to be extended to include other groups of pathogens, particularly Gram-positive bacteria such as *S. aureus* that are likely to pose the greatest risk to the crew of opportunistic infection.

We have initiated an investigation of the effect of LSMMG on the parameters of antibiotic susceptibility and virulence in clinical isolates of the Gram-positive pathogen *S. aureus*. Some of this data has been presented at the 57th International Astronautical Congress, Valencia, October 2-6th 2006. The methicillin susceptible *S. aureus* (MSSA) isolates RF1, RF6 and RF11, obtained from clinical samples processed at the Royal Free Hospital in

London, were grown in Müller-Hinton (MH) broth using Synthecon RCCS-1 HARVs housed in a 37°C incubator at a rotation of 25 rpm. The bioreactors were orientated to grow cells under conditions of LSMMG (rotation about the horizontal axis) or NG (rotation about the vertical axis). To determine growth rates, samples were removed *via* the sampling port and the OD_{600} measured. In all other experiments, cells were harvested after 24 h incubation.

Changes in growth kinetics induced by LSMMG may impact on other parameters of interest. We therefore compared the growth rates of *S. aureus* grown under NG and LSMMG (Figure 2). No differences in growth kinetics over the 24h culture period were seen with any of the three clinical isolates examined. The Gram-negative bacteria *S. enterica* [46] and *E. coli* [47] appear to show increased rates of division and reach a higher cell density under LSMMG compared to NG, although such effects could not be demonstrated with *P. aeruginosa* [48]. These published studies were undertaken using minimal salts medium. When *S. enterica* was examined following growth in complex Luria-Bertani broth, the growth kinetics of LSMMG-grown cells were identical to those obtained under NG conditions [46]. As staphylococci have complex nutritional requirements, our experiments could not be undertaken using chemically defined minimal salts media. Thus, our data obtained using nutritionally complex MH broth lend some support to the view that the combination of minimal medium and LSMMG induce increased metabolic activity that drives growth [40].

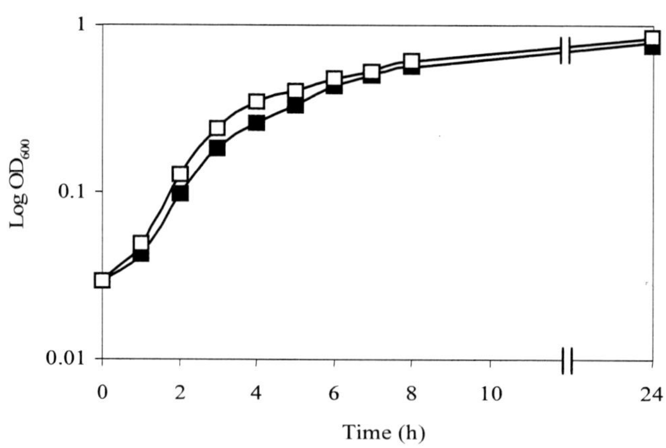

Figure 2. Growth kinetics of *S. aureus* RF1 under normal gravity (■) and simulated microgravity (□). Similar growth curves were obtained for isolates RF6 and RF11.

In marked contrast to the images obtained from the cells cultured aboard Salyut 7 (Figure 1), there were no discernable differences in staphylococcal cell morphology as revealed by scanning electron microscopy (Figure 3) and transmission electron microscopy (Figure 4). In contrast to reports of in-flight investigations conducted in nutritionally complex medium with the Crétien *S. aureus* isolate [31, 33], we found no changes in the susceptibility of RF1, RF6 and RF11 to the macrolide antibiotic erythromycin, the β-lactam flucloxacillin and the glycopeptide vancomycin (Table 1). These antibiotics were selected for investigation because they interact with different targets in the bacterial cell: erythromycin interrupts the early stages of protein synthesis by binding to the 50S ribosomal subunit whereas flucloxacillin and vancomycin interfere with different steps in cell wall peptidoglycan synthesis.

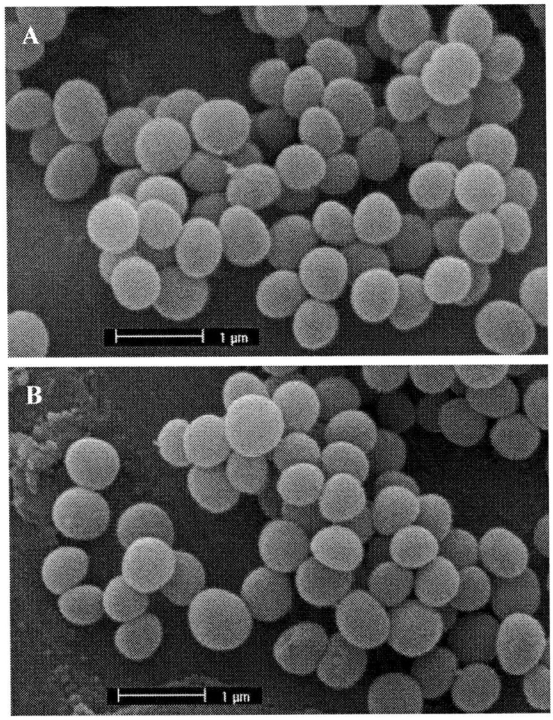

Figure 3. Scanning electron microscopy images of *S. aureus* RF6 grown under normal gravity (A) and simulated microgravity (B). Bacteria were recovered by centrifugation, washed twice with phosphate buffered saline (PBS), fixed in 1.5% w/v glutaraldehyde for at least 2 h at room temperature and washed once with 70% ethanol and twice in 100% ethanol. Air-dried, gold-coated preparations were examined using a FEI XL30 scanning electron microscope.

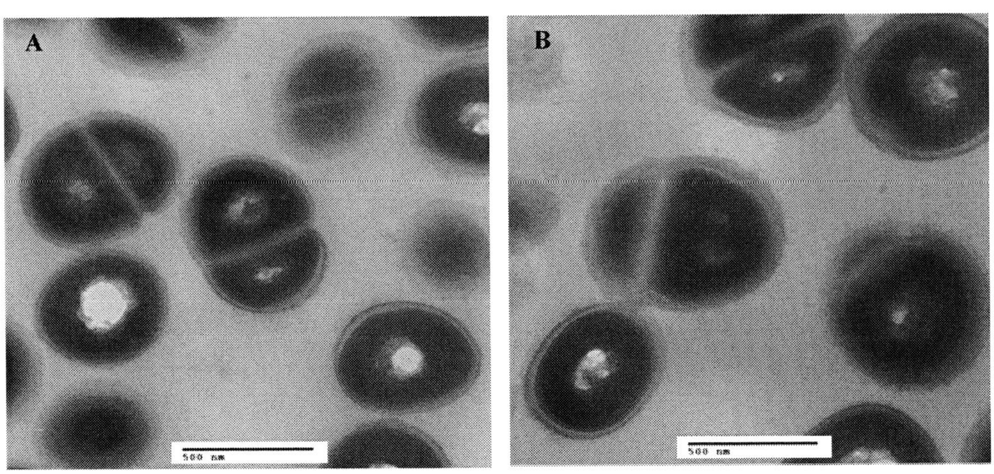

Figure 4. Transmission electron microscopy images of sections of *S. aureus* RF6 grown under normal gravity (A) and simulated microgravity (B). Bacteria were recovered by centrifugation and washed twice in PBS. Bacteria were fixed in 1.5% w/v glutaraldehyde for at least 2 h at room temperature, treated with osmium tetroxide and embedded in epoxy resin. Sectioning and staining with uranyl acetate was followed by Reynolds' lead citrate. Ultra-thin sections were viewed and photographed using a Philips 201 transmission electron microscope.

Table 1. Influence of simulated microgravity on the susceptibility of S. aureus isolates RF1, RF6 and RF11 to antibiotics (NG, normal gravity; LSMMG, low-shear modelled microgravity). MICs were determined by the broth dilution method using serial twofold dilutions of the antibiotics. Assays were performed in 96-well microtiter trays with an inoculum of 10^4 CFU in μ100 l of MH broth. *S. aureus* ATCC29213 was used as a reference strain to ensure comparability with published data. MIC values were obtained after incubation at 37°C for 24 h. The MIC was determined to be the lowest antibiotic concentration at which no visible bacterial growth occurred.

Antimicrobial Agent	Minimum inhibitory concentration (μg. ml^{-1})					
	RF1		RF6		RF11	
	NG	LSMMG	NG	LSMMG	NG	LSMMG
Erythromycin	1	1	0.25	0.12	>128	>128
Flucloxacillin	0.25	0.12	0.25	0.25	0.12	0.12
Vancomycin	1	1	0.5	0.5	1	1

Growth in a microgravity environment had an impact on a number of factors associated with the virulence of *S. aureus*. Most *S. aureus* strains produce carotenoid pigments that impart a golden hue to colonies of this pathogen. The main pigment is staphyloxanthin, a triterpenoid esterified with a C15 fatty acid [49]. A recent study has linked carotenoid pigment synthesis to staphylococcal virulence; mutants unable to synthesise these molecules have impaired survival within neutrophils and are less pathogenic in a mouse subcutaneous abscess model [50]. All three strains that we examined produced less staphyloxanthin when grown under simulated microgravity compared to NG-grown cells (Figure 5).

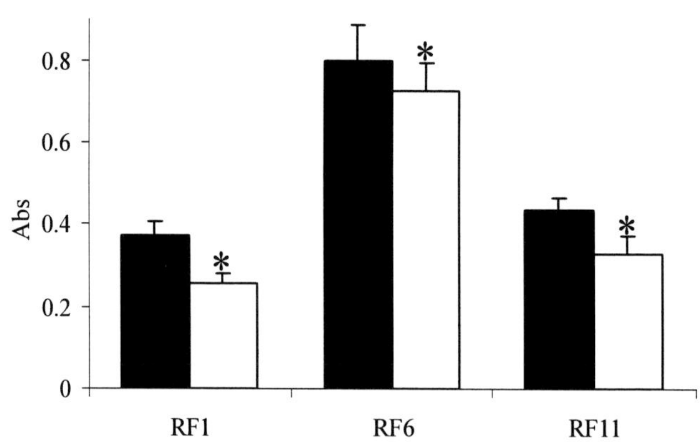

Figure 5. Production of staphyloxanthin by *S. aureus* isolates grown under normal gravity (■) and simulated microgravity (□). Error bars represent 1SD; n=6. *Paired Student's *t* test, p<0.01. Staphyloxanthin and intermediate carotenoids were extracted and quantified according to Wieland et al. [49]. Bacteria were harvested by centrifugation (3,000 *g*, 10 min) and washed twice with PBS. The packed cells were either extracted immediately or maintained at -20°C for up to 3 months without affecting their carotenoid content. Cells (10 mg dry weight) were suspended in 500 μl of methanol, held at 55°C for 5 minutes, cooled and extracts obtained by centrifugation. The extraction was repeated twice more. Carotenoids were estimated quantitatively by measuring absorbance of solutions at 462nm, the absorbance maximum of the primary carotenoid pigment staphyloxanthin.

Staphylococcal pathogenesis is multifactorial and involves secreted proteins that include cytotoxins, tissue-degrading enzymes and superantigens [23]. In addition, cell surface-bound proteins that recognize adhesive matrix molecules such as fibronectin- and collagen-binding protein, along with cell surface components such as the capsule and peptidoglycan, play a key role in determining the capacity of staphylococci to cause disease in humans. Secreted proteins enable the organism to attack the cellular and structural elements of diverse tissues and organs of the infected host and contribute to the spread of the pathogen and its ability to produce symptoms of disease. We have therefore examined some of these key secreted virulence determinants. The three isolates secrete a wide array of proteins into the surrounding medium and there are clear differences in protein secretion patterns between these strains (Figure 6). Microgravity cultures secreted less protein than NG-cultured bacteria and the quantity of some individual proteins recovered from the growth medium was markedly reduced. The largest reduction in protein secretion occurred with RF6 (Figure 6). Interestingly, the secretion of a few proteins was increased under LSMMG: for example, RF6 and RF11 appeared to increase the secretion of a protein with a molecular mass of ca 24kDa. We are in the process of identifying the nature of this protein.

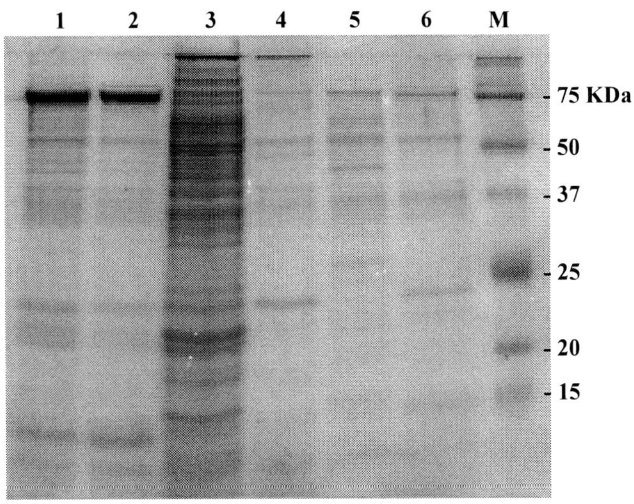

Figure 6. Sodium dodecyl sulphate polyacrylamide gel electrophoresis (SDS-PAGE) of proteins secreted by *S. aureus* isolates visualised with Comassie blue. Cells were grown under normal gravity (NG) and simulated microgravity (LSSMG). Lanes: (1) RF1 NG; (2) RF1 LSSMG; (3) RF6 NG; (4) RF6 LSSMG; (5) RF11 NG; (6) RF11 LSSMG; (7) Molecular weight marker. Concentration of acrylamide: 10% w/v. Proteins were visualised by staining with Coomassie blue.

S. aureus isolates secrete one or more cytotoxic proteins that have the capacity to lyse red blood cells; α-toxin, the *hla* gene product, is one of the most potent bacterial toxins known and contributes significantly to morbidity and mortality in human infections [23]. A significant number of *S. aureus* strains produce additional hemolysins such as β-toxin, δ-toxin, γ-toxin and other less well-characterised biocomponent toxins [23]. The three isolates that we have examined secreted toxins that lysed erythrocytes from sheep blood (Figure 7a); all the major membrane-active staphylococcal exotoxins possess lytic activity against erythrocytes of this animal species [23, 52]. Growth under LSMMG reduced the expression of sheep erythrocyte haemolysins in all isolates; with RF6 the effect was marked, with toxin secretion almost completely absent. As the β-hemolysin, in contrast to the α- and δ- forms, is unable to lyse rabbit erythrocytes [52], we examined the capacity of culture supernatants to lyse red cells from this animal species (Figure 7b). LSMMG-grown bacteria possessed less hemolytic activity and the largest difference was again found with strain RF6; comparison of Figure 7a and 7b indicates that β-toxin activity made a significant contribution to the overall lytic profile of RF6 against sheep erythrocytes.

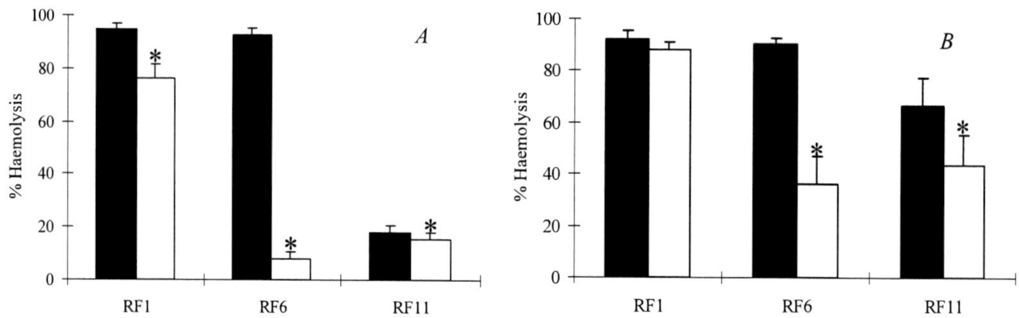

Figure 7. Secretion of sheep (A) and rabbit (B) erythrocyte haemolysins by *S. aureus* strains grown under normal gravity (■) and simulated microgravity (□). Error bars represent 1SD; n=5-6. *Paired Student's *t* test, $p<0.01$. Defibrinated sheep or rabbit blood (10 ml) was centrifuged (1,000 *g*, 15 min), the erythrocytes washed three times with PBS and suspended in cold sterile PBS (1% v/v). A normalized volume of culture supernatant was added to aliquots of this suspension to give a combined final volume of 1.5 ml; mixtures incubated at 37°C for 30 min, followed by 25 min at 4°C. After centrifugation to remove intact cells, the degree of lysis was determined by absorption at $A_{540\ nm}$.

These studies strongly suggest that *S. aureus*, a problematical Gram-positive opportunistic pathogen in the context of the theme of this contribution, is rendered less virulent following growth in modelled microgravity. This is in marked contrast to the situation with the Gram-negative pathogen *S. enterica* serovar Typhimurium that has been rigorously examined by Nickerson and colleagues [40, 44] and found to possess increased virulence when cultured under LSMMG. The separation of bacteria on the basis of their Gram-staining characteristics represents a fundamental division within the prokaryotic kingdom and it is possible that the two bacterial groups interpret the microgravity signal in entirely different ways. However, it is clear that more research to establish ground rules in the area is badly needed. Very little effort has been expended on examination of the effect of microgravity on the phenotype of Gram-positive bacteria. Impaired immune function provides an opportunity for bacteria of moderate virulence to more easily overcome innate and adaptive host defense mechanisms than is the case in fully immunocompetent individuals. Thus, a microgravity-induced decrease in virulence, in particular with pathogens such as *S. aureus*, is welcome news for long-term space travel. Conversely, increases in the capacity of bacteria to cause infection in an unpredictable way may put great strain on the onboard pharmacy, which will of necessity be of limited in the range of drugs carried. In this context, agents such as the catechin gallates [36], which compromise the virulence properties and antibiotic resistance profiles of *S. aureus*, may be valuable additions to the complement of onboard therapeutics [21].

Implications for Control of Infection during Mars-Bound Flight

Manned flight to Mars is a logical next step in the evolution of space flight. Manned flight progressed rapidly from sub-orbital and short orbital flights to lunar missions. The launch of the first Soviet orbiting space station occurred in April 1971 and was followed by the NASA Skylab Space Station in 1973. The current era of international cooperation has seen the construction of the ISS. The ISS will enable the elucidation of the biomedical risks and hazards relating to space habitation and will play a key role in the launch of projected missions to Mars. However, extended travel through deep space combined with a prolonged stay on the Martian surface will expose astronauts and equipment to a myriad of novel, and in some cases, unknown risks. The impact of a journey into deep space on the behavior of the fellow-travelling microbial population will also be difficult to predict with any certainty.

The effects of microgravity on bacterial physiology have been well studied in comparison to other parameters associated with space flight, although this body of work sheds little light on possible microbial responses to gravitational fluctuations that will be encountered on a Mars mission: a period of microgravity followed by zero gravity, then a significant interval under the influence of the Martian gravitational field of 0.38 g before a repetition of the flight conditions on the return transit. The risks to the crew from a continuous flux of high-energy, ionizing cosmic ray nuclei (HZE) – leftovers from collapsing stars and supernova explosions – and occasional but intense fluxes of solar energetic particles, primarily protons and alpha particles, have been well recognized [53, 54], but they may also affect microbes in ways that are barely understood. This gap in our knowledge is unlikely to be filled by conducting radiation exposure experiments aboard the ISS, as the station sits inside Earth's magnetic field. The thin Martian atmosphere ensures that there will be little respite from radiation exposure, including HZE, on the surface of the Red Planet [55], although bombardment will at least be unidirectional. In addition to radiation effects on the properties of microorganisms that may impact on astronaut health, bacteria will be essential for the recycling of waste and production of food in-flight and on the Martian base, so it is essential that the impact of radiation on microbial physiology is better understood before embarking on interplanetary exploration. After a 40-day flight aboard Mir, mutation rates for a cloned bacterial gene carried by a yeast strain were two to three times higher compared to the ground control [56]; as the emergence of bacterial drug resistance and adaptation to a more virulent form may arise through gene mutation, there may be potential for the generation of more harmful bacterial genotypes through increased mutation rates.

Current measures to reduce levels of microbial contamination [11] combined with good housekeeping practise, such as the use of UV light and a range of disinfectants and other sanitizing materials, should be adequate to restrict the potential for cross-contamination and fortuitous infection of wounds, cuts and abrasions although concerns have been raised regarding the impact of flight-induced changes to the bacterial phenotype on the efficacy of disinfectants [57]. It has been demonstrated that biofilms, communities of bacteria that grow on surfaces and characterized by the elaboration of a protective adhesive matrix, appear to form readily in space flight [58]. These complex aggregations are associated with biofouling and are far more resistant to both antibiotics and disinfectants than their free-floating (planktonic) counterparts. There are clear implications here for onboard hygiene and sanitation as well as infection control.

The stability of pharmaceutical products is a complex issue and many factors impact on the changes in the composition of the finished product [59]. The ability of a pharmaceutical product to retain its properties within specified limits determines its shelf life, the period of time during which a product is expected, if stored correctly, to comply with the specification. Physical (e.g., dissolution, moisture content, pH, package integrity), chemical (degradation products, loss of potency and alterations to excipients) and microbiological (proliferation, sterility, preservative efficacy) changes will contribute to the determination of shelf life. For most products in tablet, capsule or liquid form a shelf life of 24 to 60 months will ensure the viability of the product in the commercial marketplace. Once released, the product will be outside the control of the manufacturer and may be dispatched to regions of the globe where storage conditions will be less than ideal; the recommended storage life should be reduced accordingly in such situations [60]. Clearly, the stability of medicines will be central to the maintenance of an effective onboard pharmacy assembled for missions to Mars. There are indications that the radiation environment in space has a deleterious effect on the stability of pharmaceutical preparations [61]. Data gathered from Space Shuttle missions suggest that medicines and vitamins undergo decreases in physical stability such that bioavailibility and potency are affected after even short periods of time. Significant degradation of the active ingredients in the antibiotic preparations Augmentin and Bactrim are particularly worrisome for the effective treatment of infections that are likely to arise in the weightless environment on board a Mars voyager due to injury and trauma, such as lacerations and open fractures. Augmentin is a broad-spectrum combination agent active against a wide range of bacteria and could form a key component of the onboard pharmacy. This important area is under-researched. The Stability project, begun during ISS Expedition 13, should go some way to correcting this: the in-flight stability of the antibiotics Azithromycin, Augmentin, Bactrim, Cefadroxil, Ciprofloxacin and Clotrimazole (an antifungal) are being evaluated, in addition to a number of other pharmaceutical products.

There are significant hurdles that need to be overcome before the astronautical community can be confident that adequate measures are in place to cover the eventuality of serious infection during a 1,000-day Martian mission. A weakened immune system will provide opportunities for bacteria to cause serious infections that may be difficult to treat if reduced gravity and other factors unique to the space environment conspire to increase bacterial virulence and reduce the susceptibility of pathogens to conventional antibiotics, which may themselves be rendered less effective by radiation bombardment. We continue to rely on antibacterial drugs that were identified and developed during the "Golden Age" of antibiotic discovery from around 1945 to 1970, and their utility is being constantly eroded by the emergence and dissemination of bacterial drug resistance genes. There has been much recent interest in unconventional approaches to the treatment of infection [62-66]; not all of these rely on destruction of target pathogens, but on their modification such that they are less "fit" – less able to survive within the body and cause damage. This approach may circumvent the emergence of resistant forms by Darwinian natural selection, as modifying agents do not remove the overwhelming number of drug-susceptible bacteria and allow the few pre-existing variants to predominate [64]. These alternative approaches should be considered for use on board a Mars bound mission [21]: some of the potential therapeutics in this area are derived from natural sources and could be generated during flight to become components of a biopharmacy helping to sustain permanent human colonies beyond Earth's boundaries.

CONCLUSION

The opportunistic bacterial pathogen *Staphylococcus aureus*, regularly found as a contaminant aboard the ISS and with the potential to cause infections during Mars expeditions, shows significant reductions in the expression of secreted virulence-associated proteins when grown under modelled microgravity. The three staphylococci examined showed no increases in resistance to widely used antibiotics, suggesting that, in contrast to earlier studies, microgravity does not act as an environmental trigger for the expression of a more robust phenotype. These studies highlight the need for further investigations to determine the nature of risk factors impacting on the expression of antibiotic resistance and virulence in the deep space environment.

REFERENCES

[1] Ball, J.R., & Evans, C.H. (2001). *Safe Passage: Astronaut Care for Exploration Missions*. Washington DC: National Academy Press.

[2] Grigoriev, A.I., Kozlovskaya, I.B., & Potapov, A.N. (2002). Goals of biomedical support of a mission to Mars and possible approaches to achieving them. *Aviat. Space Environ. Med., 73*, 379-384.

[3] Wheatcroft, W. (1989). Effects of simulated Skylab missions on the oral health of astronauts. *J. Gr. Houston Dental Soc., 61*, 7.

[4] Thomas, T.L., Hooper, T.I., Camarca, M., Murray, J., Sack, D., Mole, D., Spiro, R.T., Horn, W.G., & Garland, F.C. (2000). A method for monitoring the health of U.S. Navy submarine crewmembers during periods of isolation. *Av. Space Environ. Med., 71*, 699-705.

[5] Taylor, D.M., & Gormly, P.J. (1997). Emergency medicine in Antarctica. *Emerg. Med. 9*, 237-245.

[6] Taylor, G.R. (1993). Immune changes during short-duration missions. *J. Leukoc. Biol. 54*, 202-208.

[7] Taylor, G.R. (1974). Recovery of medically important microorganisms from Apollo astronauts. *Aerospace Med. 45*, 824-882.

[8] Sonnenfeld, G., & Shearer, W.T. (2002). Immune function during space flight. *Nutrition 18*, 899-903.

[9] La Duc, M.T., Nicholson, W., Kern, R., & Venkateswaran, K. (2003). Microbial characterization of the Mars Odyssey spacecraft and its encapsulation facility. *Environ. Microbiol. 5*, 977-985.

[10] La Duc, M.T., Kern, R., & Venkateswaran, K. (2004). Microbial monitoring of spacecraft and associated environments. *Microb. Ecol. 47*, 150-158.

[11] Pierson, D.L. (2001). Microbial contamination of spacecraft. *Gravit. Space Biol. Bull. 14*, 1-6.

[12] Castro, V.A., Thrasher, A.N., Healy, M., Ott, C.M., & Pierson, D.L. (2004). Microbial characterization during the early habitation of the International Space Station. *Microb. Ecol. 47*, 119-126.

[13] Salisbury, F.B. (1999). Growing crops for space explorers on the Moon, Mars or in space. *Adv. Space Biol. Med. 7*, 131-162.

[14] Yamashita, M., Ishikawa, Y., Kitaya, Y., Goto, E., Arai, M., Hashimoto, H., Tomita-Yokotani, K., Hirafuji, M., Omori, K., Shiraishi, A., Tani, A., Toki, K., Yokota, H., & Fujita, O. (2006). An overview of challenges in modelling heat and mass transfer for living on Mars. *Ann. N.Y. Acad. Sci. 1077*, 232-241.

[15] Ott, C.M., Bruce, R.J., & Pierson, D.L. (2004). Microbial characterization of free floating condensate aboard the Mir Space Station. *Microb. Ecol. 47*, 133-136.

[16] Novikova, N., De Boever, P., Poddubko, S., Deshevaya, E., Polikarpov, N., Rakova, N., Coninx, I, & Mergeay, M. (2006). Survey of environmental biocontamination on board the International Space Station. *Res. Microbiol. 157*, 5-12.

[17] Guarner, F., & Malagelada, J.R. (2003). Gut flora in health and disease. *Lancet 361*, 512-519.

[18] Lencner, A.A., Lencner, C.P., Mikelsaar, M.E., Tjuri, M.E., Toom, M.A., Valjaots, M.E., Silov, V.M., Liz'ko, N.N., Legenkov, V.I., & Reznikov, I.M.. (1984). Die quantitative Zusammensetzung der Lactoflora des Verdauungstrakt vor und nach kosmischen Flügen unterschiedlicher Dauer. *Nahrung 28*, 607-613.

[19] Nefedov, Y.G., Shilov, V.M., Koustantinova, I.V., & Zaloguev, S.N. (1971). Microbiological and immunological aspects of extended manned space flights. *Life Sci. Space Res. 9*, 11-16.

[20] Ilyin, V.K. (2005). Microbiological status of cosmonauts during orbital spaceflights on Salyut and Mir orbital stations. *Acta Astronaut. 56*, 839-850.

[21] Taylor, P.W., & Sommer, A.P. (2005). Towards rational treatment of bacterial infections during extended space travel. *Int. J. Antimicrob. Agents 26*, 183-187.

[22] Decelle, J.G., & Taylor, G.R. (1976). Autoflora in the upper respiratory tract of Apollo astronauts. App. *Environ, Microbiol. 32*, 659-665.

[23] Fischetti, V.A., Novick, R.P., Ferretti, J.J., Portnoy, D.A., & Rood, J.I. (2006). *Gram-positive pathogens*. Washington DC: ASM Press.

[24] Mishra, S.K., & Pierson, D.L. (1992). Space flight, effects on microorganisms. In J. Lederberg (Ed.), *Encyclopedia of Microbiology* (vol. 4, pp53-60). San Diego: Academic Press.

[25] Klaus, D.M. (2002). Space microbiology: microgravity and microorganisms. In G. Britton (Ed.), *Encyclopedia of Environmental Microbiology* (vol. 6, pp. 2996-3004). New York: John Wiley.

[26] Kacena, M.A., Leonard, P.E., Todd, P., & Luttges, M.W. (1997). Low gravity and inertial effects on the growth of *E. coli* and *B. subtilis*. *Aviat. Space Environ. Med. 68*, 1104-1108.

[27] Mennigmann, H.D., & Lange, M. (1988). Growth and differentiation of *Bacillus subtilis* under microgravity conditions. In N. Longdon, & V. David (Eds.), *Biorack on Spacelab D1: an overview of the first flight of Biorack, an ESA facility for Life Sciences Research in Microgravity.* (ESA SP-1091, pp. 37-44). Noordwijk: Kluver.

[28] Klaus, D., Simske, S., Todd, P., & Stodieck, L. (1997). Investigation of space flight effects on *Escherichia coli* and a proposed model of underlying physical mechanisms. *Microbiology 143*, 449-455.

[29] Kacena, M.A., Merrell, G.A., Manfredi, B., Smith, E.E., Klaus, D.M., & Todd, P. (1999). Bacterial growth in space flight: logistic growth curve parameters for *Escherichia coli* and *Bacillus subtilis*. *Appl. Microbiol. Biotechnol. 51*, 229-234.

[30] Tixador, R., Richoilley, G., Gasset, G., Templier, J., Bes, J.C., Moatti, N., & Lapchine, L. (1985). Study of minimal inhibitory concentration of antibiotics on bacteria cultured *in vitro* in space (Cytos 2 experiment). *Aviat. Space Environ. Med. 56*, 748-751.

[31] Tixador, R., Richoilley, G., Gasset, G., Planel, H., Moatti, N., Lapchine, L., Enjalbert, L., Raffin, J., Bost, R., Zaloguev, S.N., Bragina, M.P., Moroz, A.F., Antsiferova, N.G., & Kirilova, F.M. (1985). Preliminary results of Cytos 2 experiment. *Acta Astronaut. 12*, 131-134.

[32] Andrews, J. (2001). The development of the BSAC standardized method of disc diffusion testing. *J. Antimicrob. Chemother. (Suppl. 1) 48*, 29-42.

[33] Lapchine, L., Moatti, N., Gasset, G., Richoilley, G., Templier, J., & Tixador, R. (1985). Antibiotic activity in space. *Drugs Exptl. Clin. Res. 12*, 933-938.

[34] Sieradzki, K., Pinho, M.G., & Tomasz, A. (1999). Inactivated *pbp4* in highly glycopeptide-resistant laboratory mutants of *Staphylococcus aureus*. *J. Biol. Chem. 274*, 18942-18946.

[35] Sieradzki, K., & Tomasz, A. (2003). Alterations of cell wall structure and metabolism accompany reduced susceptibility to vancomycin in an isogenic series of clinical isolates of *Staphylococcus aureus*. J. Bacteriol. 185, 7103-7110.

[36] Taylor, P.W., Hamilton-Miller, J.M.T., & Stapleton, P.D. (2005). Antimicrobial properties of green tea catechins. *Food Sci. Technol. Bull. 2*, 71-81.

[37] Hamilton-Miller, J.M.T., & Shah, S. (1999). Disorganisation of cell division of methicillin-resistant *Staphylococcus aureus* by a component of tea (*Camellia sinensis*): a study by electron microscopy. *FEMS Microbiol. Lett. 176*, 463-469.

[38] Khandelwal, P., & Banerjee-Bhatnagar, N. (2003). Insecticidal activity associated with the outer membrane vesicles of *Xenorhabdus nematophilus*. *Appl. Environ. Microbiol. 69*, 2032-2037.

[39] Tixador, R., Gasset, G., Eche, B., Moatti, N., Lapchine, L., Woldrigh, C., Toorop, P., Moatti, J.P., Delmotte, F., & Tap, G. (1994). Behavior of bacteria and antibiotics under space conditions. *Aviat. Space Environ. Med. 65*, 551-556.

[40] Nickerson, C.A., Ott, C.M., Wilson, J.W., Ramamurthy, R., & Pierson, D.L. (2004). Microbial responses to microgravity and other low-shear environments. *Microbiol. Molec. Biol. Rev. 68*, 345-361.

[41] Kacena, M.A., & Todd, P. (1999). Gentamicin: effect on E. coli in space. *Micrograv. Sci. Technol. 12*, 135-137.

[42] Hammond, T.G., & Hammond, J.M. (2001). Optimized suspension culture: the rotating-wall vessel. *Am. J. Physiol. Renal Physiol. 281*, F12-25.

[43] Nickerson, C.A., Ott, C.M., Wilson, J.W., Ramamurthy, R., LeBlanc, L., Höner zu Bentrup, K., Hammond, T., & Pierson, D.L. (2003). Low-shear modeled microgravity: a global environmental regulatory signal affecting bacterial gene expression, physiology, and pathogenesis. *J. Microbiol. Meth. 54*, 1-11.

[44] Nickerson, C.A., Ott, C.M., Mister, S.J., Morrow, B.J., Burns-Keliher, L., & Pierson, D.L. (2000). Microgravity as a novel environmental signal affecting *Salmonella enterica* serovar Typhimurium virulence. *Infect. Immun. 68*, 3147-3152.

[45] Wilson, J.W., Ramamurthy, R., Porwollick, S., McClelland, M., Hammond, T., Allen, P., Ott, C.M., Pierson, D.L, & Nickerson, D.L. (2002). Microarray analysis identifies *Salmonella* genes belonging to the low-shear modelled microgravity regulon. *Proc. Natl. Acad. Sci. USA 99*, 13807-13812.

[46] Wilson, J.W., Ott, C.M., Ramamurthy, R., Porwollick. S., McClelland, M., Pierson, D.L, & Nickerson, C.A. (2001). Low-shear modeled microgravity alters the *Salmonella enterica* serovar Typhimurium stress response in a RpoS-independent manner. *Appl. Environ. Microbiol. 68*, 5408-5416.

[47] Fang, A., Pierson, D., Mishra, S., Koenig, D., & Demain, A. (1997). Effect of simulated microgravity and shear stress on microcin B7 production by *Escherichia coli* and on its excretion into the medium. *Appl. Environ. Microbiol. 63*, 4090-4092.

[48] Guadarrama, S., Pulcini, E.D., Broadaway, S.C., & Pyle, B.H. (2005). *Pseudomonas aeruginosa* growth and production of exotoxin A in static and modified microgravity environments. *Gravit. Space Biol. 18*, 85-86.

[49] Pelz, A., Wieland, K-P., Putzbach, K., Hentschel, P., Albert, K., & Götz, F. (2005). Structure and biosynthesis of staphyloxanthin from *Staphylococcus aureus*. *J. Biol. Chem. 280*, 32493-32498.

[50] Lui, G.Y., Essex, A., Buchanan, J.T., Datta, V., Hoffman, H.M., Bastian, J.F., Fierer, J., & Nizet, V. (2005). *Staphylococcus aureus* golden pigment impairs neutrophil killing and promotes virulence through its antioxidant activity. *J. Exp. Med. 202*, 209-215.

[51] Wieland, B., Feil, C., Gloria-Maercker, E., Thumm, G., Lechner, M., Bravo, J-M., Poralla, K., & Götz, F. (1994). Genetic and biochemical analyses of the biosynthesis of the yellow carotenoid 4,4'-diaponeurosporene of *Staphylococcus aureus*. *J. Bacteriol. 176*, 7719-7726.

[52] Marks, J., & Vaughan, A.C.T. (1950). Staphylococcal δ-haemolysin. *J. Path. Bact. 62*, 597-615.

[53] Curtis, S.B., & Letaw, J.R. (1989). Galactic cosmic rays and cell-hit frequencies outside the magnetosphere. *Adv. Space Res. 9*, 293-298.

[54] Todd, P., Pecaut, M.J., & Fleshner, M. (1999). Combined effects of space flight factors and radiation on humans. *Mutat. Res. 430*, 211-219.

[55] Zeitlin, C., Cleghorn, T., Cucinotta, F., Saganti, P., Andersen, V., Lee, K., Pinsky, L., Atwell, W., Turner, R., & Badhwar, G. (2004). Overview of the Martian radiation environment experiment. *Adv. Space Res. 33*, 2204-2210.

[56] Fukada, T., Fukada, K., Takahashi, A., Ohnishi, T., Nakano, T., Sato, M., & Gunge, N. (2000). Analysis of deletion mutations of the *rpsL* gene in the yeast *Saccharomyces cerevisiae* detected after long-term flight on the Russian space station Mir. *Mutat. Res. 470*, 125-132.

[57] Leys, N.M.E.J., Hendrickx, L., de Boever, P., Baatout, S., & Mergeay, M. (2004). Space flight effects on bacterial physiology. *J. Biol. Reg. Homeostat. Agents 18*, 193-199.

[58] McLean, R.J.C., Cassanto, J.M., Barnes, M.B., & Koo, J.H. (2001). Bacterial biofilm formation under microgravity conditions. *FEMS Microbiol. Lett. 195*, 115-119.

[59] Carstensen, J.T., & Rhodes, C.T. (2000). *Drug Stability: Principles and Practices.* Marcel Dekker: New York.

[60] Ballereau, F., Prazuck, T., Schrive, I, Lafleuriel, M.T., Rozec, D., Fisch, A., & Lafaix, C. (1997). Stability of essential drugs in the field: result of a study conducted over a two-year period in Burkina Faso. *Am. J. Trop. Med. Hyg. 57*, 31-36.

[61] Robinson, J.A., Thumm, T.L., & Thomas, D.A. (2006). NASA utilization of the International Space Station and the Vision for Space Exploration. *Proceedings of the 57th International Astronautical Federation Congress* IAC-06-A1.7/A2.7.06.

[62] Tan, Y.-T., Tillett, D.J., & McKay, I.A. (2000). Molecular strategies for overcoming antibiotic resistance in bacteria. *Molec. Med. Today 6*, 309-314.

[63] Alksne, L.E., & Projan, S.J. (2000). Bacterial virulence as a target for antimicrobial chemotherapy. *Curr. Opin. Biotechnol. 11*, 625-636.

[64] Taylor, P.W., Stapleton, P.D., & Luzio, J.P. (2002). New ways to treat bacterial infections. *Drug Disc Today 7*, 1086-1091.

[65] Matsuzaki, S., Rashel, M., Uchiyama, J., Sakurai, S., Ujihara, T., Kuroda, M., Ikeuchi, M., Tani, T., Fujieda, M., Wakiguchi, H., & Imai, S. (2005). Bacteriophage therapy: a revitalized therapy against bacterial infectious disease. *J. Infect. Chemother. 11*, 211-219.

[66] González, J.E., & Keshavan, N.D. (2006). Messing with bacterial quorum sensing. *Microbiol. Molec. Biol. Rev. 70*, 859-875.

In: Planet Mars Research Focus
Editor: L. A. Costas, pp. 187-215
ISBN: 978-1-60021-826-2

Chapter 7

THE CASE FOR A COLD AND WET MARS

Alberto G. Fairén[1,2], Alfonso F. Davila[1], Darlene Lim[1], Esther R. Uceda[1], Jhony Zavaleta[1], Ricardo Amils[2,3] and Chris P. McKay[1]

[1] NASA Ames Research Center, Space Science and Astrobiology Division, Moffett Field, 94035 CA, USA.

[2] Centro de Biología Molecular, CSIC-Universidad Autónoma de Madrid, 28049-Cantoblanco, Madrid. Spain.

[3] Centro de Astrobiología (CSIC-INTA), 28850-Torrejón de Ardoz, Spain.

ABSTRACT

Today we have compelling geomorphologic and geochemical evidence for the presence in the past of vast quantities of liquid water on Mars. The numerous and diverse erosive, depositional and chemical features that characterize the Martian surface include ocean-related landforms, massive layered outcrops, valley networks, dendritic gullies, anastomosing and meandering rivers, deltas and evaporitic paleolake deposits. These features are the remnant testimony of an era when the atmosphere, surface and subsurface of Mars were linked by a complex hydrological cycle, in which water melting, freezing, evaporation and sublimation processes were largely controlled by the chemical and physical characteristics of the Martian atmosphere and surface.

Traditionally, the presence of liquid water at any given time on Mars has been explained by invoking two controversial and mutually exclusive models: the "hot and wet" Mars, where an atmosphere rich in greenhouse-gases enabled surface temperatures close or above the freezing point of water, and the "cold and dry" Mars, with a hydrological cycle mainly controlled by thermal energy sources that locally and episodically enabled the presence of liquid water on the planet's surface. Both models have difficulties to explain the timing, diversity and number of erosive and depositional features that characterize the Martian atmosphere, and have been contradicted by geochemical data and atmospheric models of Mars.

Based on the observation that weathering fluids loaded with high abundances of diverse salts can remain in the liquid state for temperatures well below 0 °C in the presence of a thick CO_2 atmosphere, we envision an alternative model that accounts for the origin and fate of water throughout the entire history of Mars. During the Noachian (>3.7 Gyr. ago) Mars was a cold and wet planet with a planetary hydrological cycle that witnessed the formation of glaciers, lakes, rivers and oceans. In this "cold and wet" Mars, massive glaciers and ice-rich regolith accumulated vast amounts of ice water in the southern highlands. The seasonal partial melting of the glaciers and regolith at lower latitudes, resulted in fluvial-lacustrine systems in which meltwater was transported through large river valleys and accumulated in crustal aquifers, confined lakes and ponds. This process originated the mature, terrestrial-like, dendritic valley networks observed in orbital explorations. The chemical reequilibration of glacier meltwater with iron- and sulfur-enriched soils produced high ionic level of superloaded acidic waters, with elevated concentrations of Fe, SO_4, Br, Na, Cl, Ca, and other ions. Such solutions would not require above-freezing temperatures to flow. Additionally, water filled the Martian lowlands forming extended and long-lasting acidic oceans and seas, partially covered by sea-ice at higher latitudes. During the local summer, periods of maximum volcanic activity and/or variations in planetary orbital parameters, ponded water would evaporate to the atmosphere, and rock and water glaciers would undergo sublimation and wasting. Water accumulated in the atmosphere would then precipitate almost exclusively in the form of snowfall, because of the low temperatures. The net accumulation of snow and its metamorphosis to ice form glaciers in the highlands again would close the hydrogeological cycle.

The proposed hydrological cycle was active only during episodes of dense atmosphere in the Noachian. During the Hesperian epoch (3.7-3.1 Gyr ago), Mars underwent a series of dramatic changes that resulted in the partial loss of its atmosphere, accompanied by a loss of vast quantities of its water reservoir and by progressive crustal assimilation of H_2O. The thinning of the atmosphere in a water depleted planet likely triggered the partial freezing of the ocean and other bodies of standing waters, resulting in an ice-covered ocean on the northern plains, and ice covered lakes and ponds in the southern highlands. This ultimately led to the interruption of the hydrological cycle and represents the onset of the colder and dryer conditions that characterize Mars today. The Hesperian is also characterized by the occurrence of catastrophic flow events.

The most recent Martian epoch, the Amazonian (3.1-present), has witnessed brief and intriguing wet episodes, in which the climate was relatively warm and moist, until geologically very recent times. Multiple glacial landscapes formed and phases of tropical mountain glaciation occurred even within the past few tens of millions of years, including water runoff during the past decade.

THE EVIDENCE FOR WATER ON MARS

Understanding the evolution of the Martian hydrosphere is of paramount importance to elaborate reliable models for the evolution of the planet and to better determine if life may have originated on Mars, and how Martian organisms could have evolved and adapted to the extreme climate changes that occurred throughout the planet's history.

Today Mars is a cold and extremely dry desert. At low latitudes the daily temperatures range from about -100 °C to +17 °C, and the average is -60 °C. The atmosphere is very thin, and exerts a mean pressure of 5.6 millibars on the surface of the planet, less than one hundredth that of Earth. The atmosphere is composed of Carbon Dioxide (95%), Nitrogen (2.7%), Argon (1.6%), Oxygen (0.13%) and minute traces of other gases. Water is still present on the planet, mainly in the form of water ice and trapped in the North and South Poles and in the regolith.

The most relevant and enigmatic feature of the Martian topography is arguably the dichotomy which divides the planet's surface into the ancient, heavily cratered southern highlands, and the younger, smooth, northern lowlands. An abrupt elevation change of several kilometers seems to occur at the boundary. Both the highlands and the lowlands are carved by erosive features that have been interpreted as the result of liquid or ice water flows. The number and scale of fluid related geomorphologic features are a testimony of an era when liquid water covered large portions of the planet, likely in amounts that sufficed to host for extended periods of time a high latitude ocean in the northern hemisphere, and a large number of lakes, ponds, rivers and crustal aquifers in the cratered lands of the southern hemisphere.

Evidences of liquid water flowing and pounding on the surface of the planet include ocean-like shorelines (Parker et al., 1993; Head et al., 1998; Clifford and Parker, 2001; Fairén et al., 2003), massive layered outcrops usually occurring within impact craters (Malin and Edgett, 2000a), valley networks with associated fluvial redistribution of sediment (Craddock and Howard, 2002; Mangold et al., 2004; Bhattacharya et al., 2005), anastomosing and meandering rivers and deltas (Malin and Edgett, 2003) and cross-stratification in rock outcrops (Squyres et al., 2004). Mineralogical and geomorphological data obtained by the recent Mars Rovers Missions have yielded unambiguous evidence for the presence of ancient bodies of standing water in crater impacts such as Eagle Crater (Squyres et al., 2004). Outcropped rocks analyzed by the Mars Exploration Rover Opportunity have revealed layered deposits that contain 30–40 wt. % of Ca-, Mg-, and Fe-sulfates (jarosite), with imbedded hematite spherules. Along with these minerals, the outcrop rocks also contain a large fraction of siliciclastic materials likely of a basaltic nature. Altogether the exposed rocks suggest a series of episodic inundations of the crater by shallow surface water, followed by evaporation and desiccation (Squyres et al., 2004).

The surface of Mars also presents clear indications of ice water flows and ice-water deposits. Today both the Martian North Pole and South Pole are covered by massive ice caps, and evidence suggests that these ice caps have been part of the Martian landscape for most of the planet's history. A large portion of the surface of Mars is covered by ice-rich regolith, with thicknesses approaching 7 km near the poles and 1-3 km near the equators (Fanale et al., 1986). Giant polygons can also be found in several locations on the planet, likely formed due to cooling and dessication of wet sediments, and some curvilinear ridges in mid and high latitudes provide striking morphological evidence for ancient esker-like glaciation processes.

LIQUID WATER ON MARS

Any model attempting to describe the evolution of water on Mars, has to consider the fact that some 4.5 Gyr ago the Sun was cooler and smaller by about 10% its present size, resulting in a luminosity of about 70% of that of the present Sun (Sagan and Chyba, 1997). This resulted in much cooler temperatures on early Mars than today. To explain the presence of liquid water on Mars, two competing and mutually exclusive models have been traditionally proposed: the "Hot and Wet" Mars and the "Cold and Dry" Mars.

The "Hot and Wet Mars" Model

This model argues that the early Martian climate was substantially different from that at present due to the presence in the atmosphere of greenhouse gases. In this scenario the planet would be capable of sustaining long-term warm periods, resulting in an Earth-like hydrological cycle with large lakes or oceans as the evaporative sources. The proposed greenhouse species acting as warming agents include CO_2 injected periodically in the atmosphere by the thermal decomposition of carbonate rocks (Pollack et al., 1987), and reduced greenhouse gases such as NH_3 (Sagan and Mullen, 1972) and CH_4-NH_3 (Sagan and Chyba, 1997) produced via extensive volcanism (Dohm et al., 2001; Phillips et al., 2001). According to this model, the increasing amounts of greenhouse gases in the atmosphere resulted in the episodic melting of surface ice deposits and the formation of oceans and large bodies of standing waters (Baker et al., 1991). Yet, greenhouse model has been difficult to reconcile with precise reconstructions of the atmospheric conditions in early Mars. A putative greenhouse atmosphere produced by the CO_2-H_2O combination would have saturated at temperatures well below 0 °C, resulting in CO_2 condensation and the formation of high altitude ice clouds that strongly increase the albedo and preclude further warming of the planet's surface (Kasting, 1991). Some authors have argued, however, that CO_2 ice clouds in the troposphere could be responsible instead for further greenhouse warming due to the scattering of infrared radiation from the planet's surface (Forget and Pierrehumbert, 1997). Arguably, the role of CO_2 clouds as warming or cooling agents on early Mars remains controversial as atmospheric models are very sensitive to input variables (Mischna et al., 2000; Colaprete and Toon, 2003). Furthermore the amount of CH_4 required to increase the surface temperatures of Mars to near the melting point of water would imply a terrestrial-like biological source (Kasting, 1997), which has not yet been identified, and NH_3 would be consumed by UV photolysis in less than 10 years (Kuhn and Atreya, 1979). For these reasons greenhouse models need to incorporate additional mechanisms for planetary warming. A suggested mechanism is the orbital variations of the planet's spin axis obliquity and/or eccentricity (Ward, 1992; Touma and Wisdom, 1993; Mischna et al., 2003); however, currently there is lack of evidence for orbital forcing of the Martian climate before the Amazonian time.

The "Cold and Dry Mars" Model

In the "Cold and Dry" model, liquid water-related features on Mars have been explained as a result of spatially- and temporally-localized thermal energy sources in a cold planet. Advocates of this model argue that, due to the reduced solar luminosity during the Noachian, not even 5 bar of CO_2 in the atmosphere would have been enough to rise the temperatures above the freezing point of pure water (Colaprete & Toon, 2003). In such enduring and intensely cold desert conditions, only punctual events such as hydrothermally-released groundwater (Griffith and Shock, 1997; Cabrol and Grin, 1997) or meteoritic impacts (Segura et al., 2002) triggered liquid water release to the surface. Valley networks would then have been carved by groundwater circulation (Gulick, 2001), or under a protective ice cover (Wallace and Sagan, 1979; Carr, 1983). A number of observations seem to reinforce the notion of a dry and cold Mars, including the absence of evident rain erosion (Gulick, 2001), the widespread geomorphic evidence of ancient massive ice sheets (Gaidos and Marion, 2003; Head et al., 2004), and the detection of olivine in precise locations (Hoefen et al., 2003; Mustard et al., 2005) along with the paucity of secondary minerals that would otherwise be typical in a warmer and wetter environment (Gaidos and Marion, 2003; Poulet et al., 2005). However the notion of punctual events resulting in liquid water on the Martian surface or groundwater circulation, does not account for the large amount of recharge necessary for maintaining the valley network systems formed during the Noachian, and fails to explain the mechanisms by which water occupied the Martian regolith in the first place. Furthermore the spatial patterns of erosion and crater degradation in the highlands (Craddock and Howard, 2002), the absence of obvious association between many flow features and geothermal heat sources, and the formation of evaporites on Mars, which would require stable bodies of standing liquid water, are difficult to reconcile with the "Cold and Dry" model.

Hence it appears that when Mars was wet, it was also very cold. The question therefore reduces to envision how liquid water could be stable on the surface of the planet with a mean temperature of roughly -25 °C and for relatively extended periods of time. The model also needs to account for the observed surface geochemistry and mineralogy of Mars, and also be consistent with the sequence and time of deposition of the different mineralogical units.

Water on Early Mars was a Cold, Hypersaline and Acidic Solution

The freezing point of an aqueous solution is a function of pressure and chemical composition. The partial pressure of CO_2 on early Mars, approximately accounting for 95% of the Martian atmospheric pressure, likely ranged between 1.5-2 bars (Kasting, 1991; Phillips et al., 2001; Fairén et al., 2004). At these atmospheric pressures pure water would remain in the solid state for temperatures below 0 °C. The dissolution of specific compounds can however lower the freezing point of water to the point that liquid water may be stable at the above mentioned atmospheric pressures and temperatures. In the Antarctic Dry Valleys, for instance, water in contact with soils with up to 3,000 ppm salt concentrations remains liquid at temperatures as low as -56 °C (Mahaney et al., 2001).

The possibility that liquid, hyperloaded water solutions may have been present on Mars at temperatures below 0 °C has been considered in a number of pioneering studies (i.e., Clark and Van Hart, 1981; Kuzmin and Zabalueva, 1998; Haberle et al., 2001; Mahaney et al., 2001; Knauth and Burt, 2002; Burt and Knauth, 2003; Jean et al., in press). Melting temperatures estimated for model Martian solutions go as low as -55 °C (see Table 1), and several of the various ternary salt waters have freezing points below -48 °C, well below the estimated mean temperature for early Mars of -25 °C. Sulfates, the most abundant salts identified so far on the Martian surface, are poor freezing point depressors and none prevent the formation of ice below ~ -10 °C. Yet other antifreeze compounds have melting temperatures down to -28 °C when dissolved in pure water; and when mixed with sulfate salts could greatly lower its freezing point while raising its boiling point (Jean et al., in press). Clark and Van Hart (1981) argued that Ca- and Fe-chloride salts, with typically large freezing point depressions, could react with soluble sulfates in the soil to form less effective "antifreeze" compounds, which would still maintain liquid water stable well below 0°C. Interestingly, the formation of hyperloaded solutions in a long-term cold Mars would also considerably increase the stability of liquid water on the surface by decreasing the evaporation rate by orders of magnitude (Sears and Chittenden, 2005). The stability against freezing of several solutions modeled in the last decades is summarized in Table 1. These water solutions represent a first approximation of the type of aqueous solutions that were likely present on the Martian surface.

Table 1. Very low melting temperatures of usual examples of putative Martian simple solutions and antifreezers.

Type	Substance	°K
Solution	$H_2SO_4 \cdot 9H_2O$	199
Soil	Up to 3,000 ppm salt concentrations (Antarctic Dry Valles)	217
Solution	$CaCl_2$ brine, $CaCl_2$-$MgCl_2$ and $FeCl_3$	218
Solution	$CaCl_2$-$NaCl$-H_2O	221
Solution	$CaCl_2$-rich	223
Salt	Various ternary salt waters	225
Solution	$MgBr_2$ eutectic	230
Solution	$MgCl_2$ eutectic	238
Solution	$NaClO_4$ eutectic	240
Antifreezer	1,3-propylene glycol + pure water	245
Solution	NaCl eutectic	252
Antifreezer	Ethylene glycol + pure water	260
Salt	Sulfate salts (maximum)	263
Antifreezer	Diethylene glycol + pure water	265
Antifreezer	Triethylene glycol + pure water	266

Geochemical and mineralogical data collected in the past four decades support the notion that water on Mars was indeed a hyperloaded solution. However these data also suggest that liquid water on Mars was a multicomponent solution much more complex than those included in the models mentioned above. In the five widely separated sites visited by Mars landers, the data gathered is consistent with environments where moderate to extreme acidic pH and high abundances of Fe, Mg, and the salt-forming elements S, Cl and Br prevailed. These observations have recently been confirmed by the discovery of ancient acidic brines (Squyres et al., 2004), derived from the alteration of the mafic-ultramafic crust, which favored the formation of sulfate minerals relative to carbonates and other salts, as advanced by Fairén et al. (2004). The discovery of widespread sulfate deposits with high chloride and bromide concentrations, as well as of iron-oxide in outcrop materials in Meridiani Planum (Rieder et al., 2004, Gendrin et al., 2005) are further indicators of the presence in the past of salt rich water solutions on the planet's surface. These salts are likely the dissolution and re-precipitation products of olivine-bearing basalts and iron-sulfur minerals, rocks which are found in the soil throughout the planet's surface, and play an important role in the soil geochemistry.

The identification in a few locations of early Noachian deposits of Fe- and Al-rich phyllosilicates, minerals that commonly form in alkaline environments, suggests that other environments with different geochemical characteristics also existed on the planet, albeit in a more reduced temporal and spatial scale. Based on these observations, Poulet et al. (2005) and Bibring et al. (2006) proposed that Mars witnessed a sharp change in the geochemical conditions of standing bodies of water on its surface between the Early Noachian and the Late Noachian-Early Hesperian epochs, the first characterized by depositional environments with alkaline type waters, and the later characterized by depositional environments with acidic type waters. The almost complete absence of carbonate deposits on the Martian surface, and the relative scarcity of the Fe- and Al-rich phyllosilicates, is however difficult to reconcile with models that advocate for an early Noachian alkaline dominated geochemistry. An alternative scenario can instead be envisioned in which both major water-alteration products were deposited contemporarily but in different locations, characterized by diverse geochemical conditions during the wet Noachian/Hesperian times. This hypothesis is consistent with published geochronologies for the occurrence of sulfate deposition, which indicate that most of them are Noachian to Hesperian in age (Squyres et al., 2004; Squyres and Knoll, 2005). Interestingly, the later model is in stark contrast to present day conditions on Earth. Here, water chemistry is generally characterized by alkaline or circumneutral conditions and acidic environments are considered rare and extreme. Instead, when liquid water was present on the surface of Mars in large quantities, the scenario was likely the opposite: predominantly acidic water conditions, with rare and extreme neutral to alkaline environments. Irrespectively of the speculated geochemical conditions during the early Noachian era, there is common agreement now that a surface water geochemistry characterized by acidic conditions prevailed on Mars since the Late Noachian to the present, a time span than encompasses approximately 3.5 Gyr.

All these observations seem to support the idea that early Mars was characterized by intense cold temperatures and yet large amounts of stable, hypersaline liquid water systems on its surface. In addition to the early liquid, hyperloaded water models mentioned above, we propose four additional considerations that further strengthen the value of the antifreeze solution models. First, taking into account the composition of the evaporitic deposits studied in settings such as Eagle crater, we have modeled the water deposits in early Mars as very complex multicomponent solutions, composed of mixtures of Fe-Mg-(Ca)-SO_4-(Cl)-rich salts. These solutions are more efficient in depressing the freezing point of water than those considered in earlier relatively simple models. Second, our multicomponent model implies that water solutions on Mars were not dense liquids, thick eutectic brines with a high viscosity, as usually proposed by the classical models; these properties are inconsistent with the shape and morphology of the valley networks, outflow channels and gullies, which suggest rapid fluvial flow rather than the slow, halting movement of low temperature cool molasses. Third, the above summarized geochemical models have mainly been applied to the present climate conditions on Mars, as it is commonly accepted that the surface temperatures on early Mars were close to 0 °C, and therefore antifreeze solution models would not be required to explain the presence of liquid water on the planet's surface; instead, our model is based on conditions of intense cold temperatures in the past and increased atmospheric pressure. And finally, antifreeze solution models were never followed by a comprehensive explanation of its significance in a complete hydrological cycle like the one presented here.

The "Cold and Wet Mars" Model

Following the geochemical analyses summarized above, a global hydrological cycle can be proposed to explain the long-term presence of open bodies of liquid water on a basically cold Mars. Depending on input variables, climatic models considering the addition of greenhouse gases to the atmosphere offer a wide range of mean surface temperatures. For 2 bars of CO_2 and a solar luminosity 75% that of the current value, temperatures are estimated to be in an ample range between -45 to 25 °C (Figure 1). The more conservative values result from standard radiative-convective models, considering only thermodynamic arguments and not cloud optical properties (Kasting, 1991; Colaprete and Toon, 2003). The upper values consider a near-100% cover of clouds with a high optical depth to scattering (Forget and Pierrehumbert, 1997). For this range of temperatures, some of the acidic saline supercold solutions described above may have remained liquid in an open acid-sulfate weathering regime for most of the Martian history (Table 1).

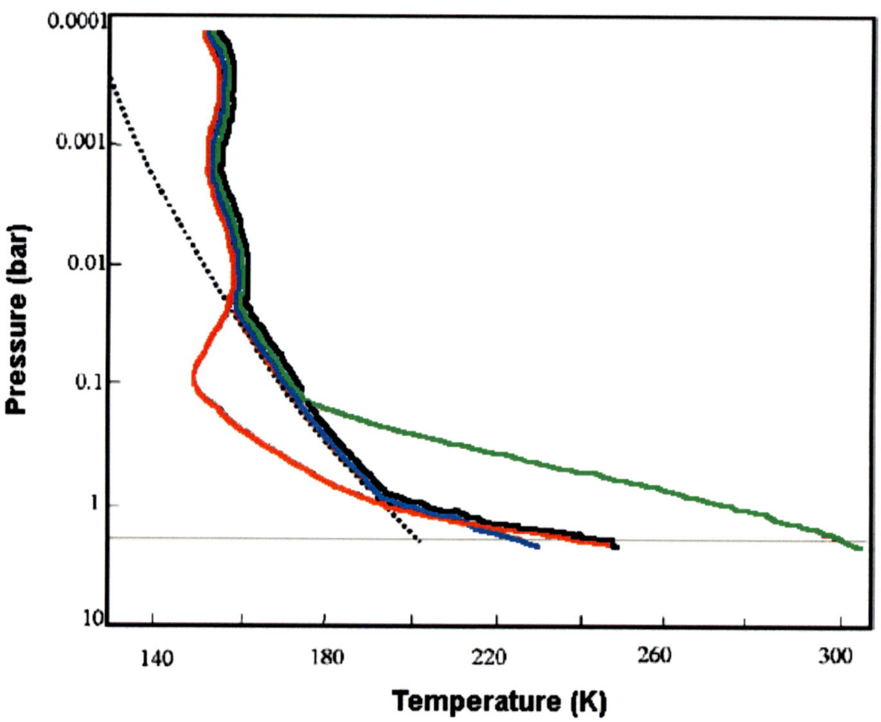

Figure 1: Temperature profiles on the surface of Mars plotted for 2 bars of CO_2 (horizontal line) and a solar luminosity 75% that of the current value. Blue) Radiative-Convective model by Kasting (1991), considering the effect of CO_2 clouds; Red) Standard Radiative-Convective model by Pollack et al. (1987); Black) Temperature profiles in Colaprete and Toon (2003); Green) Radiative-Convective model considering cloud scattering, after Forget and Pierrehumbert (1997). Dotted line indicates CO_2 saturation. For most modelled atmospheric conditions, multicomponent saturated solutions as those described in Table 1 can remain in liquid state on the surface of Mars.

The Noachian (>3.7 Gyr Ago): The Onset of the Martian Hydrological Cycle

In a generally supercold climate produced by a solar value 75% that at present and a 2-bar CO_2-H_2O-CH_4 atmospheric composition, resulting in a greenhouse warming of the Martian surface of about -25° C, the hydrological cycle of Mars was likely characterized by the presence of vast ice deposits generating widespread, permanent and massive rock glaciers in the Martian highlands of the southern hemisphere. Most of the glacially-derived geomorphologies on Mars, including ridges, lobes and outlets, show distinctive similarities to terrestrial moraines and rock glaciers (Head et al., 2004). These features are found throughout almost the entire highland terrains, from the South Pole to near the equator.

At the bottom of the glacier deposits, water ice melted due to geothermal heating and/or conductive and advective heat loss from magma reservoirs (Fassett and Head, 2006). This promoted cation exchange between the meltwater, the salt solutions and the bed rock. The chemical reequilibration of melt water with iron- and sulfur-enriched soils, resulted in superloaded acidic waters, with elevated concentrations of Fe, SO_4, Br, Na, Cl, and Ca ions, and a total salt content over 10%. Similar processes occur on Earth in arctic glaciers, where groundwater emerges to the surface in the form of hyperloaded springs (Andersen et al., 2002). Due to the antifreeze effects of the dissolved ions, such solutions would not require above-freezing temperatures to flow, and thus formed confined crustal aquifers enclosing substantial meltwater.

Basal drainage of hyperconcentrated meltwater beneath the ice sheets likely created a significant volume of circulating groundwater under the rock glaciers, which evolved into a subterranean system of aquifers in diverse locations of the periglacial environment. This system lead in some locations to the formation of cold, acidic and salty springs and glacial outwashes, as is also found on Earth in the Arctic (Grasby et al., 2003) and Antarctica (Lyons et al., 2003; Mikucki and Priscu, 2004) (Figure 2). The presence of this type of spring system on early Mars can also be inferred from the very low Cl/Br ratios (~20 to 80) in salts from Meridiani and Gusev rock rinds, which have been interpreted as the evaporitic result of hyperconcentrated subsurface Martian waters. In addition, sulfate minerals are scattered over a wide range of elevations on Mars (Gendrin et al., 2005), and not merely as playa deposits, suggesting spring discharge at different hillslope heights. Further evidence for landforms that support a link between the development of ice sheets and the flow of water is provided in Gaidos and Marion (2003). Other studies have suggested the presence of lakes or piedmont glaciers just at the outlet of dense water-worn features (Quantin et al., 2004).

The melting of ice and accumulated snowpacks was likely rapid, as can be inferred from the observed Martian channels length and depth, and the estimated discharges and flood rates, which are equivalent to mean annual floods in terrestrial drainage basins of similar size (Howard et al., 2005). Once on the surface, water must have remained in liquid state for a relatively long time, in order to carve out the drainage networks, some of which are several thousands of kilometers long. Runoff channels and valley networks would then have developed for millions of years on the Martian highlands, transporting acidic waters to ponds and lakes formed in topographical lows such as impact craters (e.g, Argyre, Hellas), where alluvial fans and deltas occasionally evolved.

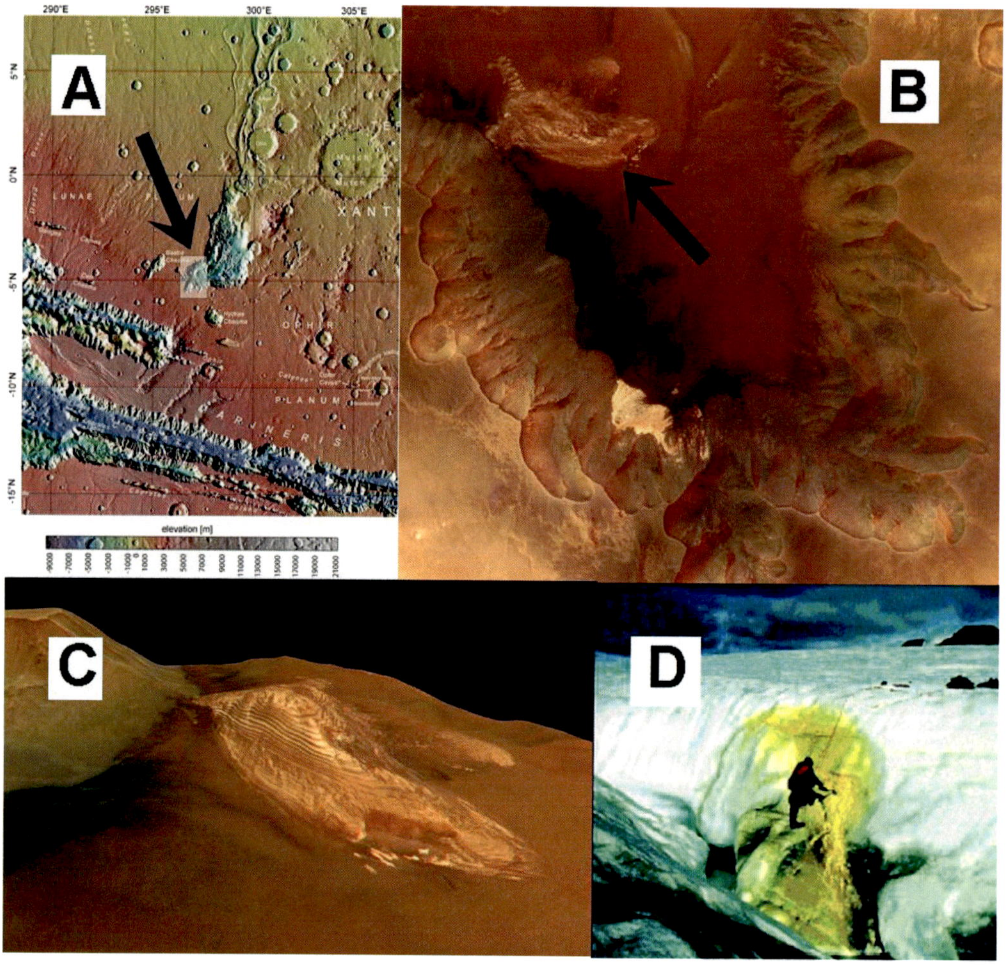

Figure 2: **A)** Location of Juventae Chasma cutting into the plains of Lunae Planum, MOLA elevation data. Arrow indicates the location of a huge sulfate outcrop. **B)** Arrow marks the location of a mountain composed of bright and layered material, likely sulfates, outcropping from the valls of Juventae. Image from HRSC (ESA/DLR/FU Berlin/G. Neukum). **C)** Perspective view of the mountain in Juventae Chasma, thought to be composed of sulfates, looking Northeast. This mountain is approximately 2.5 kilometres high, 59 kilometres long and up to 23 kilometres wide, thus in this HRSC image the vertical profile is substantially exaggerated. (ESA/DLR/FU Berlin/G. Neukum). **D)** Dispersed native sulfur at the outlet on the wall of a incised supraglacial melt water channel in the Canadian High Arctic. A similar process of basal drainage and discharge of subterranean cold, acidic and salty aquifers in periglacial environments could have originated the sulfate mountain in Juventae Chasma. (After Grasby et al., 2003).

The presence on the Martian surface of two basic types of geographic features similar to river valleys on Earth further supports the hypothesis that long-lived rivers occurred in the past on the planet. The occurrence of mature terrestrial-like dendritic valleys with inner channels (Mangold et al., 2004; Bhattacharya et al., 2005; see Figure 3a) indeed requires the presence of liquid water flows over geological time scales. Similarly, long valleys which extend for hundreds of kilometers, with few and small tributaries and rather homogeneous widths and shapes, were likely carved by liquid water flowing on the surface for extended periods of time (see Figure 3b). It is important to note here that a significant number of ancient valleys and rivers may remain buried under the Martian surface. Recent radar imaging of the Martian lowlands revealed a large amount of craters buried under layers of eolic and volcanic deposits (Frey et al., 2002). Similarly complex drainage networks in the Sahara desert, consisting of ancient river channels and lakes, can be observed by radar imaging buried under layers of sand (McCauley et al., 1982; Figure 4). The extreme aridity and winds on the Martian surface in later periods, likely facilitated the burying processes of small and reduced geomorphologies. On Earth, low-frequency synthetic aperture radar (SAR) has demonstrated its capabilities to reveal subsurface images of dust-covered geomorphologies, especially in arid regions. The radar imaging systems incorporated in future Mars missions will definitively unveil the complete fluvial history of the planet, revealing where the channels originate, where they debouch, and what was the actual extent of the Martian fluvial system.

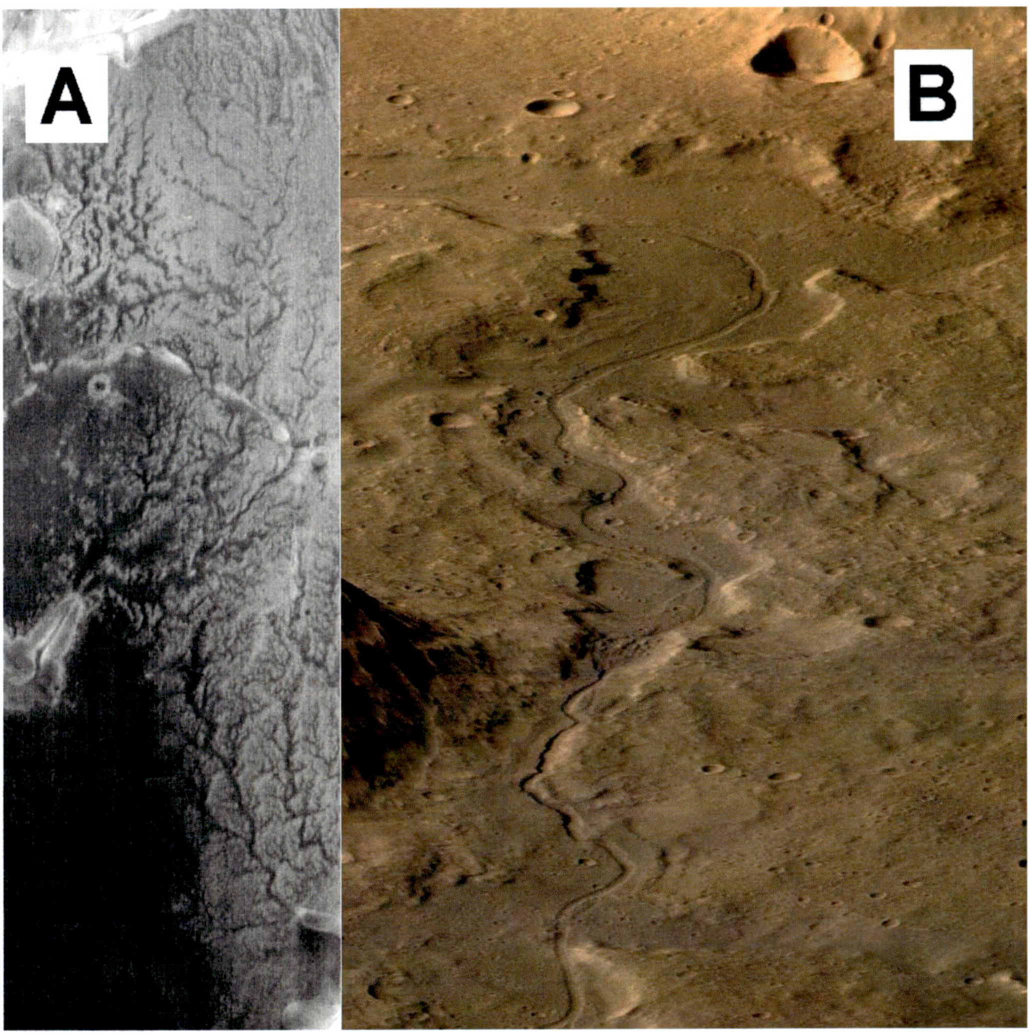

Figure 3: **A)** Portion of THEMIS nighttime infrared image I02119003, showing ancient (Hesperian) dendritic fluvial channels on the mesa west of Echus Chasma. The image is 32 km wide, centered 0.5° S and 279° E. (THEMIS Public Data Releases, Arizona State University). **B)** Portion of a HRSC image of a modern (Amazonian) valley on Lybia Montes, with no tributaries and constant widht and shape. Image centered 2° N and 81° E, (ESA/DLR/FU Berlin/G. Neukum).

Figure 4: An ancient river channel buried under layers of sand in the Sahara Desert, southeast Lybia (23.3° N, 22.9° E). The buried river channel was revealed in images taken by the Spaceborne Imaging Radar SIR-C/X-SAR that flew twice on the Space Shuttle Endeavour in 1994. The radar images were processed at JPL and the University of Texas at Dallas. (From http://www.op.dlr.de/ne-hf/SRL-2/p45719_kufra.html).

This complex network of river valleys transported melt water from the highland glaciers to lakes and ponds and finally to the Martian lowlands, forming extended and long-lasting acidic ultracold oceans and seas. Similar to hypersaline lakes in Antarctica (Burton, 1988; Haberle et al., 2001), the waters in the lowlands oceans and seas may have never frozen completely under average temperatures well below the freezing point of pure water. Both rivers and seas likely carried seasonal floating ice floes, and possibly icebergs (Woodworth-Lynas and Guigné, 2004; Murray et al., 2005) as in the Arctic and Southern Oceans on Earth. The bedrock shorelines identified in the Martian lowlands may therefore be related to these ice-covered oceans (Fairén et al., 2003; Kraal et al., 2006; see Figure 5). The acidic nature of the waters would explain the paucity of carbonates observed on the surface of Mars (Fairén et al., 2004), and would also have contributed to the enrichment in CO_2 of the Martian atmosphere. This in turn helped to keep surface temperatures over -25 ºC. During the local summer, rock glaciers and sea ice would undergo sublimation and wasting, and small fractions of pounded waters would evaporate to the atmosphere. Complementary to the seasonal cycles, periods of maximum volcanic activity and/or spin axis obliquity and orbit eccentricity would have enhance these processes. Water vapor accumulated in the atmosphere would then precipitate mainly in the form of snowfall at high latitudes and on top of elevated volcanoes. The metamorphosis of the seasonal snows into ice would close the hydrological cycle as schematically represented in Figure 6a.

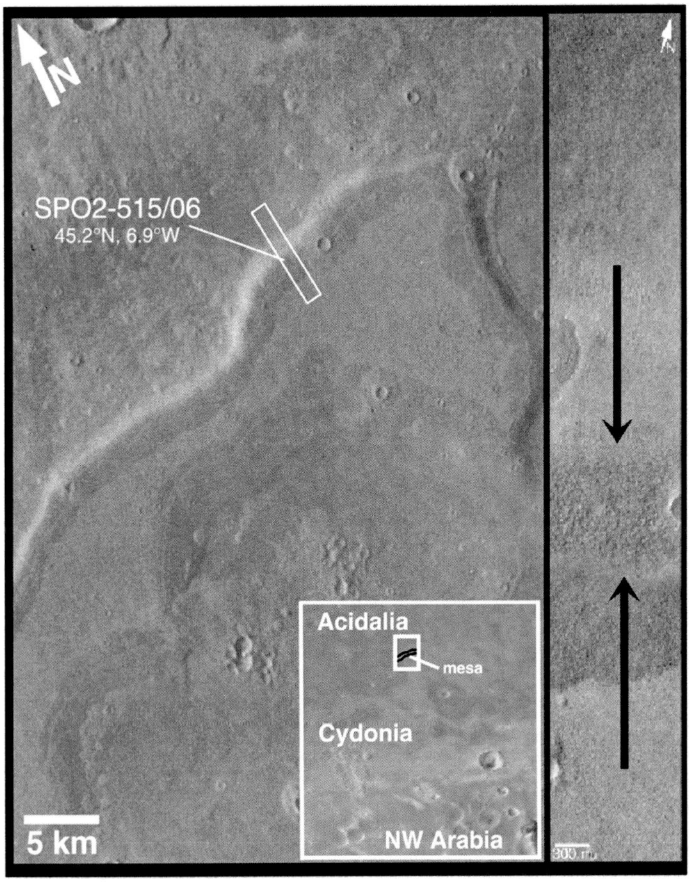

Figure 5: 026A72 Viking image (left) and subframe of SPO2-515/06 MGS-MOC image, evidencing Artic and Antartic-like shore morphologies, such as lines of boulders, scour marks and gravel bars (indicated by arrows in the MOC image) on the Acidalia Plains (45º N, 7º W). The flat-topped mesa has been interpreted as a possible island emerging in a northern plains latitudinally and/or seasonally ice-covered ocean.

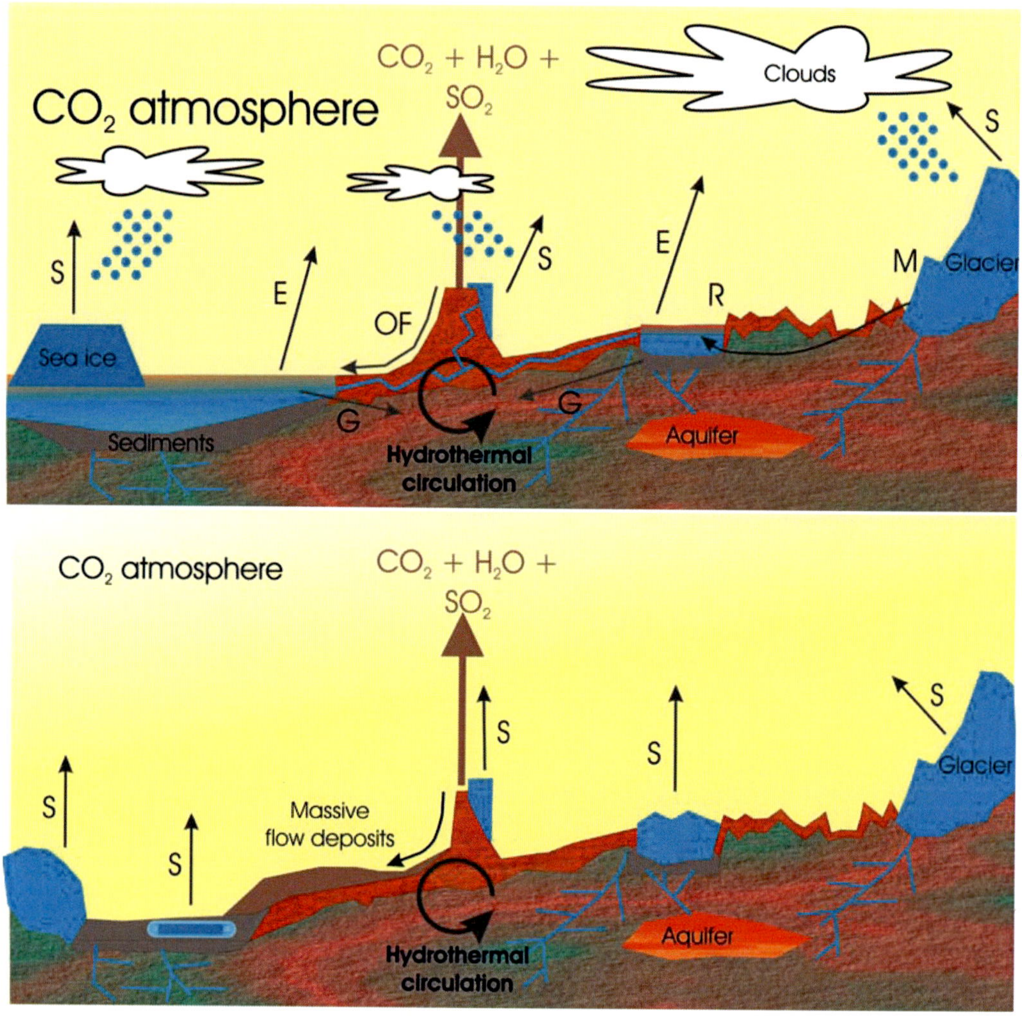

Figure 6: **a)** Schematic representation of land–ocean–atmosphere interactions associated with the presence of a long-term hydrological cycle on Mars in a very cold Noachian climate. S: Sublimation; E: Evaporation; M: Meltwater; R: Runoff; OF: Outburst Flooding; G: Groundwater flow systems recharge. For the putative Martian aqueous solutions modelled here, the cycle is active with an atmospheric pressure of about 2 bar and associated greenhouse effect raising temperatures over 250 K. Similarities to the seasonal water cycle in the terrestrial Antarctic's Dry Valles are evident; **b)** The hydrological cycle during the Hesperian, dominated by the activity of outflow channels and by the thinning of the atmosphere and the hydrosphere; **c)** Amazonian episodic wet environments are dispersed on the Martian surface. H_20-H_2O_2 mixtures provide potential for the water to be in the liquid state, under a milibar-CO_2 atmosphere.

The hydrological cycle in a cold and wet Mars could have been active only in periods of dense atmosphere, as having a minimum water vapor partial pressure is essential for water to flow, and temperatures of at least -25 ºC are required to trigger evaporation and snowfall. Such conditions may have prevailed during the Noachian (Phillips et al., 2001), when the melting of water ice and snow was responsible for both surface valley erosion and groundwater recharge. According to our model, the formation of valley networks in the southern highlands (e.g., Bhattacharya et al., 2005), would have been particularly efficient at low paleolatitudes, due to the higher mean temperatures in the Martian tropics. Hemispheric size oceans of water and ice likely formed on the northern lowlands (Clifford and Parker,

2001) with increasing quantities of ice towards the North Pole and a vast northern polar cap almost the size of the Vastitas Borealis. Figure 7 summarizes some of the photogeological evidence of Noachian hyperloaded water-related landforms.

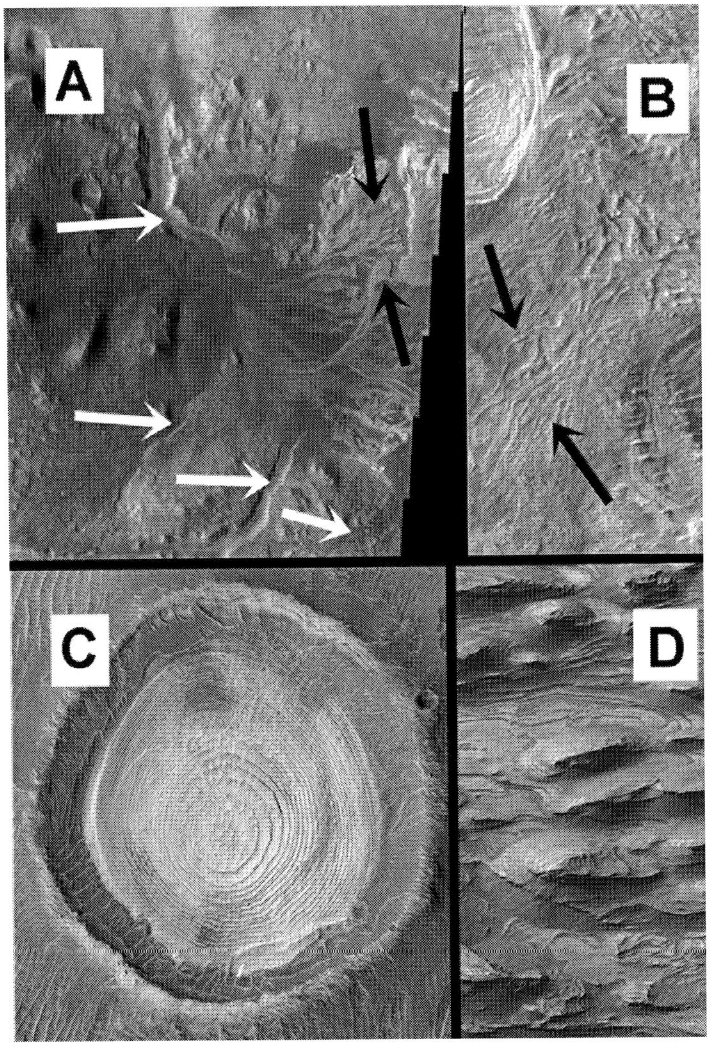

Figure 7: It is impossible to date Martian surfaces from impact craters in either an absolute or relative sense, given the problems of burial and exhumation, but the images presented here have been interpreted as most likely Noachian features. All them are aptly explained by deposition of salts and sediments in craters and valley floors by hyperloaded circulating waters, causing inverted reliefs by differential erosion of the less coarsed and weakly cemented sorrounding materials.
A: Portion of THEMIS V08790003 image probing persistent hyperloaded water flow over long periods of time, that created this delta located in a crater at 23.5° S, 33.5° W. White arrows mark source channels; black arrows indicate cut-off meanders and crossing channels, all showing inverted reliefs.
B: Portion of THEMIS V10963002 image showing a system of channels with inverted relief, marked with blak arrows, in Ganges Chasma (5.2° S, 76° W).
C: MOC2-403 image of an impact crater (0.9° S, 346.2° W) on the northwestern floor of the 2.3 km across Schiaparelli Basin. The constant thickness, uniform physical properties and bedding of the sedimentary layers indicate a recurrent process, likely driven by the climatic modulation of water deposition in a lake. Note the inverted relief in the mesa at the bottom of the crater.
D: Subframe of MOC image FHA-01278, in Candor Chasma (5° S, 74° W). The large number and thickness uniformity of the layers, about 10 meters, indicate regular deposition of materials in hyperloaded waters.

The Hesperian (3.7-3.1 Gyr Ago): Switching Off the Planetary Hydrological Cycle

It is commonly accepted that towards the end of the Noachian, Mars underwent a series of dramatic changes that resulted in the partial loss of its atmosphere (Melosh and Vickery, 1989; Brain and Jakosky, 1998; Lundin et al., 2004), accompanied by a loss of vast quantities of its water reservoir through thermal and nonthermal escape processes, chemical weathering of the surface soil (Lammer et al., 2003), and by progressive crustal assimilation of H_2O (Clifford and Parker, 2001). This ultimately led to the interruption of the hydrological cycle. The thinning of the atmosphere in a water depleted planet likely triggered the partial freezing of the ocean and other bodies of standing waters, resulting in an ice-covered ocean on the northern plains, and ice covered lakes and ponds in the southern highlands. These environmental changes also facilitated the entrapment of frozen water in the Martian regolith. The ultimate fate of any Martian ocean or big lake was probably the complete freezing, and later sublimation, as evaporation is a very ineffective process in hyperloaded waters (Sears and Chittenden, 2005) and at very low ambient temperatures.

This super-cold and dry scenario is supported by the large concentration of highly hydrated salts in the dominant layers of basin sequences, which is consistent with a cold depositional environment, and the negligible quantities of anhydrous or less hydrated salts, that would be expected in an evaporitic scenario at higher temperatures (Kargel, 2004). The presence on the Martian surface of large scale erosive features, likely due to catastrophic flows, has been interpreted as evidence for the presence during this epoch of large amounts of ice-deposits, and further supports the idea that conditions during the Hesperian hindered the occurrence of stable bodies of liquid water on the surface. The Hesperian witnessed intense volcanism, which may have been related to the episodic melting of ice deposits and the subsequent large scale catastrophic flows. Due to the continuous escape of water to the atmosphere through the sublimation of ice, liquid water solutions in the subsoil became more concentrated and dense, resulting in the formation of a highly saline cryolithosphere. Brine migration likely occurred within the Martian subsoil leading to the widespread formation of highly saline lenses of cold acidic brines (cryopegs/taliks). The Hesperian hydrological system is schematically represented in Figure 6b.

Currently, there is some evidence pointing to the occurrence of a transient wet period in the Late Hesperian-Early Amazonian, possibly due to a relatively denser atmosphere, which may have resulted in the formation of minor seas (Fairén et al., 2003). Evidence includes glacier type erosive features and deposits (Head et al., 2004), dendritic valley formation (Mangold et al., 2004; Quantin et al., 2004, 2005; Pondrelli et al., 2005) and repeated deposition of layered sediments in lacustrine environments (Glotch and Christensen, 2005; Quantin et al., 2005) found in different near-equatorial locations and formed during the same time period (Figure 8). Also, the Hesperian is the time when most highlands' outflow channels developed (Tanaka et al., 2003). It has been possible to assign well-defined potential source of fluids to most outflow channels, such as large enclosed basins, a region of chaos, or major fissures; however the origin of some outflows with multiple sources is not so obvious, as is the case for Mawrth Valles and some other valleys in the Arabia Terra region (Howard & Moore, submitted). A compelling explanation for the origin of the later is the local occurrence of aquifer springs consisting of cold, acidic and salty waters, remnants of earlier Noachian times: the continuous accumulation of hyperloaded waters could have occurred progressively under large water-ice deposits, until they reached their maximum water retention capacity and the subsequent discharge of vast amounts of water in brief lapses of time. Hyperloaded water derived from such events debouched in the northern lowlands, which were already cemented by thick deposits of salts derived from the disappearance of the Noachian oceans, contributing to the formation of small, acidic and saturated Hesperian-Early Amazonian seas. Geologic and mineralogical data strongly support this interpretation, as Hesperian and Amazonian deposits cover the Noachian basement of the Martian northern lowlands (Fairén and Dohm, 2004) with relatively recent secondary, water-related minerals, such as sulfur-rich basalts, relatively enriched in bromine relative to chlorine, and with anhydrous nanophase ferric oxides. Unaltered mafic materials such as olivine or pyroxene, which are generally related to water-free environments, are totally absent in these regions (Bibring et al., 2006; McSween *et al.,* in press; see Figure 9).

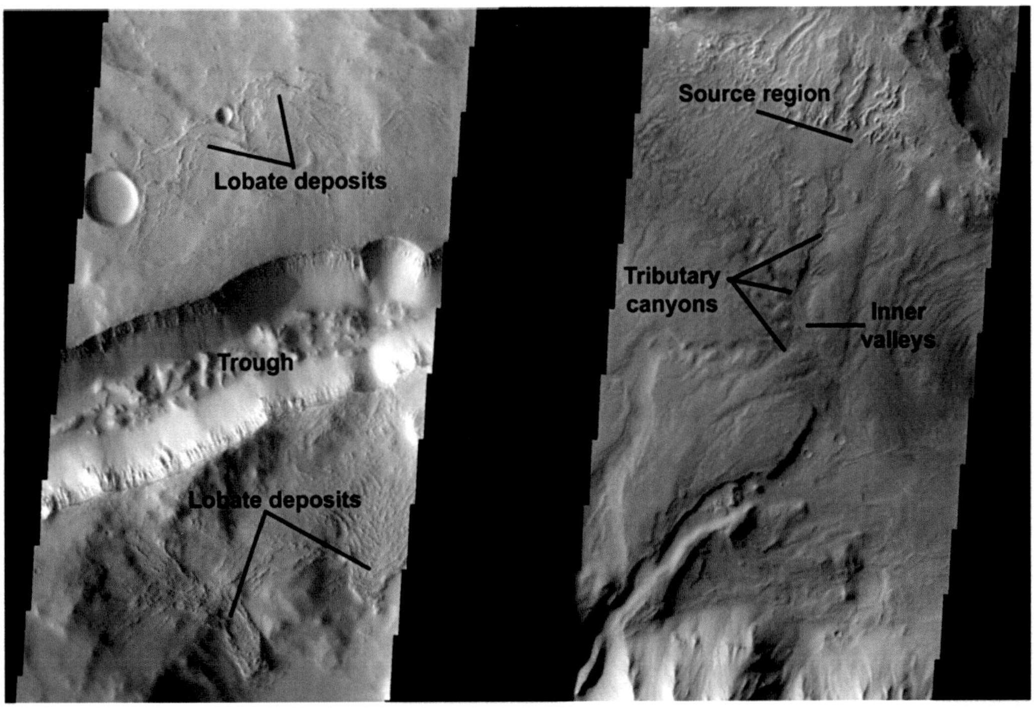

Figure 8: THEMIS visible-light images of Late Hesperian – Early Amazonian units on equatorial Mars; **Left**: Portion of THEMIS image V06597003 of ridged deposits and lobate features on the Northern trough rim of the Mangala Valles source region (149.3°W, 18.2°S), interpreted to be formed by accumulations of snow and ice on the graben rim; **Right**: Portion of THEMIS image V3249001 of a drainage basin showing dense valley networks, tributary canyons and meandering valleys on East Melas Chasma inner plateau (77.5°W, 10°S).

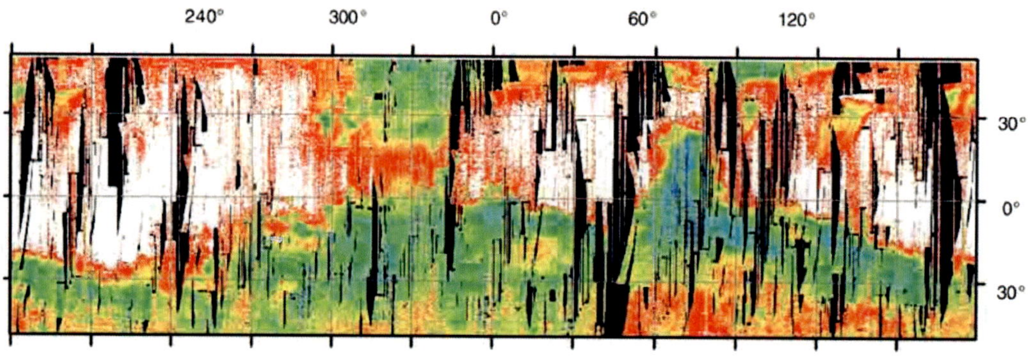

Figure 9: Global map of anhydrous nanophase ferric oxides (code color red to white) and mafic materials mainly representing pyroxene content (code color blue to green), after Bibring et al. (2006). Distribution indicates global dissolution of iron phases into acidic waters across the southern highlands, massive fluvial transfer of soluble ferrous iron into northern plains oceans (in the form of both long-term sustained flow of groundwater and surface valley systems during the Noachian and episodic massive flood discharges by outflow channels during the Hesperian), precipitation of ferrous iron, and photolytic oxidation of ferrous to ferric iron.

The Amazonian (3.1 Gyr Ago-Present): Transient Ponds in a Cold and Dry Planet

Conditions for liquid water became suitable again during brief and intriguing episodes throughout the Amazonian, in which the climate was intermittently relatively warm and moist, until geologically very recent times (Fairén et al., 2003). The geologic record of that period confirms enhanced periodic activity in the Tharsis and Elysium volcanic regions, coupled with hydrothermal mobilization of water and the subsequent development of glaciers (Neukum et al., 2004). Multiple glacial landscapes formed and phases of tropical mountain glaciation occurred even within the past few tens of millions of years (Kargel and Strom, 1992; Neukum et al., 2004; Shean et al., 2005; Head et al., 2005; Milkovich et al., 2006; Forget et al., 2006). Minor amounts of water also flowed between rocks and filled in closed depressions to form lakes (Figure 10a). These lakes rested on top of highly saline terrains formed after the freezing of Noachian-Hesperian brine solutions, and were composed of highly saline, dense and acidic waters. The presence of erosive and depositional features such as deltas, shorelines and terraces, Amazonian in age, confirms the occurrence of these wet episodes over extended periods of time (Cabrol and Grin, 2001, see Figure 6c).

The presence of very young gullies on impact crater walls and cliff sides has been explained as a result of recent melting snow deposits in a cold climate (Christensen, 2003; Figure 10b). The Mars Exploration Rovers Missions have shown aqueous alteration in present-day Mars by transient acidic water at Gusev crater (Haskin et al., 2005; Cabrol et al., 2006), with surface and subsurface soils dated at less than 2 Gy. (Cabrol et al., 1998), suggesting wetter conditions sometime in the Amazonian. The Opportunity Rover explorations at Meridiani Planum also provided evidence of very recent episodes of transient liquid water (Squyers et al., 2004, Chavdarian and Sumner, 2006). Finally, a recent event of surface water runoff has been documented by the Mars Global Surveyor Orbiter. During its mission to document new deposits formed by downslope movement of material in mid-latitude gullies, the Mars Global Surveyor Orbiter Camera found evidence for the formation of new gully deposits since August 1999 (Figure 11). This observation has been interpreted as evidence that liquid water flowed on the surface of Mars during the past decade. Despite these episodes of transient liquid water on the surface of Mars, any post-Hesperian chemical weathering has likely remained spatially limited (Christensen et al., 2003; Hoefen et al., 2003; Hamilton and Christensen, 2004; Rogers et al., 2005; Bibring et al., 2006), and atmospheric alteration characterizes Amazonian geochemistry.

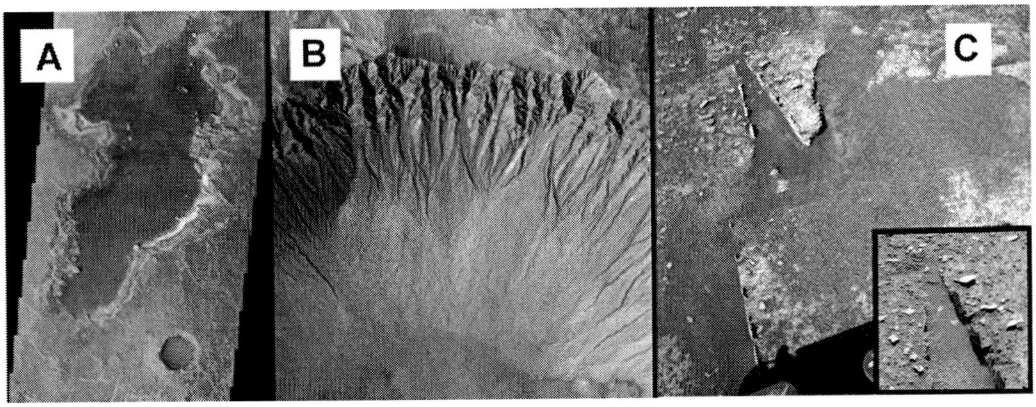

Figure 10: **A:** Subframe of THEMIS image V07889029 showing a very young dark unit embaying underlying terrain, visible in the main body and in the crater southeast. The dark unit likely has a water-related origin. Image centered 1.5° N, 354.4° W; **B:** Very recent water erosion after melting of ice and/or snowpacks create gullies in the north wall of this crater located at 38 °S, 167.2° W. Image MOC2-1292; **C:** *Opportunity* picture 1N143189015EFF3243P1961L0M1, obtained on sol 169 inside Endurance crater. The channeled features may have been formed by liquid water flowing through fractures. Centimeter-sized pointy features, highlighted in the inset, seem to be so delicate that their presence suggests a very recent epoch of formation.

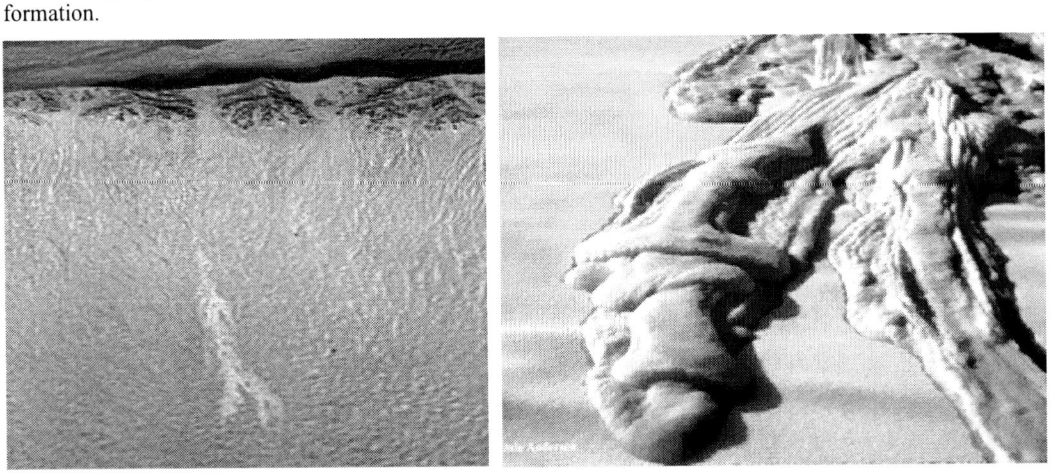

Figure 11: **Left**: MGS MOC2-1619 image of gully deposits in the slope of a crater in the Centauri Montes region, highlighted in September 2005 and not observed in previous MOC images of the same crater in August 1999, suggesting contemporaneous water flow on Mars. The new deposits are light toned and exhibit attributes expected from emplacement aided by a fluid with the properties of liquid water: relatively long, extended, digitate distal and marginal branches, diversion around obstacles, and low relief. (MSSS); **Right**: Icy brine flowing in low temperature in Axel Heiberg island, Canadian high Arctic. The brine flows for some minutes and freezes over when the flowing velocity reduces. A complete video can be shown at: http://dale andersen.seti.org/Dale%20Andersen/ Home.html. The same phenomena may have created the gully deposits on Mars. (Courtesy of Dale Andersen).

In the coldest periods of thin atmosphere, water vapor partial pressure is the limiting factor for water to flow, and temperatures are close to the freezing point for much of the analyzed solutions (Table and Figure 1), and consequently only transient liquid water can potentially exist on most of the Martian surface (Richardson and Mischna, 2005; Heldmann et al., 2005; Cabrol et al., 2006). These conditions are prevailing today, with atmospheric pressure of 6.2 mbar and mean temperatures about -50 °C, although a maximum atmospheric temperature of -8 °C was measured in daylight at both Pathfinder (15° N) and Opportunity (2° S) landing sites; and up to 15 °C in soil temperature in the Opportunity landing site. In such conditions, water is forming ice in the residual polar caps (Bibring et al., 2005), interstitial ice and adsorbed water in the regolith (Boynton et al., 2002), gullied terrains (Malin and Edgett, 2000b; Heldmann and Mellon, 2004) as young as less than 250,000 years (Heldmann et al., 2005), or lakes beneath a protective ice sheet (Murray et al., 2005), possibly remaining in liquid state at depth because the combination of insulation from the atmosphere, chemical composition, pressure and geothermal heat flux.

Astrobiological Implications

The main constrains for the apparition and evolution of life on Mars were probably the availability of diverse energy sources and the presence of standing bodies of water for considerable periods of time. Furthermore, if life ever originated on Mars on the first place, then Martian organisms would have needed to adapt to dramatic environmental changes, and evolve physiological and ecological adaptations to harmful solar radiation, dissecation, extremely low temperatures, highly oxidizing environments saline and acidic waters, and ionic H_2O_2 strength, among other environmental challenges.

Mars possessed in the past the same energy sources that organisms use on Earth, including light, hydrothermal and volcanic activity or the presence of reduced inorganic compounds, which could sustain a photoautotroph and/or chemoautotroph ecology. Early in the planet's history, the presence of standing bodies of liquid water provided stable, albeit challenging, environmental conditions for the apparition and evolution of life. The physico-chemical characteristics of rivers, lakes, seas and aquifers on Mars, were similar to environments on Earth which are considered extreme, yet host abundant and diverse communities of organisms (e.g., Amils et al., 2007). Therefore Mars likely possessed early in its history the necessary ingredients for the apparition of life. Taking the evolution of life on Earth as an example, life could have reached relatively complex levels of organization before Mars underwent the most dramatic environmental changes, around 3.5 Gy ago, and Martian organisms could have evolved similar physiological adaptations to challenging environmental conditions as extremophiles on Earth.

Our model of the Martian hydrological cycle could arguably accommodate a variety of terrestrial organisms, and thus offers new insights for the planning and design of future life-search missions to Mars. A putative Martian biosphere would be dominated by halophilic, acidophilic and psychrotolerant organisms, likely with physiological and ecological adaptations that further optimized their survivability. Such organisms abound on Earth, in environments that here are considered extreme. Active communities of psychrophilic bacteria thrive within liquid inclusions in frozen environments at temperatures as low as -25 °C both in Siberia, Greenland and the Arctic (Rivkina et al., 2000; Bakermans et al., 2003; Junge et al., 2004; Tung et al., 2006) and in Antarctica (Priscu et al., 1998; Mahaney et al., 2001; de Angelis et al., 2005). These organisms derive energy from iron and sulfur compounds in areas covered by glaciers and thick permafrost in or near the Arctic (Grasby et al., 2003; Tung et al., 2006) and the Antarctic (Mikucki and Priscu, 2004). Such ecosystems could be considered as models for Mars.

The detection recently of small amounts of methane in the Martian atmosphere has intensified the debate about the possibility that extant organisms may still be present on the planet. After models advocating for a volcanic source of the methane have been severely discussed (Ryan et al., 2006), a subsurface microbial metabolism coupled to sulfide weathering at very cold temperatures is the most likely candidate to explain the origin of the methane. Such microorganisms could form communities of acidophilic methanogenic psychrofiles, existing contemporarily and/or in the wetter past. Sulfate mineral matrices can shield these microorganisms from UV radiation (Martínez-Frías et al., 2006) and also preserve biological compounds for geologically long periods of time (Aubrey et al., 2006); consequently, sulfate deposits are prime candidate sites to search for extinct and/or extant life forms on Mars.

Conclusion: The Case for a "Cold and Wet" Mars

The hydrogeological cycle presented here reconciles accumulating evidence for long lasting liquid water on the surface of Mars concurrently with widespread glacier formation, and is in accordance with existing climatic models which predict serious obstacles to warm Martian climate, offering a compelling and self-consistent theory to account for a complete Martian hydrological cycle. Assuming that surface temperatures have always been maintained below 0 °C, Mars can be considered a "cold and wet" planet for its entire geological history. Indeed, all the factors previously suggested to warm the Martian atmosphere over the freezing point of pure water, as well as punctual events releasing water to the surface and groundwater circulation, could be important contributors to firmly assure the presence of substantial quantities of liquid water in different moments on the surface of a really alien, Martian Mars.

References

R. Amils, E. González-Toril, D. Fernández-Remolar, F. Gómez, Á. Aguilera, N. Rodríguez, M. Malki, A. García-Moyano, A. G. Fairén, V. de la Fuente, J. L. Sanz Extreme environments as Mars terrestrial analogs: the Rio Tinto case. *Planetary and Space Science*, submitted.

D. T. Andersen, W. H. Pollard, C. P. McKay, J. Heldmann. Cold springs in permafrost on Earth and Mars. *J. Geophys. Res.* 107, doi: 10.1029/2000JE001436 (2002).

V. Ansan and N. Mangold. New observations of Warrego Valles, Mars: Evidence for precipitation and surface runoff. *Planet Space Sci.*, 54, 219-242 (2006).

R. E. Arvidson, F. Poulet, J.P. Bibring, M. Wolff, A. Gendrin, R. V. Morris, J. J. Freeman, Y. Langevin, N. Mangold, G. Bellucci. Spectral reflectance and morphologic correlations in eastern Terra Meridiani, Mars. *Science* 307, 1591-1594 (2005).

Andrew Aubrey, H. James Cleaves, John H. Chalmers, Alison M. Skelley, Richard A. Mathies, Frank J. Grunthaner, Pascale Ehrenfreund, Jeffrey L. Bada. Sulfate minerals and organic compounds on Mars. *Geology*: Vol. 34, No. 5, pp. 357–360, 2006.

J.L. Bada and A. Lazcano. Some like it hot, but not the first biomolecules. *Science* 296, 1982-1983 (2002).

Baker, V.R., Strom, R.G., Gulick, V.C., Kargel, J.S., Komatsu, G., Kale, V.S. Ancient oceans, ice sheets and the hydrological cycle on Mars. *Nature* 352, 589-594 (1991).

C. Bakermans, A. I. Tsapin, V. Souza-Egipsy, D. A. Gilichinsky and K. H. Nealson. Reproduction and metabolism at − 10°C of bacteria isolated from Siberian permafrost. *Environmental Microbiology*, 5, 321-326, 2003.

Bandfield, J. L., Glotch, T.D.&Christensen, P. R. Spectroscopic identification of carbonateminerals in the Martian dust. *Science* 301, 1084–1087 (2003).

Bhattacharya, J. P., T. H. D. Payenberg, S. C. Lang, and M. Bourke. Dynamic river channels suggest a long-lived Noachian crater lake on Mars. *Geophys. Res. Lett.* 32, L10201, doi:10.1029/2005GL022747 (2005).

J.P. Bibring, Y. Langevin, A. Gendrin, B. Gondet, F. Poulet, M. Berthé, A. Soufflot, R. Arvidson, N. Mangold, J. Mustard, P. Drossart, and the OMEGA team. Mars surface diversity as revealed by the OMEGA/Mars Express observations. *Science* 307, 1576-1581 (2005).

J.P. Bibring, Y. Langevin, J. F. Mustard, F. Poulet, R. Arvidson, A. Gendrin, B. Gondet, N. Mangold, P. Pinet, F. Forget, and the OMEGA team. Global mineralogical and aqueous Mars history derived from OMEGA/Mars Express data. *Science* 312, 400-404, 2006.

Boynton, W.V., Feldman, W.C., Squyres, S.W., Prettyman, T.H., Brückner, J., Evans, L.G., Reedy, R.C., Starr, R., Arnold, J.R., Drake, D.M., Englert, P.A.J., Metzger, A.E., Mitrofanov, I., Trombka, J.I., d'Uston, C., Wänke, H., Gasnault, O., Hamara, D.K., Janes, D.M., Marcialis, R.L., Maurice, S., Mikheeva, I., Taylor, G.J., Tokar, R., Shinohara C. Distribution of hydrogen in the near surface of Mars: Evidence for subsurface ice deposits. *Science* 297, 81-85 (2002).

Brain, D. A. & Jakosky, B. M. Atmospheric loss since the onset of the Martian geologic record: combined role of impact erosion and sputtering. *J. Geophys. Res.* 103, 22689-22694 (1998).

Brocks Jochen J., Gordon D. Love, Roger E. Summons, Andrew H. Knoll, Graham A. Logan, Stephen A. Bowden, Biomarker evidence for green and purple sulphur bacteria in a stratified Palaeoproterozoic sea, *Nature* 437, 866-870 (06 Oct 2005).

Burt, D. M., and L. P. Knauth. Electrically conducting, Ca-rich brines, rather than water, expected in the Martian subsurface. *J. Geophys. Res.*, 108(E4), 8026, doi:10.1029/2002JE001862 (2003).

Burton, H.R. The annual cycle of heat content and mechanical stability of hypersaline Deep Lake, Vestfold Hills, Antarctica. *Hydrobiol.* 165, 115-128 (1988).

Cabrol, N. A., E. A. Grin, G. Dawidowicz. A model of outflow generation by hydrothermal underpressure drainage in volcano–tectonic environment, Shalbatana Vallis (Mars). *Icarus* 125, 455-464 (1997).

Cabrol, N. A., E. A. Grin, R. Landheim, R. O. Kuzmin, R. Greeley. Duration of the Ma'adim Vallis/Gusev crater hydrogeologic system, Mars. *Icarus* 133, 98-108 (1998).

Cabrol, N. A., and E. A. Grin. The evolution of lacustrine environments on Mars: Is Mars only hydrologically dormant. *Icarus* 149, 291–328 (2001).

Cabrol, N. A., J. D. Farmer, E. A. Grin, L. Richter, L. Soderblom, R. Li, K. Herkenhoff, G. A. Landis, and R. E. Arvidson (2006). Aqueous processes at Gusev crater inferred from physical properties of rocks and soils along the Spirit traverse, *J. Geophys. Res.*, 111, E02S20, doi:10.1029/2005JE002490.

Carr, M. H. Stability of streams and lakes on Mars. *Icarus* 56, 476–495 (1983).

Chavdarian, G. V., and D. Y. Sumner (2006), Cracks and fins in sulfate sand: Evidence for recent mineral-atmospheric water cycling in Meridiani Planum outcrops? *Geology*, 34, 229–232.

Christensen, P. R. Formation of recent Martian gullies through melting of extensive water-rich snow deposits. *Nature*, 422, 45-48 (2003).

Christensen, P. R., et al. Morphology and composition of the surface of Mars: Mars Odyssey THEMIS results. *Science* 300, 2056– 2061 (2003).

Clark, B.C. Chemical and physical microenvironments at the Viking landing sites. *J. Mol. Evol.* 14, 13–31 (1979)

Clark, B.C. On the non-observability of carbonates on Mars. In *Fifth Conference on Mars, Lunar and Planetary Institute*, Houston, p. 6214 (1999).

Clark, B.C. and D.C. Van Hart. The salts of Mars. *Icarus* 45, 370– 378 (1981).

Clifford, S. M. and Parker, T. J. The evolution of the Martian hydrosphere: Implications for the fate of a primordial ocean and the current state of the northern plains. *Icarus* 154, 40-79 (2001).

Colaprete, A., and O. B. Toon. Carbon dioxide clouds in an early dense Martian atmosphere, *J. Geophys. Res.*, 108, 5025, doi:10.1029/2002JE001967, (2003).

Craddock, R.A., Howard, A.D. The case for rainfall on a warm, wet early Mars. *J. Geophys. Res.* 107. 10.1029/2001JE001505 (2002).

de Angelis, M., M.C. Morel-Fourcade, J.M. Barnola, J. Susini, and P. Duval. Brine micro-droplets and solid inclusions in accreted ice from Lake Vostok (East Antarctica). *Geophys. Res. Lett.*, 32, doi:10.1029/2005GL022460 (2005).

Dohm, J.M., Ferris, J.C., Baker, V.R., Anderson, R.C., Hare, T.M., Strom, R.G., Barlow, N.G., Tanaka, K.L., Klemaszewski, J.E., Scott, D.H. Ancient drainage basin of the Tharsis region, Mars: potential source for outflow channel systems and putative oceans or paleolakes. *J. Geophys. Res.* 106, 32,943-32,958 (2001).

Elwood Madden, M.E., Bodnar, R.J., Rimnstidt, J.D. Jarosite as an indicator of water limited chemical weathering on Mars. *Nature* 431, 821-823 (2004).

Fairén, A. G. Does olivine indicates dry conditions on Mars? *Lunar Planet. Sci. Conf.* XXXVII, #1645 (2006).

Fairén, A. G., Dohm, J.M., Baker, V.R., de Pablo, M.A., Ruiz, J., Ferris, J., Anderson, R. Episodic flood inundations of the northern plains of Mars. *Icarus* 165, 53-67 (2003).

Fairén, A. G.; Dohm, J. M. (2004). Age and origin of the lowlands of Mars. *Icarus*, 168, 277-284.

Fairén, A. G., Fernández-Remolar, D., Dohm, J.M., Baker, V.R., Amils, R. Inhibition of carbonate synthesis in acidic oceans on early Mars. *Nature* **431**, 423-426 (2004).

Fairén, A. G., Dohm, J.M., Uceda, E. R., Rodríguez, J. A. P., Baker, V.R., Fernández-Remolar, D., Schulze-Makuch, D., Amils, R. Prime candidate sites for astrobiological exploration through the hydrogeological history of Mars. *Planetary and Space Science* 53, 1355-1375 (2005).

Caleb I. Fassett, James W. Head. Valleys on Hecates Tholus, Mars: origin by basal melting of summit snowpack. *Planetary and Space Science*, 54 (2006) 370–378.

Forget, F. & Pierrehumbert, R. T. Warming early Mars with carbon dioxide clouds that scatter infrared radiation. *Science* **278**, 1273-1276 (1997).

F. Forget, R. M. Haberle, F. Montmessin, B. Levrard, J. W. Head. Formation of glaciers on Mars by atmospheric precipitation at high obliquity. *Science* 311, 368-371, 2006.

Formisano, V., Atreya, S., Encrenaz, T., Ignatiev, N., and Giuranna, M. Detection of methane in the atmosphere of Mars. *Science* 306, 1758-1761 (2004).

Frey, H.V., J. H. Roark, K. M. Shockey, E. L. Frey and S. E. H. Sakimoto, Ancient Lowlands on Mars, *Geophys. Res. Lett.* 29, 10.1029/2001GL013832, 2002.

Gaidos, E., and G. Marion. Geological and geochemical legacy of a cold early Mars. *J. Geophys. Res*., 108, doi:10.1029/2002JE002000 (2003).

R. Gellert, R. Rieder, R. C. Anderson, J. Brückner, B. C. Clark, G. Dreibus, T. Economou, G. Klingelhoöfer, G. W. Lugmair, D. W. Ming, S. W. Squyres, C. d'Uston, H. Wänke, A. Yen, J. Zipfe. Chemistry of rocks and soils in Gusev Crater from the Alpha Particle X-ray Spectrometer. *Science* 305, 829-832 (2004).

Gendrin, N. Mangold, J.Bibring, Y. Langevin, B. Gondet, F. Poulet, G. Bonello, C. Quantin, J. Mustard, R. Arvidson, S. LeMouélic. Sulfates in Martian layered terrains: the OMEGA/Mars Express view. *Science* 307, 1587-1591 (2005).

Glotch, T. D., and P. R. Christensen (2005), Geologic and mineralogic mapping of Aram Chaos: Evidence for a water-rich history, *J. Geophys. Res*., 110, E09006, doi:10.1029/2004JE002389.

S. E. Grasby, C. C. Allen, T. G. Longazo, J. T. Lisle, D. W. Griffin, B. Beauchamp. Supraglacial sulfur springs and associated biological activity in the Canadian High Arctic—Signs of life beneath the ice. *Astrobiology* 3, 583-596 (2003).

Griffith, L.L., and Shock, E.L.. Hydrothermal hydration of Martian crust: Illustration via geochemical model calculations. *J. Geophys. Res*. 102, 9135-9143 (1997).

Gulick, V.C. Origin of valley networks on Mars: a hydrological perspective. *Geomorphology* 37, 241-268 (2001).

Haberle, R. M., C. P. McKay, J. Schaeffer, N. A. Cabrol, E. A. Grin, A. P. Zent, and R. Quinn. On the possibility of liquid water on present day Mars. *J. Geophys. Res*. 106(E10), 23,317–23,326 (2001).

L. A. Haskin, A. Wang, B. L. Jolliff, H. Y. McSween, B. C. Clark, D. J. Des Marais, S. M. McLennan, N. J. Tosca, J. A. Hurowitz, J. D. Farmer, A. Yen, S.W. Squyres, R. E. Arvidson, G.Klingelhöfer, C.Schröder, P. A. de Souza Jr, D. W. Ming, R. Gellert, J. Zipfel, J.Brückner, J. F. Bell III, K.Herkenhoff, P. R. Christensen, S. Ruff, D. Blaney, S. Gorevan, N.A. Cabrol, L.Crumpler, J. Grant, L. Soderblom. Water alteration of rocks and soils on Mars at the Spirit rover site in Gusev crater. *Nature* 436, 66-69 (2005).

Head, J.W., Kreslavsky, M., Hiesinger, H., Ivanov, M.A., Pratt, S., Seibert, N., Smith, D.E., Zuber, M.T. Oceans in the past history of Mars: test for their presence using Mars Orbiter Laser Altimeter (MOLA) data. *Geophys. Res. Lett.* 25, 4401–4404 (1998).

J.W. Head III, D.R. Marchant, G.J. Ghatan. Glacial deposits on the rim of a Hesperian-Amazonian outflow channel source trough: Mangala Valles, Mars. *Geophys. Res. Lett.* 31, doi:10.1029/2004GL020294 (2004).

J. W. Head, G. Neukum, R. Jaumann, H. Hiesinger, E. Hauber, M. Carr, P. Masson, B. Foing, H. Hoffmann, M. Kreslavsky, S. Werner, S. Milkovich, S. van Gasselt & The HRSC Co-Investigator Team. Tropical to mid-latitude snow and ice accumulation, flow and glaciation on Mars. *Nature* 434, 346-351 (2005).

Heldmann, J.L. and Mellon, M.T. Observations of Martian gullies and constraints on potential formation mechanisms. *Icarus* 168, 285-304 (2004).

Heldmann, J. L., O. B. Toon, W. H. Pollard, M. T. Mellon, J. Pitlick, C. P. McKay, and D. T. Andersen. Formation of Martian gullies by the action of liquid water flowing under current Martian environmental conditions. *J. Geophys. Res.*, 110, doi:10.1029/2004JE002261 (2005).

T. M. Hoefen, R. N. Clark, J. L. Bandfield, M. D. Smith, J. C. Pearl, P. R. Christensen. Discovery of olivine in the Nili Fossae region of Mars. *Science* 302, 627-630 (2003).

Holland, H. D. The oceans: a possible source of iron in iron-formations. *Econ. Geol.* 68, 1169-1172 (1973).

Howard, A. D., J. M. Moore, and R. P. Irwin III (2005), An intense terminal epoch of widespread fluvial activity on early Mars: 1. Valley network incision and associated deposits, *J. Geophys. Res.*, 110, E12S14, doi:10.1029/2005JE002459.

Howard, A. D. & Moore, J. M. A geomorphic transect across the Martian highlands-lowlands boundary near the Prime Meridian: Evidence for a sedimentary platform graded to a deep ocean. Submitted.

Hynek, B.M. Implications for hydrologic processes on Mars from extensive bedrock outcrops throughout Terra Meridiani. *Nature* 431, 156-159 (2004).

Jiin-Shuh Jean, Ming-Jer Lee, Ming-Hung Chien, Ming-Kuo Lee. Potential antifreeze compound(s) in the present-day Martian seepage groundwater. *Hydrogeology Journal*, submitted.

K. Junge, H. Eicken, and J. W. Deming. Bacterial activity at -2 to -20°C in Arctic wintertime sea ice. *Applied and Environmental Microbiology* 70, 550-557 (2004).

Johnson, D.B. and Hallberg, K.B. The microbiology of acidic mine waters. *Research in Microbiology* 154, 466-473 (2003).

Kargel, J. S. & Strom, R. G. Ancient glaciation on Mars. *Geology* 20, 3–7 (1992).

Kargel, J. S. "*Mars, a warmer wetter planet*". Springer-Verlag, Berlin (2004).

Kasting, J. F. CO_2 condensation and the climate of early Mars. *Icarus* 94, 1-13 (1991).

J. F. Kasting. Warming early Earth and Mars. *Science* 276, 1213-1215 (1997).

Knauth, L. P. Salinity history of the Earth's early ocean. *Nature* 395, 554–555, (1998).

Knauth, L. P., and D. M. Burt. Eutectic brines on Mars: Origin and possible relation to young seepage features. *Icarus* 158, 267– 271 (2002).

Kraal, E. R., E. Asphaug, J. M. Moore, and R. D. Lorenz (2006), Quantitative geomorphic modeling of Martian bedrock shorelines, *J. Geophys. Res.*, 111, E03001, doi:10.1029/2005JE002567.

Kuhn, W.R. and Atreya, S.W. Ammonia photolysis and the greenhouse effect in the primordial atmosphere of the Earth. *Icarus* 37, 207-213 (1979).

Kuzmin, R. O. and Zabalueva E. V. On salt solutions in the Martian criolithosphere. *Solar System Research*, 32, 187-197 (1998).

H. Lammer, H. I. M. Lichtenegger, C. Kolb et al. Loss of water from Mars: Implications for the oxidation of the soil. *Icarus* 165, 9-25 (2003).

Linke, W. *Solubilities of inorganic and metal organic compounds*, Vol. I, 3rd edn., Am. Chem. Soc., Washington DC (1958).

R. Lundin, S. Barabash, H. Andersson, M. Holmström, A. Grigoriev, M. Yamauchi, J.-A. Sauvaud, A. Fedorov, E. Budnik, J.-J. Thocaven, D. Winningham, R. Frahm, J. Scherrer, J. Sharber, K. Asamura, H. Hayakawa, A. Coates, D.R. Linder, C. Curtis, K.C. Hsieh, B.R. Sandel, M. Grande, M. Carter, D. H. Reading, H. Koskinen, E. Kallio, P. Riihela, W. Schmidt, T. Säles, J. Kozyra, N. Krupp, J. Woch, J. Luhmann, S. McKenna-Lawler, R. Cerulli-Irelli, S. Orsini, M. Maggi, A. Mura, A. Milillo, E. Roelof, D. Williams, S. Livi, P. Brandt, P. Wurz, and P. Bochsler. Solar wind-induced atmospheric erosion at Mars: first results from ASPERA-3 on Mars Express. *Science* 305, 1933-1937 (2004).

Lyons, W.B., Welch, K., McKnight, D., and Priscu, J. Blood Falls: a frozen saline discharge, a strange curiosity, or an important clue to history of Taylor Valley. *Seattle Annual Meeting*, Paper No. 190-9 (2003).

W. C. Mahaney, J. M. Dohm, V. R. Baker, Horton E. Newsom, D. Malloch, R. G. V. Hancock, Iain Campbell, D. Sheppard, and W. M. Milner. Morphogenesis of Antarctic paleosols: Martian analogue. *Icarus*, 154, 113-130 (2001).

M. C. Malin and K. S. Edgett. Sedimentary rocks of early Mars. *Science* 290, 1927-1937 (2000a).

M. C. Malin and K. S. Edgett. Evidence for recent groundwater seepage and surface runoff on Mars. *Science* 288, 2330-2335 (2000b).

M. C. Malin and K. S. Edgett. Evidence for persistent flow and aqueous sedimentation on early Mars. *Science* 302, 1931-1934 (2003).

Mangold, N., Quantin, C., Ansan, V., Delacourt, C., Allemand, P. Evidence for precipitation on Mars from dendritic valleys in the Valles Marineris area. *Science* 305, 78-81 (2004).

J. Martínez-Frías, G. Amaral, L. Vázquez. Astrobiological significance of minerals on Mars surface environment. *Reviews in Environmental Science and Bio/technology*, in press.

McCauley, J.F., G.G. Schaber, C.S. Breed, M.J. Grolier, C.V. Haynes, B.Issawi, C. Elachi, and R. Blom, 1982, Subsurface valleys and geoarchaeology of the Eastern Sahara revealed by Shuttle Radar: *Science*, v. 218, n. 4516, p. 1004-1020.

McSween, H.Y.; Wyatt, M.B.; Gellert, R.; Bell III, J.F.; Morris, R.V.; Herkenhoff, K.E.; Crumpler, L.S.; Milam, K.A.; Stockstill, K.R.; Tornabene, L.L.; Arvidson, R.E.; Bartlett, P.; Blaney, D.; Cabrol, N.A.; Christensen, P.R.; Clark, B.C.; Crisp, J.A.; Des Marais, D.J.; Economou, T.; Farmer, J.D.; Farrand, W.; Ghosh, A.; Golombek, M.; Gorevan, S.; Greeley, R.; Hamilton, V.E.; Johnson, J.R.; Joliff, B.L.; Klingelhoeffer, G.; Knudson, A.T.; McLennon, S.; Ming, D.; Moersch, J.E.; Rieder, R.; Ruff, S.W.; Schroeder, C.; de Souza, Jr., P.A.; Squyres, S.W.; Waenke, H.; Wang, A.; Yen, A.; Zipfel, J. Characterization and petrologic interpretation of olivine-rich basalts at Gusev Crater, Mars. *JGR* in press 2006.

Melosh, H. J. & Vickery, A. M. Impact erosion of the primordial atmosphere of Mars. *Nature* 338, 487-489 (1989).

J.A. Mikucki and J.C.Priscu. Microbial life in Blood Falls: an ancient Antarctic ecosystem. *2nd Conference on early Mars*, Abstract #8023 (2004).

Sarah M. Milkovich, James W. Head and David R. Marchant. Debris-covered piedmont glaciers along the northwest flank of the Olympus Mons scarp: Evidence for low-latitude ice accumulation during the Late Amazonian of Mars. *Icarus* 181, 388-407 (2006).

Mischna, M.A., Kasting, J.F., Pavlov, A., Freedman, R. Influence of carbon dioxide clouds on early Martian climate. *Icarus* 145, 546-554 (2000).

Mischna, M.A., Richardson, M.I., Wilson, R.J., McCleese, D.J. On the orbital forcing of Martian water and CO_2 cycles: a general circulation model study with simplified volatile schemes. *J. Geophys. Res.* 108, doi:10.1029/2003JE002051 (2003).

J.B. Murray, J.P. Muller, G. Neukum, S.C. Werner, S. van Gasselt, E. Hauber, W.J. Markiewicz, J.W. Head III, B.H. Foing, D. Page, K.L. Mitchell, G. Portyankina & The HRSC Co-Investigator Team. Evidence from the Mars Express High Resolution Stereo Camera for a frozen sea close to Mars' equator. *Nature* 434, 352-356 (2005).

J. F. Mustard, F. Poulet, A. Gendrin, J.-P. Bibring, Y. Langevin, B. Gondet, N. Mangold, G. Bellucci, F. Altieri. Olivine and pyroxene diversity in the crust of Mars. *Science* 307, 1594-1597 (2005).

G. Neukum, R. Jaumann, H. Hoffmann, E. Hauber, J. W. Head, A. T. Basilevsky, B. A. Ivanov, S. C. Werner, S. van Gasselt, J. B. Murray, T. McCord & The HRSC Co-Investigator Team. Recent and episodic volcanic and glacial activity on Mars revealed by the High Resolution Stereo Camera. *Nature*, 432, 971-979, 2004.

Parker, T.J., Gorsline, D.S., Saunders, R.S., Pieri, D.C., Schneeberger, D.M. Coastal geomorphology of the Martian northern plains. *J. Geophys. Res.* 98, 11061–11078 (1993).

R.J. Phillips, M.T. Zuber, S.C. Solomon, M.P. Golombek, B.M. Jakosky, W.B. Banerdt, D.E. Smith, R.M.E. Williams, B.M. Hynek, O. Aharonson, and S.A. Hauck, II. Ancient geodynamics and global-scale hydrology on Mars. *Science* 291, 2587-2591 (2001).

Pollack, J.B., Kasting, J.F., Richardson, S.M., Poliakoff, K. The case for a warm wet climate on early Mars. *Icarus* 71, 203-224 (1987).

Pondrelli, M., A. Baliva, S. Di Lorenzo, L. Marinangeli, and A. P. Rossi. Complex evolution of paleolacustrine systems on Mars: An example from the Holden crater. *J. Geophys. Res.* 110, doi:10.1029/2004JE002335 (2005).

Poulet, F., J.-P. Bibring, J. F. Mustard, A. Gendrin, N. Mangold, Y. Langevin, R. E. Arvidson, B. Gondet, C. Gómez & the Omega Team (2005), Phyllosilicates on Mars and implications for early Martian climate, *Nature*, 438, 623–627.

John C. Priscu, Christian H. Fritsen, Edward E. Adams, Stephen J. Giovannoni, Hans W. Paerl, Christopher P. McKay, Peter T. Doran, Douglas A. Gordon, Brian D. Lanoil, and James L. Pinckney. Perennial Antarctic lake ice: an oasis for life in a polar desert. *Science*, 280, 2095-2098, 1998.

C. Quantin, P. Allemand, N. Mangold, G. Dromart, C. Delacourt. Paleolake and dense valley networks originated by precipitations within Melas Chasma, Valles Marineris. *2nd Conference on early Mars,* Abstract #8010 (2004).

Quantin, C., P. Allemand, N. Mangold, G. Dromart, and C. Delacourt (2005). Fluvial and lacustrine activity on layered deposits in Melas Chasma, Valles Marineris, Mars, *J. Geophys. Res.*, 110, E12S19, doi:10.1029/2005JE002440.

Richardson, M. I., and M. A. Mischna. Long-term evolution of transient liquid water on Mars. *J. Geophys. Res.*, 110, E03003, doi:10.1029/2004JE002367 (2005).

R. Rieder, R. Gellert, R. C. Anderson, J. Brückner, B. C. Clark, G. Dreibus, T. Economou, G. Klingelhoöfer, G. W. Lugmair, D. W. Ming, S. W. Squyres, C. d'Uston, H. Wänke, A. Yen, J. Zipfe. Chemistry of rocks and soils at Meridiani Planum from the Alpha Particle X-ray Spectrometer. *Science* 306, 1746-1749 (2004).

E. M. Rivkina, E. I. Friedmann, C. P. McKay, and D. A. Gilichinsky. Metabolic activity of permafrost bacteria below the freezing point. *Appl. Environ. Microbiol.* 66, 3230-3233, 2000.

Rogers, A. D., P. R. Christensen, and J. L. Bandfield. Compositional heterogeneity of the ancient Martian crust: Analysis of Ares Vallis bedrock with THEMIS and TES data. *J. Geophys. Res.* 110, doi:10.1029/2005JE002399 (2005).

Russell, M. J. & Hall, A. J. The emergence of life from iron monosulphide bubbles at a submarine hydrothermal redox and pH front. *J. Geol. Soc.* **154**, 377–402 (1997).

S. Ryan, E.J. Dlugokencky, P.P. Tans, M.E. Trudeau, Mauna Loa volcano is not a methane source: Implications for Mars, *Geophys. Res. Lett.* 33, L12301 (2006) doi:10.1029/2006GL026223.

C. Sagan and G. Mullen. Earth and Mars: evolution of atmospheres and surface temperatures. *Science* 177, 52-56 (1972).

C. Sagan and C. Chyba. The early faint Sun paradox: organic shielding of ultraviolet-labile greenhouse gases. *Science* 276, 1217-1221 (1997).

Schaefer, M. W. Aqueous geochemistry on early Mars. *Geochim.Cosmochim. Acta* **57**, 4619-4625 (1993).

Sears, D. W. G., and J. D. Chittenden (2005), On laboratory simulation and the temperature dependence of the evaporation rate of brine on Mars, *Geophys. Res. Lett.*, *32*, L23203, doi:10.1029/2005GL024154.

T. Segura, O. B. Toon, A. Colaprete, K. Zahnle. Environmental effects of large impacts on Mars. *Science* 298, 1977-1980 (2002).

Shean, D. E., J. W. Head, and D. R. Marchant. Origin and evolution of a cold-based tropical mountain glacier on Mars: The Pavonis Mons fan-shaped deposit. *J. Geophys. Res.*, 110, doi:10.1029/2004JE002360 (2005).

D. L. Shuster and B. P. Weiss. Martian surface paleotemperatures from thermochronology of meteorites. *Science* 309, 594-597 (2005).

S.W. Squyres, J.P. Grotzinger, R.E. Arvidson, J.F. Bell, III, W. Calvin, P.R. Christensen, B.C. Clark, J.A. Crisp, W.H. Farrand, K.E. Herkenhoff, J.R. Johnson, G. Klingelhöfer, A.H. Knoll, S.M. McLennan, H.Y. McSween, Jr., R.V. Morris, J.W. Rice, Jr., R. Rieder, and L.A. Soderblom. In situ evidence for an ancient aqueous environment at Meridiani Planum, Mars. *Science* **306**, 1709-1714 (2004).

S.W. Squyres, A. H. Knoll. Sedimentary rocks at Meridiani Planum: Origin, diagenesis and implications for life on Mars. *Earth and Planetary Science Letters* 240 1 –10 (2005).

Stockstill, K. R., J. E. Moersch, S. W. Ruff, A. Baldridge, and J. Farmer (2005), Thermal Emission Spectrometer hyperspectral analyses of proposed paleolake basins on Mars: No evidence for in-place carbonates, *J. Geophys. Res.*, 110, E10004, doi:10.1029/2004JE002353.

Tanaka, K. L., J. A. Skinner, T.M. Hare, T. Joyal and A. Wenker, Resurfacing history of the northern plains of Mars based on geologic mapping of Mars Global Surveyor data, *J. Geophys. Res*, 108, E4, DOI 10.1029/2002JE001908, 2003.

Touma, J. & Wisdom, J. The chaotic obliquity of. Mars. *Science* **259**, 1294–1296 (1993).

H. Trinks, W. Schröder and C. K. Biebricher. Ice and the origin of life. *Origins of Life and Evolution of Biospheres* (2005) 35: 429–445.

H.C. Tung, P.B. Price, N.E. Bramall, and G. Vrdoljak. Microorganisms metabolizing on clay grains in 3-km-deep Greenland basal ice. *Astrobiology*, 6, 69-86, 2006.

V. Vlassov, B. H. Johnston, L. F. Landweber and S. A. Kazakov. Ligation activity of fragmented ribozymes in frozen solution: implications for the RNA world. *Nucleic Acids Research*, 32, 2966-2974, 2004.

V. Vlassov, S. A. Kazakov, B. H. Johnston, L. F. Landweber. The RNA world on ice: a new scenario for the emergence of RNA information. *J. Mol. Evol.*, 61, 264–273, 2005.

Wallace, D., and C. Sagan. Evaporation of ice in planetary atmospheres: Ice covered rivers on Mars. *Icarus* 39, 385– 400 (1979).

Ward, W.R. Long-term orbital and spin dynamics of Mars. In *Mars*, edited by H. H. Kieffer et al., pp. 298– 320, Univ. of Ariz. Press, Tucson, Arizona (1992).

Woodworth-Lynas, C., Guigné, J.Y. Ancient ice-covered lakes in the Echus / Kasei Valley system, Mars. *2nd Conference on Early Mars*, Abstract #8037 (CD-Rom), Jacksonhole, Wyoming (2004).

Zolotov, M. Y. and Shock, E. L. Formation of jarosite-bearing deposits through aqueous oxidation of pyrite at Meridiani Planum, Mars. *Geophys. Res. Lett.* 32, doi:10.1029/2005GL024253, (2005).

In: Planet Mars Research Focus
Editor: L. A. Costas, pp. 217-242
ISBN: 978-1-60021-826-2

Chapter 8

RAMAN SPECTROSCOPIC ANALYSIS OF BIOGEOLOGICAL ADAPTATIONS BY EXTREMOPHILE ORGANISMS WITH RELEVANCE TO PLANETARY EXPLORATION

Susana E. Jorge Villar ♦
Area of Geodinámica, Facultad de Humanidades y Educacion, University of Burgos, Burgos, Spain

ABSTRACT

The ability of Raman spectroscopy for the detection of bio- and geosignatures, with little or no sample manipulation have led this technique to be considered by NASA and ESA for planetary exploration in future Mars missions.

Hostile environments, such as those with cold, hot, dry, acid or salty conditions, have been proposed as Martian analogues on Earth. These hazardous habitats have not been an obstacle for the adaptation and even proliferation of many different types of micro-organisms. The survival adaptations of micro-organisms in these environments extend to both inorganic and organic strategies, especially involving the use of a wide range of radiation protective pigments

The study by Raman spectroscopy of the strategies adopted by micro-organisms in extreme terrestrial environments is a pre-requisite for the search for life in other planets. Those strategies include the use of rocks as protective habitats, mineralogical changes, mobilization of minerals, as well as, the production of specific bio-molecules. Frequently, an admixture of organic and inorganic strategies has been detected.

In this work, a Raman spectroscopic characterization of geo- and bio-strategies developed by extremophile organisms in hostile environments is presented, such cold and hot deserts, acid environments or saline deposits;

The capability of Raman spectroscopy for the detection of these biological strategies is related to different spectroscopic parameters, such as laser wavelength, spectral

♦ Correspondance to author can be sent via e-mail: seju@ubu.es; susanajorgevillar@hotmail.com.

resolution and laser power. The influence of these parameters on the observed Raman spectra could compromise the molecular assignments and therefore the compound identification, for what the effect of different parameters in the Raman spectrum have been evaluated, analysing the requirements for a miniaturized Raman spectrometer for planetary exploration and predicting the limits for unambiguous molecular identification from the observed spectra.

Keywords: Raman spectroscopy, planetary exploration, extremophiles, terrestrial analogues, biomarkers, geo-markers, astrobiology.

INTRODUCTION

There is life outside the Earth? This has been during centuries a key and difficult question asked by the scientist community along the world (1A), but there is not answer yet. In the last decades a significant number of planetary missions, particularly to Mars [1-12], carried out mainly by NASA but also by ESA have revived the hopes to give an answer to that significant question. If any kind of life signals, either extant or extinct, are found anyplace in the Universe, the extremely high repercussions will reach not only the scientist world, given clues for understanding the origin of life or its evolution, but also philosophy or religion doctrines can revolutionize the understanding of the world, the Earth creation and the humanity appearance [13-17].

The sights are, at the moment, centred on Mars, although there are also moons, like Europe (Jupiter moon), and Titan (Saturnian moon), of interest for the looking for life signals[18-28]. Why is it focussed on Mars? There are several practical and scientific reasons. Firstly, the distance between Mars and Earth is short enough to be "easily" technologically reached by planetary missions and, for that, new missions will be sending there in the next years; secondly, different satellites are spinning around Mars and have sent, and are still sending, very useful and precise information, such as atmospheric or geological data [29-36]. Thirdly, the engineering ability of miniaturize analytical instrumentation in landing rovers has allowed to achieve more direct data on the Mars surface [39-45].

Scientific reasons are based on the similarity of origin, structure and composition of both planets. Differences related to magnetic field, geothermal heat, atmosphere and atmospheric pressure as well as the presence of superficial water are related to different planet size and its later evolution. The geomorphologic signals of the presence of liquid water in the Martian past [46-50] opens a significant hope to the idea that life could also have started on Mars at the same time than started on Earth.

Liquid water looks to be the indispensable requirement for the origin and continuity of life. There are ecosystems on the Earth where water is not easily available but, despite of that, life is still present. The presence of water on Mars surface [47, 51, 52] could have happened in a sequence like [51]: a) liquid water on rivers, lakes and ocean until 3.8 billion years ago; this is the period where life appears on Earth and why not on Mars?; b) from 3.8 to 3.1 billion years, the temperature decreased below cero and the water and soil were frost, only saline lakes or springs related with a high geothermal anomaly could keep liquid water. The analogue sites on Earth are related to cold desert, such as Arctic and Antarctic areas. There are many different terrestrial organisms, called "extremophiles", described in those similar extreme conditions. c) Until 1.5 billions of years, the only liquid water on Mars could be inside porous rocks and as underground waters; micro-organisms have had to adapt and migrate to into the rocks (endoliths). When no type of water remained on the Martian surface, life, if it appeared sometimes, could survive inside permafrost or in geothermal underground water.

Recently, ice water has been detected on the polar ice caps below a CO2 frost layer and could be exposed under atmospheric conditions during the polar Martian summer. Furthermore, some recent theories assume that superficial liquid water episodes could happen in recent periods over the Mars surface [53-59].

The survival capacity of extremophile micro-organisms on the Earth is related with the development of mechanisms to withstand to extreme conditions. Those strategies include biological and geological adaptations as well as mineralogical modifications on the habitat. The study of those mechanisms is of vital importance on the looking for life signals on Mars [60-68].

The fossilization process of micro-organisms requires very specific and delicate conditions that are not easily given in the nature; cells disappear as fossil morphological signals and frequently only some pigments, often degraded, can remain as biomarkers. Obviously, organic matter degrades faster than minerals and, for that, geological strategies are more easily detected. Things are not as simple as that, mineralogical changes, mobilization of mineralogical phases or migration inside a rock by micro-organisms cannot be, most of the time, allocate to extremophile activity and biological pigments could degrade until almost an unrecognizable organic residue [69]. In those cases is when the identification of bio- and geo-makers at the same time is of vital importance for the characterization of a possible signal of life.

RAMAN SPECTROSCOPY

The construction of miniaturised Raman spectrometers is an essential requirement for the inclusion of this technique as a part of the analytical instrumentation suite on robotic landers [37, 38]. Raman spectrometer for extraplanetary exploration must be light and small to achieve the necessary limits on size and weight imposed by the engineer requirements; moreover, the instrument must be robust to withstand the landing impact, low temperatures and windy dust.

The particular characteristics of Raman spectroscopy have led this analytical technique to an important place in planetary exploration [70-73]. The capability to analyze minerals in admixture with organic compounds is of vital importance in the study of extremophile organisms which, as previously has been described, use organic pigments as well as rocks as a protective habitat against stressed environmental conditions [61,74-78]. The Raman spectroscopic ability extends not only to the identification between organic and inorganic but also to the unambiguous and precise molecular identification. It is sensitive to structure and composition, for what polymorphous and isomorphic molecules are clearly differentiated in the spectrum by their Raman signatures [79,80]. In a natural sample, such as extremophiles or lichens, one analysis use to include more than one marker and then the spectrum shows the signatures of all compounds present on the studied area.

Between three to five characteristic Raman bands are necessary for unambiguous compound identification (Table 1); the presence in one specimen of a heterogeneous admixture of bio- and or geo- markers could interfere in the precise Raman band position in the spectrum, which could shift two or three numbers to lower or higher positions; the same effect can be seen related to impurities or inclusions in minerals, making the compound identification more difficult.

A chemical or physical treatment of the sample previously to the analysis by Raman spectroscopy is unnecessary, although in some cases the specimen ruptured is required, particularly in the endolith studies, to expose the bacterial colony. The size of the sample doesn't affect to the compound identification, and either micro- or macro-analyses can be carried out on micro- or macro-specimens.

Several parameters have significant influence on the Raman spectra, changing the appearance of the spectrum and making the molecular identification difficult. Those parameters are mainly spectral resolution, laser wavelength and laser power.

The worse spectral resolution is, the wider Raman bands are [79,81]; in this case, close bands, particularly doublets and triplets or even bands belonging to different compounds, could be integrated in one broad signature in which the wavenumber position isn't always an average of the wavenumber position of the integrated bands, having effects on the interpretation of the molecular data (figure 1).

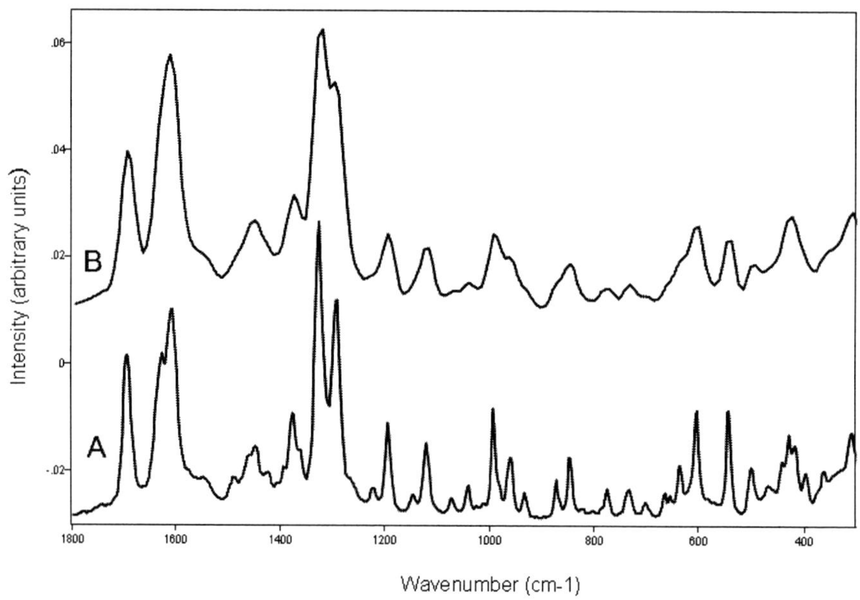

Figure 1: Raman spectra of usnic acid collected at 4 cm^{-1} (A) and 16 cm^{-1} (B) spectral resolution.

The decision about the laser wavelength more appropriated for planetary exploration has a high importance since fluorescence, molecular detection capability or relative intensity of the Raman bands are often affected by the laser excitation chosen [82]. Green lasers (514 nm or 532 nm) tends to increase fluorescence on the organic compound analyses as well as on red and dark minerals whereas near infrared have a better balance between organic molecules and minerals, although green minerals are difficult to be analyzed. Sometimes the resonant effect could increase the intensity of the Raman bands in the spectrum and, then, the marker could be easily detected even in very low concentration. That is the case of the green lasers and carotenoid pigments. By the other hand, a particular laser wavelength can be no sensitive to a compound, as happens with the green laser which, in our experiments, has not detected chlorophyll in extremophile specimens. In similar way, the relative intensity of some vibrations for a given compound can be amplified or reduced depending on the laser wavelength; this effect is seen in minor bands but also in the characteristic signatures used for the unambiguous molecular identification (figure 2).

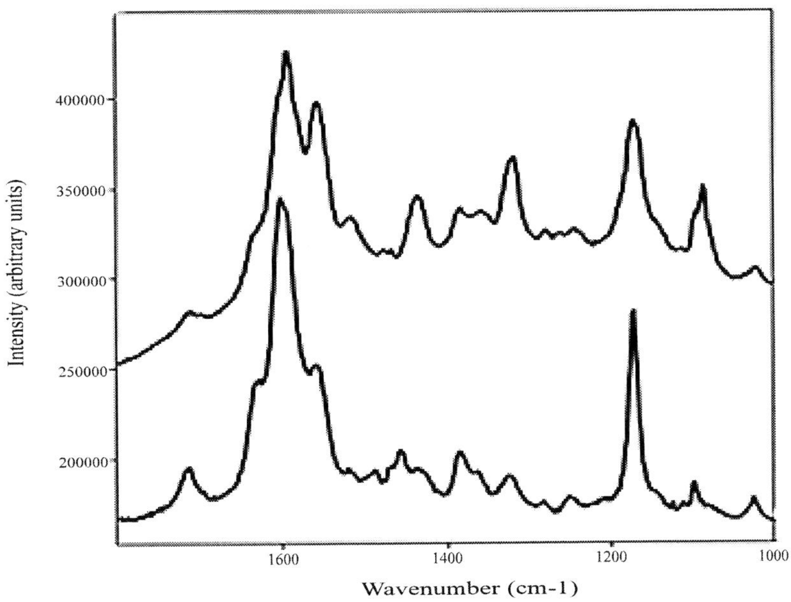

Figure 2: Raman spectra of scytonemin achieved using 785 nm (A) and 514 nm laser wavelength.

Power laser has also an enormous effect in the possible degradation of either minerals or organic molecules changing the original bio and geo-markers to other compounds or, in the worse case, could be burnt. This problem could lead to erroneous conclusions about the environmental conditions, micro-organisms type or extremophiles presence.

All of that makes the Raman spectrum interpretation complex and indicate the huge importance of building an extensive database of extremophile bio- and geo- strategies for being used on planetary exploration.

If a limitation of the wavenumber range in the Raman spectrum is necessary because of engineer requirements, it is important to notice that there are one characteristic region for the unambiguous identification of organics, mainly between 1650 to 900 cm^{-1}, and other for minerals, from 1100 to 100 cm^{-1}; a medium region between organic and inorganic wavenumber region is not useful for the identification of neither bio nor geo-markers (figure 3).

Raman spectroscopy has become an important tool on planetary exploration, and has been included in the analytical instrumentation suite by ESA to be sent in the next Mars mission rovers.

Table 1

ORGANIC BIOMARKERS	RAMAN WAVENUMBERS (cm^{-1}) Main bands shown in bold.
beta-carotene	**1515**, **1155**, **1004**
Zeaxanthin	**1520**, 1447, 1195, **1157**, **1007**, 963
Canthaxanthin	**1515**, 1272, 1191, **1154**, **1003**, 962
Lutein	**1523**, 1187, **1156**, **1005**, 964
'unidentified' Carotenoid	**1533, 1144**
Chlorophyl	1438, 1387, **1326,** 1287, 1067, 1048, **988, 916, 744, 517,** 351
C-phycocyanin	1655, **1638,** 1582, 1463, **1369,** 1338, **1272,** 1241, 1109, 1054, **815, 665,** 499
Chlorophyl-like	**1620**, 1594, 1540, 1514, **1462**, 1435, 1383, **1313**, 1277, 1226, 1209, 1099, 1069, 984, **498**
Scytonemin	1712, 1629, **1599**, 1556, 1485, **1435**, 1382, **1324**, 1249, **1171**, 1095, 1022, 983, 908, 779, 765, 753, 677, 659, 637, 628, 576, 494, 462, 437, 306
Parietin	**1556**, 1435, 1430, 1368, **1272**, 1132, 1073, **922**, 766, 633, 571, 398
Rhizocarpic acid	**1665,** 1610, **1595, 1518** , **1496,** 1477, 1347, 1303, **1002,** 944, 902, 787, 448
Calycin	1653, **1635, 1611, 1595, 1380,** 1344, 1240, 1155, 1034, **960,** 878, 498, 484
Usnic acid	**1694**, 1627, **1607**, 1444, 1372, **1322**, **1289**, **1192**, **1119**, **992**, 959, 871, 845, 651, **602**, **541**, 498, 426, 415, 308
Emodin	**1659,** 1607, 1577, 1557, **1298, 1281,** 942, **565, 467**
Atranorin	**1666, 1658,** 1632, **1303, 1294,** 1266, **588**
Pulvinic dilactone	**1672, 1603,** 1455, **1405,** 1311, **981, 504**
Gyrophoric acid	**1662,** 1628, 1612, 1334, 1304, **1291, 1235, 1138, 561**
Porphyrin	**1592**, **1485**, **1351**, 1307, 1216, 1194, 1108, 953, 847, 832, 779, **746**, **679**, 640, 594, **483**
Compound A	1606, 1562, 1435, **1412**, 1379, **1280**, 1214, **1208**, 1104, 1049
Compound B	**1595**, 640, 532
Compound C	1687, 1656, 1545, **1391**, 1315, 1278, **1241,1206**

MINERALS	RAMAN WAVENUMBERS (cm^{-1}) Main bands shown in bold.
Dolomite	**1098**, **723**, **300** 177
Haematite	607, 495, **407**, **291**, 222
Limonite	**399**
Rutile	**614**, **447**
Anatase	
Calcite	**1086, 712, 282,** 156
Aragonite	**1086, 704, 208,** 154
Dolomite	**1098, 725, 300,** 177
Magnesite	**1094, 738, 330,** **213,** 119
Hydromagnesite	**1119, 728,** 326, **232,** 202, 184, 147
Gypsum	**1133, 1007,** 669, 618, **492, 413**
Anhydrite	**1015,** 674, 628, 500, 416
Weddellite	1630, **1475,** 1411, **910,** 869, 597, **506, 188**
Whewellite	1629, **1490, 1463,** 1396, 942, **896,** 865, 596, 521, **504,** 223, 207, **185, 141**

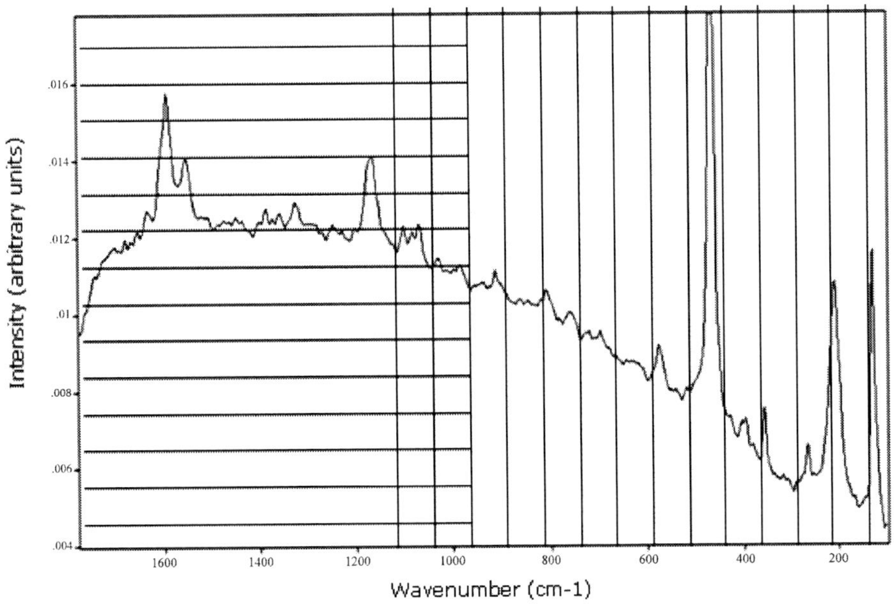

Figure 3: spectrum of scytonemin and quartz showing the Raman wavenumber region where the characteristic bands for an unambiguous identification appear.

EXTREMOPHILE ORGANISMS

If organisms on Mars surface appeared at the same time than it happened on the Earth, they should have evolved, as they did on Earth, to adapt to the new environmental conditions that, in the case of Mars, become more and more cold and dry. The ability of bacteria to adapt, in a short geological time scale, is enormous and could happen that they spread to a wide range of habitats, in a similar way that they did on Earth.

As there is no knowledge about the protective strategies of Martian bacteria (always having in mind that the premise of all theses studies are based on the theory that life could started on Mars although there is no data yet to assess it), the decision of what, where and how to look for should begin to the knowledge of where and how the terrestrial micro-organisms have adapted to Earth extreme conditions [51, 60, 63]. As Raman spectroscopy has been included as a part of the analytical instrumentation suite in Martian explorer rovers [41], mainly directed toward life signal detection, the terrestrial Raman studies should be centred on building a database necessary for answering, as much precise as possible, the questions of what, where and how to look for life on Mars.

Extreme terrestrial environments considered analogues to Mars include, for example, cold deserts (mainly Antarctic and Arctic regions), dry desert (such as Atacama), acid, basic, salty or hot waters, permafrost, ice, et cetera.

SURVIVAL STRATEGIES ADOPTED BY EXTREMOPHILE ORGANISMS

Bio- and geo- survival strategies adopted by extremophile communities can be detected by the bio- and geo- markers left on or inside rocks. Some of those markers remain on the rocks even when the micro-organisms have disappeared and can be identified using Raman spectroscopy.

Bio-Markers

Organic pigments are one of the most extended strategies [61,74,76,78,83-86] used, particularly, by organisms that don't have a rocky or mineral protection, such as lichens. Epilithic lichens use a wide range of pigments to be protected against UV-radiation, DNA damage, oxidizing molecular process, such as parietin, rhizocarpic acid, pulvinic dilactone, calycin, usnic acid, gyrophoric acid, emodin and atranorin [87,88]. Carotene and chlorophyll are also present in all lichens.

The identification of those lichen biomarkers in the Raman spectrum is made in base on the strongest Raman bands; the number of signatures depends of the compound spectrum. Only three bands are enough to detect the presence of carotenoids because of their characteristic Raman spectral shape, which has two very strong signatures (around 1515, 1155 cm^{-1}) and one medium signature close to 1000 cm^{-1} (figure 4); although there are a large number of weak to very weak peaks that, they are, frequently, no visible among other pigment bands present in natural samples. Despite of those three bands only the band placed in the highest wavenumber position is mainly used for the identification of carotenoid type. Each double C=C bond in the carotene structure shifts this band 3 cm^{-1} to lower positions; that is a general theory because the presence of different radicals in the rings, which are placed in the extreme of the chain, also affects to the position of the Raman band; for example, the signature position at 1515 cm^{-1} is characteristic of beta-carotene, 1520 cm^{-1} of zeaxanthine and 1523 cm^{-1} of lutein but beta-carotene, canthaxanthin and zeaxanthine have the same number of C=C bonds and should have the same wavenumber Raman band position; however, the strongest band located at 1515 cm^{-1} is common to beta-carotene and canthaxanthin but in the zeaxanthine Raman spectrum that signature appears at 1520 cm^{-1}; the unambiguous discrimination between beta-carotene and canthaxanthin is made by a weak band at 960 cm^{-1} which is present in canthaxanthin but not in beta-carotene (figure 4).

Although all of those carotenoids are frequent in the nature only lutein has been found in the lichens that our Raman group has studied which came from Svalbard (Arctic north), Antarctica, Atacama (Chilean desert) Hawaii (USA), Italian high altitude Alps or different areas in Spain.

Most of the organic pigments have a complex Raman spectrum and need several signatures to be identified [87,88]. Twelve bands appear in the spectrum of parietin; the strongest signatures are located at 1556, 1272, 922 cm^{-1} and can be used for the pigment identification but the use of more than only these thee signatures make the parietin assignment unambiguous (see table 1). The Raman spectrum of pure usnic acid shows twenty main signatures (figure 2) and, in this case, the use of five significant bands are required for its identification (1694, 1607, 1322, 1289, 992 cm^{-1}); the usnic acid bands at 1289, 1192,

1119, 602, 541 cm^{-1} are usually shown in the spectrum even in a mixture of bio- geo-molecules because of their medium intensity. The most common biomarkers with all Raman signatures and the most identifying Raman bands for each organic pigment (in bold) are shown in table 1.

There is no relationship between pigments and lichen genus, two species of the same genus can have different protective pigments and different genus can have the same protective pigments. One example is shown in. The distribution of pigments in lichen is not homogeneous and its profile is easily studied by micro-Raman in a lichen profile (figure 5A and 5B).

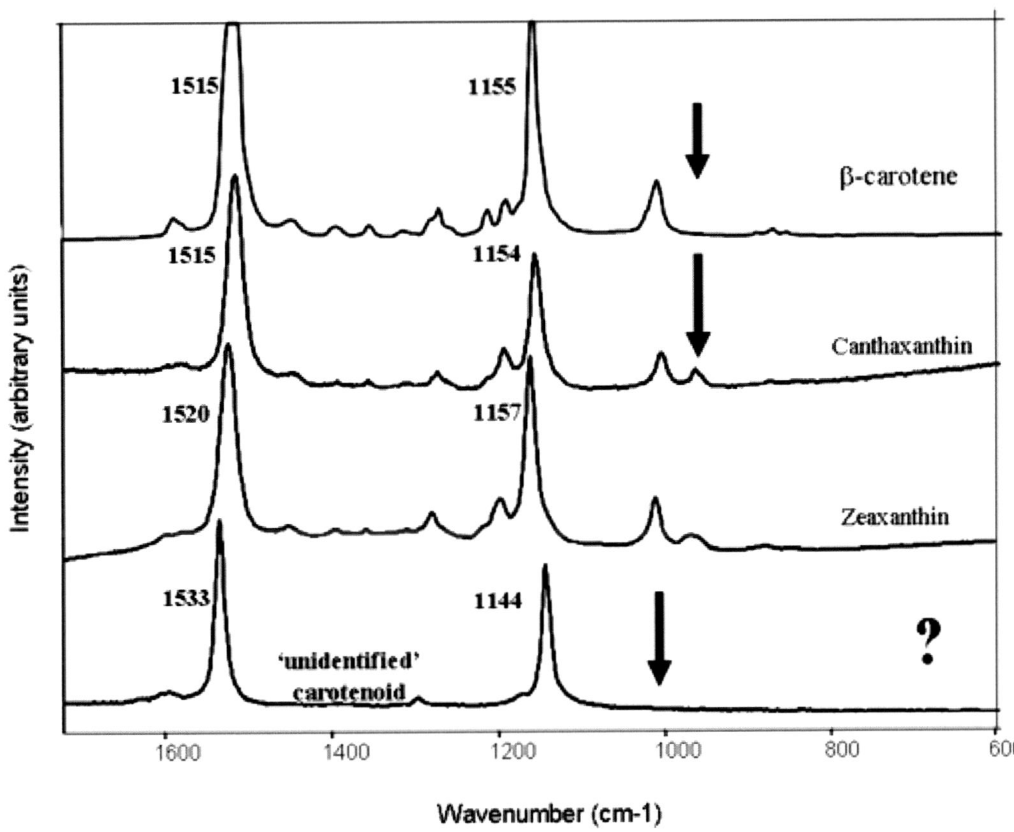

Figure 4: Raman spectrum of different carotenoids showing the characteristic Raman bands for the unambiguous identification. The spectrum with a question mark has not been identified yet but it probably has not rings in the end of the chain because the absence of the Raman band around 1000 cm^{-1}.

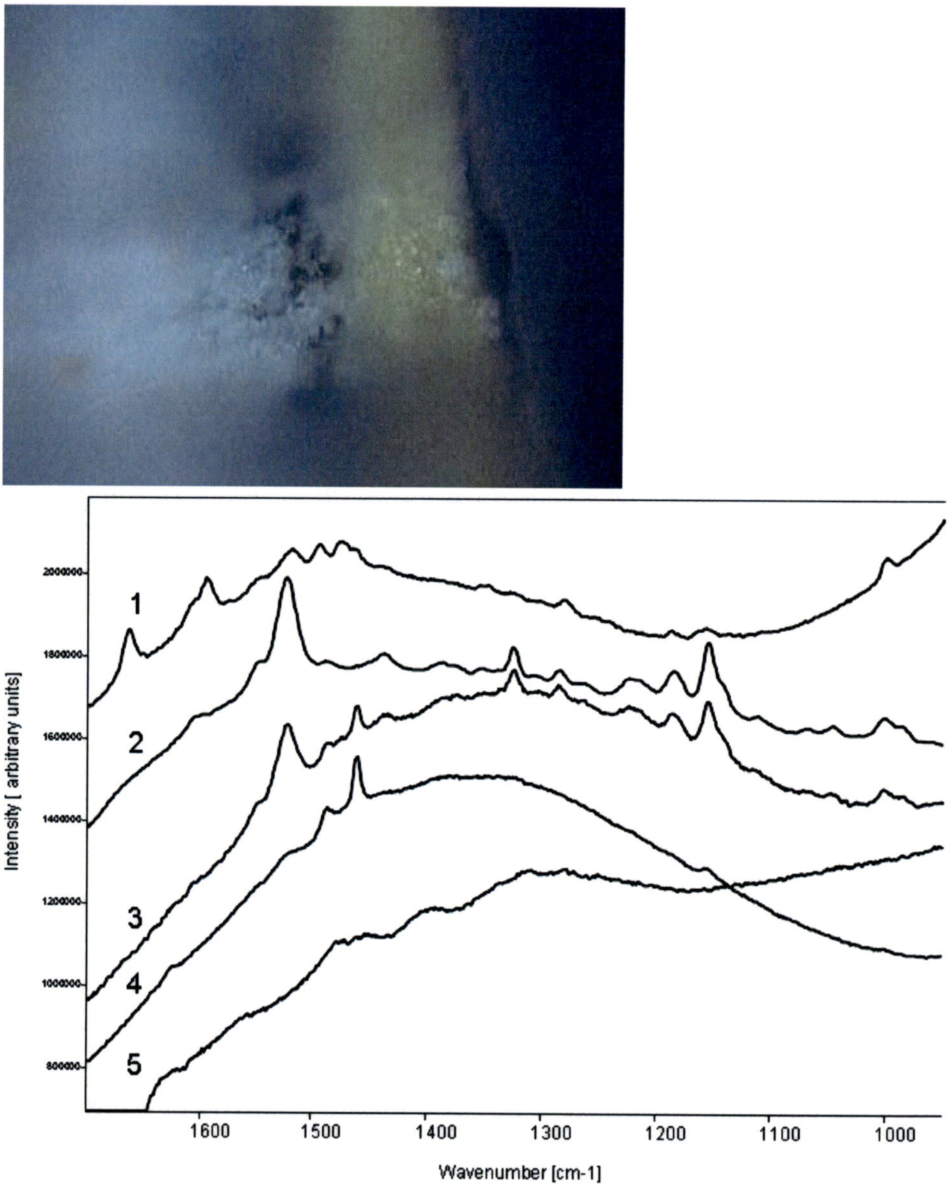

(A)

(B)

Figure 5: Micro-photo (x50) of a lichen profile showing different bands (5A). In 5B, spectra plot of different lichen layers. In number 1, external yellow band, a pigment UV radiation screen is the main marker; second band shows the carotene and chlorophyll characteristic signatures. Band number three, among of carotene and chlorophyll, bands of calcium oxalate monohydrate are visible; spectrum number 4, in contact with the rock, shows only whewellite signatures and, finally, on the fifth spectrum no organic signatures are recognized.

As previously has been notice, most of the analyses achieved on lichens but also on endoliths show the presence of more than one compound and to distinguish between those bio-markers require the valuation of the strongest Raman bands, relative peak intensity and the analysis of how many pigments could be present in the specimen at the same time, such as in the spectrum shown in figure 6 where phycocyanin, carotene, chlorophyll and calcite signatures appear. For a correct interpretation, experience is a big deal!

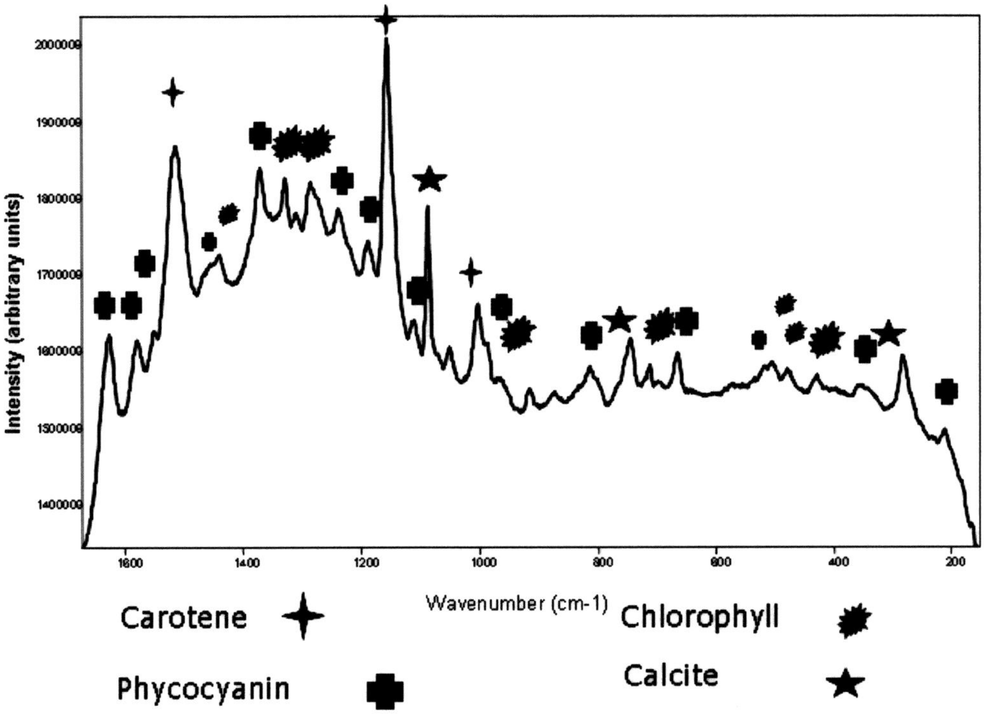

Figure 6: Raman spectrum of an endolith showing phycocyanin, carotene, chlorophyll and calcite bands.

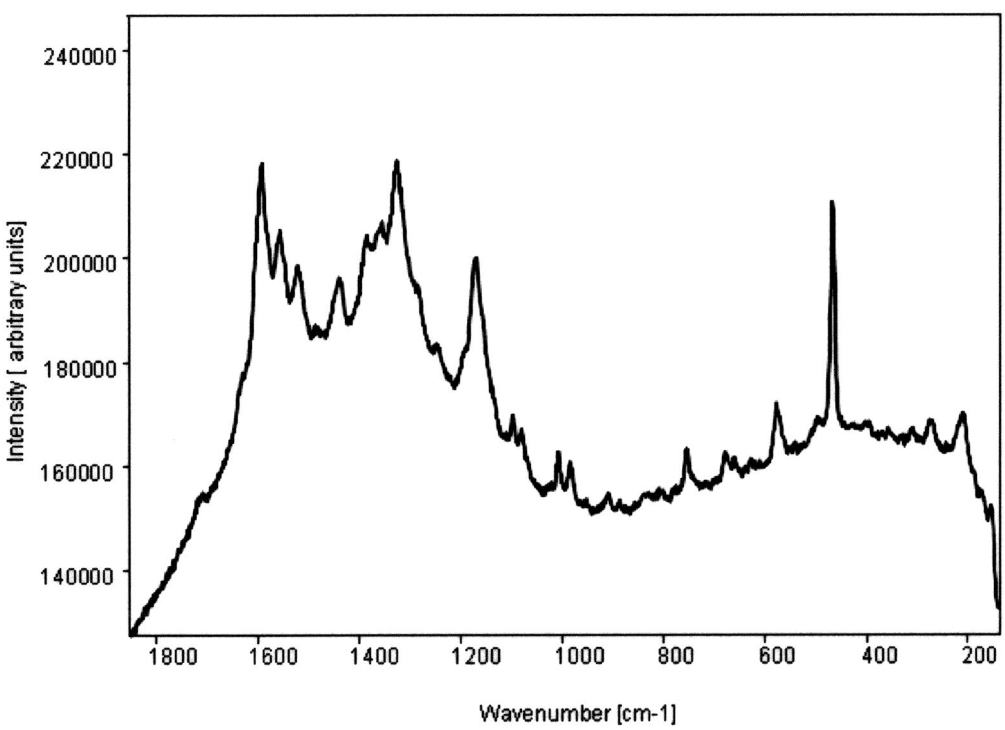

Figure 7: Raman spectrum of scytonemin, carotene and chlorophyll achieved on an endolith collected during AMASE 2004.

Bacterial colonies are the most dominant micro-organisms present in endolith communities. Scytonemin is a common biomarker used by this group of extremophiles, as well as chlorophyll and carotene. The Raman spectrum of scytonemin shows twenty six bands but the signatures used for its unambiguous assignment are only four (1599, 1435, 1324, 1171 cm^{-1}), frequently, appear together with chlorophyll and carotene Raman bands. Scytonemin is a very efficient pigment against UV-radiation, wich is very high in polar areas because of the ozone deplection [76,78]. (figure 7).

Endoliths are microorganisms living inside rock pores, the most appropiate rocks are the sedimentary group, particularly detritic rocks such as sandstones, because their higher porosity. Furthermore, the need of sunlight for the photosynthetic function is neccessary and light minerals, such as quartz, building the rock skeleton, are the most appropiate for this purpose; however, a dark basalt, without porosity but with vacules in contact with the atmosphere by an orifice, can also plays the same role. Sunlight is weaker in the bottom of a dark vacuole but extremophiles have developed mechanisms to compensate the need of light [74]; in one of our specimens collected on Svalbard, during the AMASE (Arctic Mars Analogue Svalbard Expedition) 2004 expedition, there is a clear strategy differenciation between organisms living close to the orifice, which only produced chlorophyll for photosyntetic purposes, that those colonies far away from the light entrance, which also generated phycocyanin, a harvesting sunlight compound, used to collect as much photosynthetic light as posible (figure 8). The same happens when endolith communities are located in a deeper position from the crust than use to appear (between 2 milimeters to 1 centimeter for normal endoliths), then the production of this pigment is of vital importance.

Carotenoids found in endolith and chasmolith (microorganisms living inside cracks) communities, on the contrary than in the case of lichens, are diferent from specimen to specimen; beta-carotene, lutein, zeaxanthine, canthaxanthin or astaxantine are common.

Figure 8: Black basalt showing a very small orifice (left) and the internal vacuole covered by extremophile organisms.

There are many other types of extremophiles that don´t use rocks as protective habitats for what a wider range of protective molecules are usually neccesary to withstands in such as extreme conditions. One of those examples are cryo-algae or algae living on or inside ice or snow (figure 9). Spectra of cryo-algae of the specie *chlamydomonas nivalis* show the presence of different carotenoids, such as beta-carotene, zeaxanthin, lutein, and canthaxanthin, the assignment of these carotenoids have been explaining before and are showed in Table 1. The extreme conditions which means to live in ice or in a permanent snow field, close to the poles and in an environment which is temporally frost, with no liquid water and with very limited food, drive those organisms to develop different survival strategies which can be detected as bio-markers on the Raman spectrum. Several different pigments are clearly visible (showed in table 1 as compound A, B and C) and could work, for example, as antifreeze or DNA repair (figure 10).

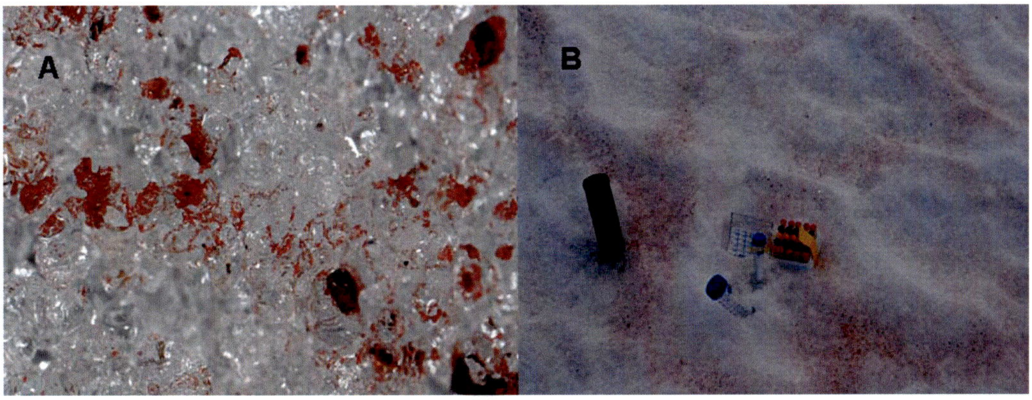

Figure 9: A) detail of red snow algae, B) Svalbard permanent snow field covered with red algae.

Raman spectroscopy (and other spectroscopic techniques) allows the possibility to relate the spectrum of a non identified compound with its structure basing on comparison with the spectrum of a well defined pigment if part of their bands appears in a close position. Example of this is shown in the figure 10 where the lowest spectrum represent an emodin-like compound and, although the unambiguous assignment has not been done yet, is possible to assess that the structure of this compound is similar to antraquinone structure.

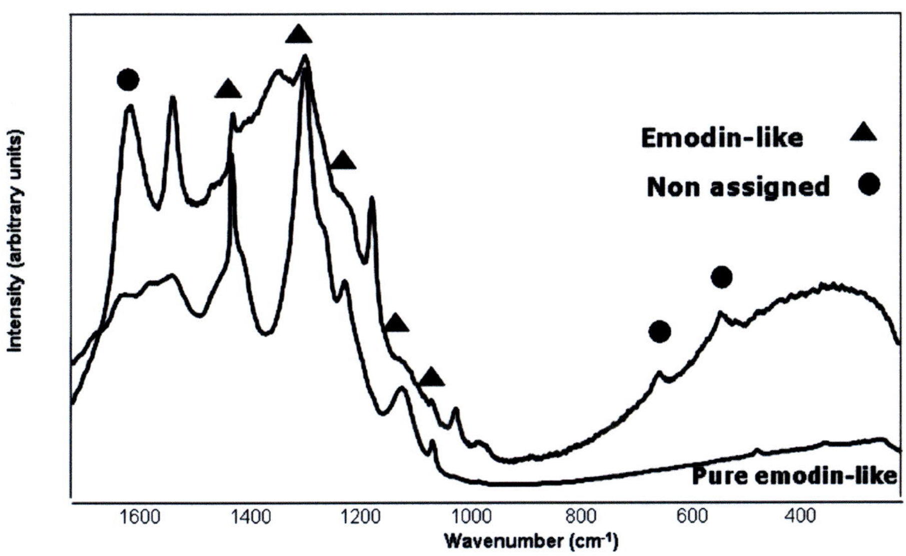

Figure 10: Spectra collected over snow algae *Chlamidomonas nivalis*, several non identified compounds appear together with carotene and chlorophyll.

Calcium oxalates, mono- or dihydrated (whewellite and wedellite respectively) are frequently found related with lichen, endolith and chasmolith communities in stressed environments. Some theories support the idea that oxalates could serve as water storage in extreme dry situations [90]. From our studies, the most common oxalate found related with lichens is whewellite (Figure 11), easily identified by its Raman bands at, 1490, 1463, 896, 504, 185, 141, cm^{-1}, particularly characteristic are the signatures at 1490, 1463 cm^{-1} which clearly differentiate this oxalate monohydrated from wedellite, which only has a band in this wavenumber region at 1475 cm^{-1} (see table 1).

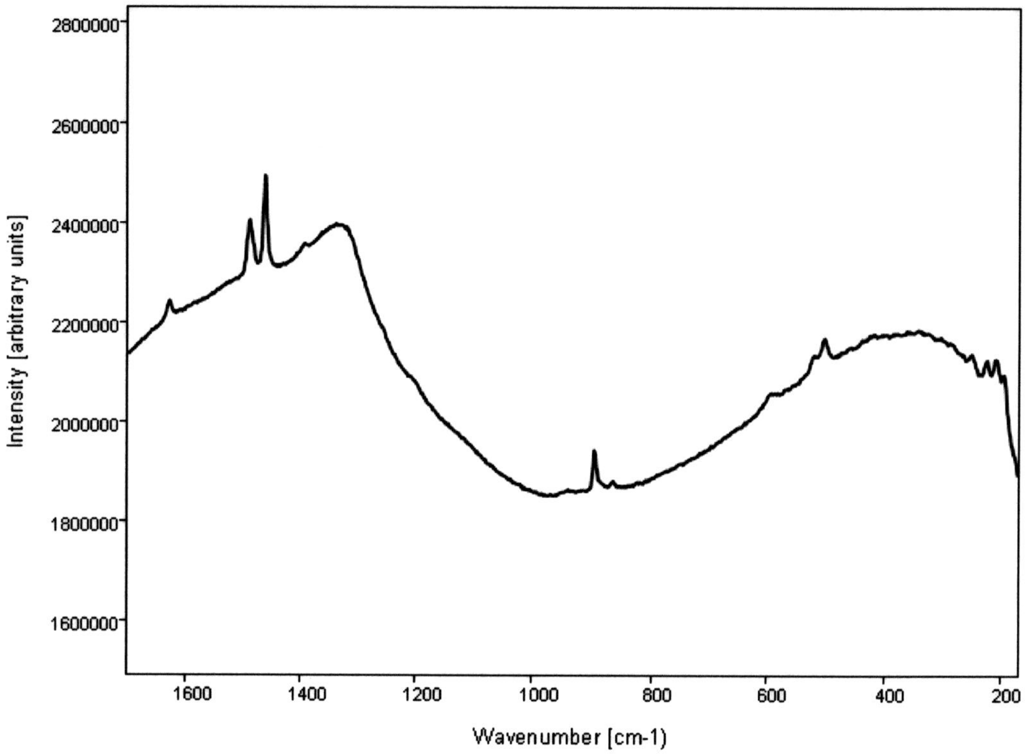

Figure 11: Raman spectrum of whewellite, a calcium oxalate monohydrate.

Geo-markers

Organisms in extreme environment migrate inside rocks to protect themselves against hazardous conditions. Apart from the production of organic pigments as protective strategies, minerals have also been used with the goal to create a more protective environment. These strategies include mineralogical changes as well as mineral mobilization [74,75,78]. The knowledge of those mineralogical strategies could be used as a life geo-marker, particularly when the bio-markers have degraded to an unidentifiable compound.

Haematite (Fe_3O_4), with Raman bands at 607, 495, 407, 291, 222 cm^{-1}, has been described as UV-radiation screen mineral. In figure 12 the red colour on the crust is owing to haematite; it is interesting to see that in the area close to the micro-organisms there is a depletion in the red colour and then on haematite, meaning that a mobilization of haematite to "clean" the organism habitat has been produced. Different strategy has been used by a chasmolith community (figure 13) from Svalbard, AMASE 2004 expedition; in this case, a light sandstone composed by quartz, feldspar, mica and ankerite presents a reddish crust because of the presence of haematite, the chasmolith community closer to the surface has an orange colour and, despite of its proximity to the haematite rich surface, this mineral has not been detected but now goethite (FeO OH) is responsible of the orange colour.

Gypsum (Ca (SO_4) 2 HO_2), with signatures at 1133, 1007, 669, 618, 492, 413 cm^{-1}, is also an UV- radiation screen mineral which can complement the protection of organic pigments. In the grey endolith shown in figure 12 left a thin layer of gypsum appears inside the porosity in the 2-3 first millimetres from the crust, over the micro-organism community. The sandstone specimen (figure 13) collected in Svalbard has also gypsum as cement and all communities can benefit of the gypsum UV protection.

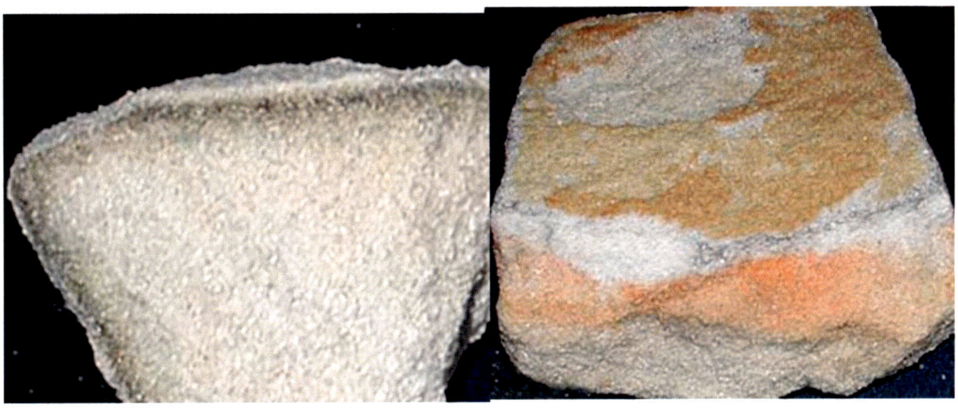

Figure 12: White sandstone with a greenish-grey endolith (left); pores on the crust are filling with gypsum, after this layer and over the organic band calcite has been detected. Orange sandstone (right), the red colour is owing to haematite; observe the haematite depletion around of endolith community.

Figure 13: Photo of a sandstone with endolith (bluish-green) and chasmolith (red and dark green) communities collected in Svalbard during the 2004 AMASE expedition.

There are mineral transformations that have been detected in relation with the micro-organism activity; in those cases, a mineral phase appears related with the rock and only on the bacterial area or very close to it a new phase or a mineral transformation is detected. Calcite (a calcium carbonate), for example, appears as a thin layer only between the gypsum and the grey organic bands in the endolith of figure 12 left; in the chasmolith community shown in figure 14, aragonite (a polymorph phase of calcite) appears as a white dust on the crack whereas calcite only appears related with the organic green layer. As Raman spectroscopy is sensitive not only to composition but also to structure the differentiation of both calcium carbonate polymorphous is possible, although they have similar spectra, The strongest band is located at 1086 cm^{-1}, a weak band appears at 712 cm^{-1} for calcite and 704 cm^{-1} for aragonite spectrum, but when there is a mixture of minerals, the use of this band for assignment can be no conclusive because of band interferences of other minerals. The unambiguous identification of both minerals is made by the medium intensity signatures at 282 and 208 cm^{-1} for calcite and aragonite respectively.

Figure 14: Chasmolith community from Antarctica; the crack is cover by a fine aragonite dust but the green areas calcite is absent and its polymorph, aragonite, has been detected.

Hydrocerussite (a lead carbonate identified by 1119, 728, 326, 232, 202, 184, 147 cm^{-1} Raman bands) appears in admixture with calcite over the endolith community shown in figure 12 left. This lead carbonate is a toxic mineral and its presence in contact with the micro-organisms could be related with a protective geo-strategy against other colonizations

Some places on Mars surface has been related with the presence of magnesite (magnesium carbonate) or hydromagnesite deposits that is why deposits of those minerals on Earth have relevance on the study of Martian analogues. In Salda Lake (Turkey) evaporitic outcrops of hydromagnesite are colonized by micro-organisms; the only areas where magnesite can be identified are related with the endolith colonization. The strongest band for hydromagnesite appears at 1119 cm^{-1} and this band shifts until 1094 cm^{-1} in the Raman spectrum of magnesite (figure 15). Table 1.

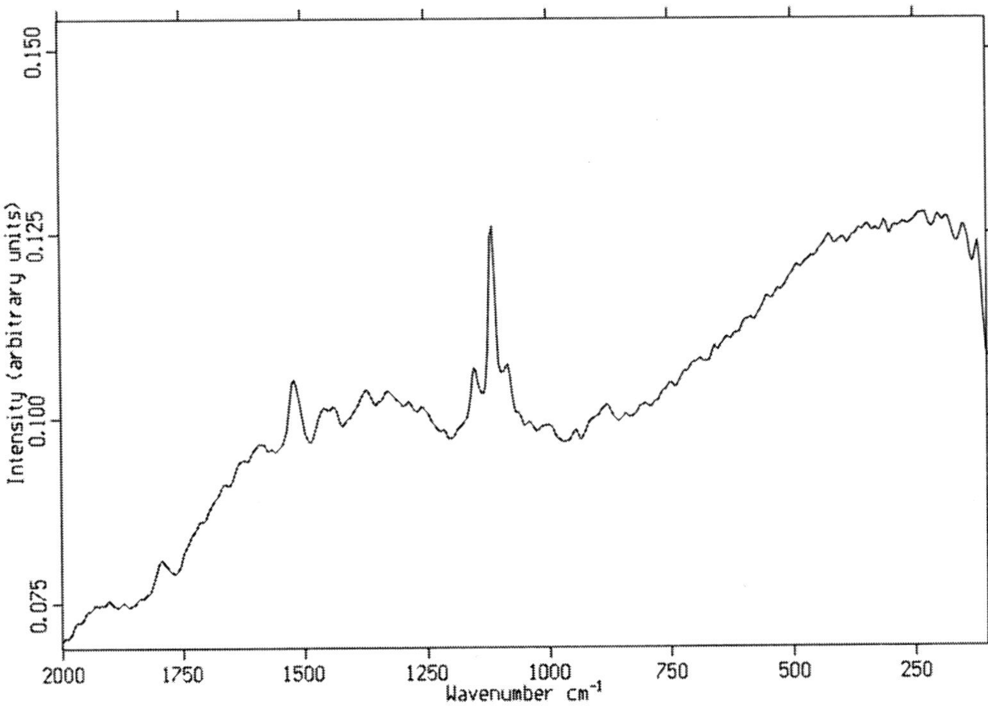

Figure 15: Raman spectrum showing bands of hydromagnesite and magnesite together with organic signatures.

Here, as in the case of calcite - aragonite or haematite – goethite minerals, there is a clear relationship between the presence of bacterial colonies and mineralogical changes

Fossil Bio-Markers

One important challenge in planetary exploration is the identification of fossil bio- and geo- markers once the micro-organisms have disappeared. Although many scientists accept that life could start on Mars surface at the same time that happens on Earth the general hypothesis is that there is no life on Mars surface at this moment. The identification of life remains is, then, one important challenge in extraplanetary exploration. Some markers degraded easily but other can remain in or on the rocks once the biological colony has disappeared.

Scytonemin has been identified by Raman spectroscopy although no bacterial morphologies were detected on the surface of hot desert rocks. In an old travertine terrace specimen, collected from a hot spring in Svalbard (79.5 degrees north), a porphyrine has been detected by its Raman bands at 1592, 1485, 1451, 1307, 1216, 1194, 1108, 953, 847, 832, 779, 746, 679, 640, 594 and 483 cm^{-1} but no micro-organisms cells have been seen using SEM (figures 16 and 17). The same Raman spectrum has been collected on a blue spot (figure 18) in a specimen from Trendal Pilbara stromatolith (3.5 billion years old although the colonization is recent). The sharp bands in both spectra indicate that the compound has not been degraded or has been in low degree.

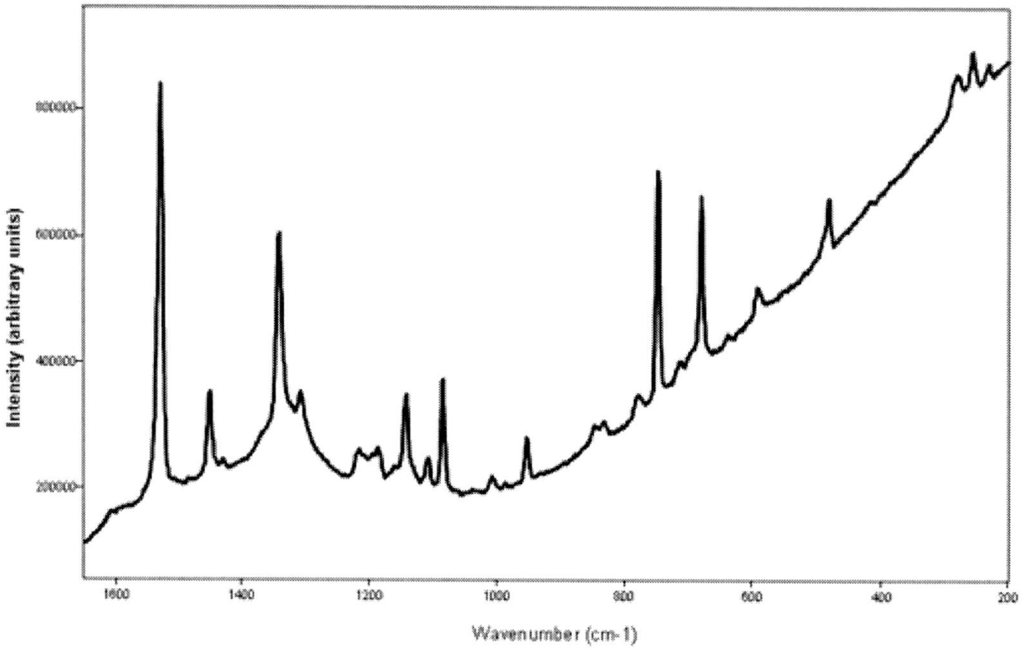

Figure 16: Raman spectrum of porphyrin achieved on a travertine specimen from Svalbard (North of Norway); no organic morphology were seen under SEM analysis.

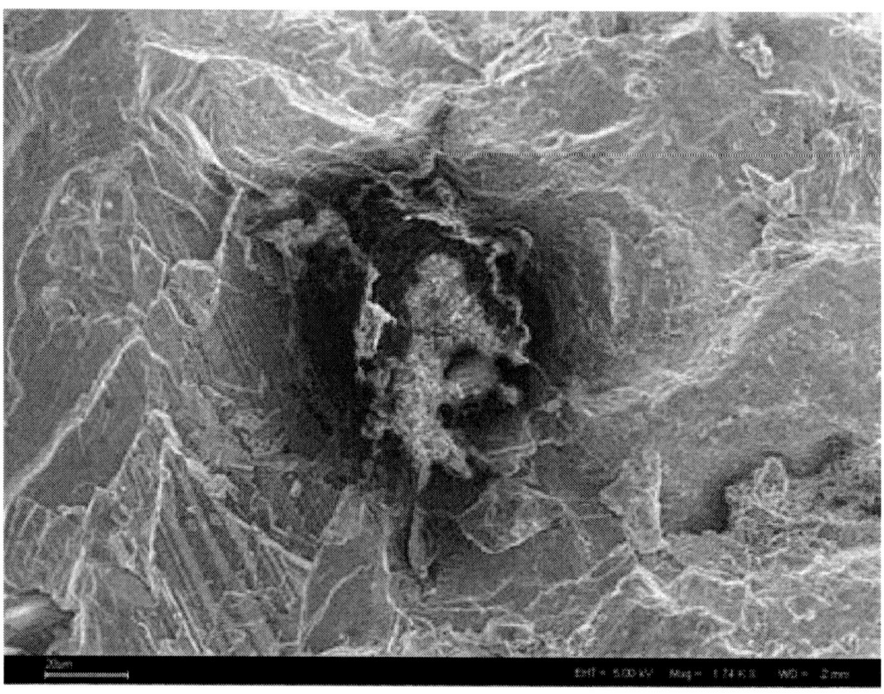

Figure 17: SEM photo collected over the area explaining in figure 16.

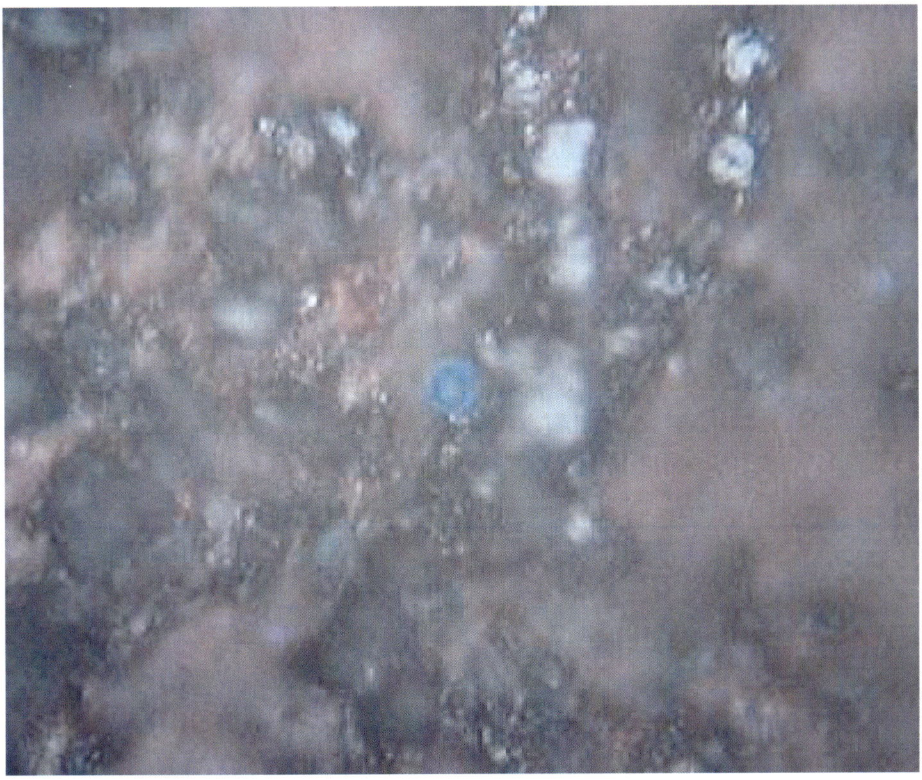

Figure 18: Blue circle in Kitty´s Gap stromatolith (Pilbara, Australia) specimen. Its Raman spectrum were assigned to the same porphyrin than in the Svalbard travertine specimen

CONCLUSIONS

The particular characteristics of Raman spectroscopy have made of this technique a valuable instrument for the study of extremophile organisms in stressed terrestrial environments analogue to Mars. The building of mini Raman spectrometers, small, light and robust allows them to take in consideration for being included as a part of the analytical suite in lander rovers for planetary exploration.

The capability of Raman spectroscopy to detect a mixture of organic and inorganic markers with no sample preparation together with the possibility of an unambiguous identification of a mixture of bio- and geo- markers makes Raman spectroscopy a valuable technique for the looking for life in other planets.

Several parameters (such as wavelength, spectral resolution laser power) could have an important influence on the Raman spectrum and make its interpretation more difficult, for what a new and good database, according to the parameter limits of the miniaturised Raman spectrometer, is an important requirement.

As Mars and Earth could have had in their beginning similar geological characteristics there are no solid arguments to think that if life started on Earth, as it is evident, life couldn't also start on Mars. The different evolution of Mars planet to harder environmental conditions makes the study of the most extreme terrestrial habitats on Earth, called terrestrial analogues to Mars, a pre-requisite before the searching for life on Mars.

Survival strategies developed by epilithic and endolithic communities in stressed environments include protective bio- molecules as well as mineral mobilizations or, even, mineral transformations; frequently, all those strategies have been used at the same time as complementary extra protection. Organic pigments are easily degraded once the micro-organisms have disappeared whereas geo markers can remain during long periods. As the possibility to find actual life on Mars surface has a little probability the need to analyse and evaluate bio and geo markers at the same time is important for a more correct interpretation of possible fossil life signals.

The unambiguous identification of a compound by its Raman spectrum needs the presence of around 3 to 5 of its strongest bands. As very often many different compounds, either organic or inorganic, appear in admixture, the Raman spectrum shows the mixture of their characteristic bands. In general, the relative band intensities are related to the bio- and geo- marker concentration but, in some cases, the laser wavelength sensitivity to one particular marker can be noticed on the Raman spectrum by very strong signatures although the molecule concentration was low, as it happens with carotene and the green laser (514 or 532 nm).

Laser wavelength affects to the Raman molecular detection because there are lasers are no very sensitive to specific. The fluorescence effect is related with the laser excitation; when it is strong could hide Raman signatures and, then, the identification of a pigment or mineral in the sample is unnoticed. Other parameter to have in mind is the spectra resolution that uses to be worse in miniaturized Raman spectrometer than in laboratory instruments. In this case, the assimilation of weaker bands by close and stronger signatures could complicate the spectrum interpretation. Laser power is also an important parameter because the heating of the sample can induce organic or mineralogical changes or even the total compound destruction.

Polymorphous (molecules or minerals with the same composition but different structure) and isomorphous (same composition but different structure) molecules can be unambiguously identified, which is an important capability of Raman spectroscopy. Polymorphs and isomorphs give similar but different signature positions and, sometimes, relative intensity. That is clear in the case of minerals, where groups of minerals, such as, for example, carbonates have the same Raman pattern and, for that, a group of signatures can be related to its mineral group.

REFERENCES

[1] Vago J; Gardini B; Kminek G; Baglioni P; Gianfiglio G; Santovincenzo A; Bayon S; Van Winnendael M. ExoMars - Searching for life on the red planet. *Esa Bulletin-European Space Agency*, 2006, 126, 16-23.

[2] Sims MR; Cullen DC; Bannister NP; Grant WD; Henry O; Jones R; McKnight D; Thompson DP; Wilson PK. The specific molecular identification of life experiment (SMILE). *Planetary and Space Science*, 2005, 53 (8), 781-791.

[3] Sims MR; Pillinger CT. Beagle 2: The astrobiology lander on ESA's Mars Express mission. *IAU Symposia*, 2002, 12, 652-652.

[4] Brooks CA; Iagnemma K. Vibration-based terrain classification for planetary exploration rovers, *IEEE Transactions on Robotics*, 2005, 21 (6), 1185-1191.

[5] Picardi G; Biccari D; Seu R; Marinangeli L; Johnson WTK; Jordan RL; Plaut J; Safaenili A; Gurnett DA; Ori GG; Orosei R; Calabrese D; Zampolini E. Performance and surface scattering models for the Mars Advanced Radar for Subsurface and Ionosphere Sounding (MARSIS), *Planetary and Space Science*, 2004, 52 (1-3), 149-156.

[6] Montmerle T; Claeys P; Gargaud, M; Lopez-Garcia P; Martin, H; Pascal R; ReisseJ; Selsis F. Life on earth ... and elsewhere?, *Earth Moon and Planets*, 2006, 98 (1-4), 299-312.

[7] Neukum, G; Jaumann, R; Hoffmann, H; Hauber, E; Head, JW; Basilevsky, AT; Ivanov, BA; Werner, SC; Van Gasselt, S; Murra, JB; McCord, T. Recent and episodic volcanic and glacial activity on Mars revealed by the High Resolution Stereo Camera, *Nature*, 2004, 432 (7020), 971-979.

[8] Squyres, SW; Arvidson, RE; Bell, JF; Bruckner, J; Cabrol, NA; Calvin, W; Carr, MH; Christensen, PR; Clark, BC; Crumpter, L; Des Marais, DJ; D'Uston, C; Economou, T; Farmer, J; Farrand, W; Folkner, W; Golombek, M; Gorevan, S; Grant, JA; Greeley, R; Grotzinger, J; Haskin, L; Herkenhoff, KE; Hviid, S; Johnson, J; Klingelhofer, G; Knoll, AH; Landis, G; Lemmon, M; Li, R; Madsen, MB; Malin, MC; McLennan, SM; McSween, HY; Ming, DW; Moersch, J; Morris, RV; Parker, T; Rice, JW; Richter, L; Rieder, R; Sims, M; Smith, M; Smith, P; Soderblom, LA; Sutlivan, R; Wanke, H; Wdowiak, T; Wolff, M; Yen, A. The Opportunity Rover's Athena science investigation at Meridiani Planum, Mars, *Science*, 2004, 306 (5702), 1698-1703.

[9] Lundin, R; Barabash, S. Evolution of the Martian atmosphere and hydrosphere: Solar wind erosion studied by ASPERA-3 on Mars Express, Planetary and Space *Science*, 2004, 52 (11), 1059-1071.

[10] Zasova, L; Formisano, V; Moroz, V; Grassi, D; Ignatiev, N; Giuranna, M; Hansen, G; Blecka, M; Ekonomov, A; Lellouch, E; Fonti, S; Grigoriev, A; Hirsch, H; Khatuntsev, I; Mattana, A; Maturilli, A; Moshkin, B; Patsaev, D; Piccioni, G; Rataj, M; Saggin, B. Water clouds and dust aerosols observations with PFS MEX at Mars, *Planetary and Space Science*, 2005, 53 (10), 1065-1077.

[11] Mase, RA; Antreasian, PG; Bell, JL; Martin-Mur, TJ; Smith, JC. Mars Odyssey navigation experience, *Journal of Spacecraft and Rockets*, 2005, 42 (3), 386-393.

[12] Silverman, S; Christensen, P. Successful mars remote sensors, MO THEMIS and MER Mini-TES, *Acta Astronautica*, 2006, 59 (8-11), 1039-1047.

[13] Cirkovic, MM. Too early? on the apparent conflict of astrobiology and cosmology, *Biology & Philosophy*, 2006, 21 (3), 369-379.

[14] BRIN, GD. The great silence - the controversy concerning extraterrestrial intelligent life, Quarterly, *Journal of The Royal Astronomical Society*, 1983, 24, 283.

[15] Staley JT. Astrobiology, the transcendent science: the promise of astrobiology as an integrative approach for science and engineering education and research, *Current Opinion in Biotechnology*, 2003, 14 (3), 347-354.

[16] Cockell CS. 'Astrobiology' and the ethics of new science, *Interdisciplinary Science Reviews*, 2001, 26 (2), 90-96.

[17] Cockell, CS. Astrobiology - a new opportunity for interdisciplinary thinking, *Space Policy*, 2002, 18 (4), 263-266.

[18] He, YQ; Geng, RX; Ming, L; Zhou, MM. Titan may hold clues to origin of life, *Origins of Life and Evolution of the Biosphere*, 2006, 36 (3), 330-331.

[19] Dick, SJ. NASA and the search for life in the universe, *Endeavour*, 2006, 30 (2): 71-75.

[20] Schulze-Makuch, D; Irwin, LN. The prospect of alien life in exotic forms on other worlds, *Naturwissenschaften*, 2006, 93 (4), 155-172.

[21] Rodier, C; Sternberg, R; Szopa, C; Buch, A; Cabane, M; Raulin, F. Search for organics in extraterrestrial environments by in situ gas chromatography analysis. space life sciences: astrobiology: steps toward origin of life and titan before cassini, *Advances In Space Research*, 2005, 36 (2)195-200.

[22] McKay, CP; Smith, HD. Possibilities for methanogenic life in liquid methane on the surface of Titan, *Icarus*, 2005, 178 (1), 274-276

[23] Irwin, LN; Schulze-Makuch, D. Mars and Titan: Assessing the plausibility of life on two worlds with similar features and exotic differences, *Geochimica et Cosmochimica Acta*, 2005, 69 (10), A529-A529.

[24] Tang, BL; Many possibilities for life's emergence, *JBIS-Journal of the British Interplanetary Society*, 2005, 58 (7-8), 218-222.

[25] Chyba, CF; Phillips, CB. Europa as an abode of life, *Origins of Life and Evolution of the Biosphere*, 2002, 32 (1): 47-68.

[26] Raulin, F. Exo-astrobiological aspects of Europa and Titan: from observations to speculations, *Space Science Reviews*, 2005, 116 (1-2), 471-487.

[27] Chyba Christopher, F; Phillips, Cynthia, B. Europa s an abode of life, *Origins of Life and Evolution of the Biosphere*, 2002, 32, 47.

[28] Jeffrey, SK; Jonathan, ZK; James, WH; Giles MM; Roger, S; James, KS; Ballesteros, Prieto;O; Steven AG; David LH. Europa's crust and ocean: Origin, composition, and the prospects for life, *Icarus*, 2000, 148, 226.

[29] Perry, RS; Hartmann, WK. Mars primordial crust: Unique sites for investigating proto-biologic properties, *Origins of Life and Evolution of the Biosphere*, 2006, 36 (5-6), 533-540.

[30] Paillou, P; Lasne, Y; Heggy, E; Malezieux, JM; Ruffie, G. A study of P-band synthetic aperture radar applicability and performance for Mars exploration: Imaging subsurface geology and detecting shallow moisture. *Journal of Geophysical Research-Planets,* 2006, 111 (E6), Art. No. E06S11.

[31] Bibring, JP; Langevin, Y; Gendrin, A; Gondet, B; Poulet, F; Berthe, M; Soufflot, A; Arvidson, R; Mangold, N; Mustard, J; Drossart, P. Mars surface diversity as revealed by the MEGA/Mars Express observations, *Science*, 2005, 307 (5715), 1576-1581.

[32] Campbell, BA. Radar backscatter from Mars: Properties of rock-strewn surfaces; *Icarus*, 2001, 150, 38.

[33] Campbell, BA; Maxwell, TA; Freeman, A. Mars orbital synthetic aperture radar: Obtaining geologic information from radar polarimetry, *Journal of Geophysical Research-Planets*, 2004, 109 (E7), Art. No. E07008.

[34] De Sanctis, K; Ferretti, R; Visconti, G. Preliminary results of a Mesoscale Model for Nuovo Cimento della Societa Italiana di Fisica, *C-Geophysics and Space Physics*, 2006, 29 (6), 641-649.

[35] Esposito, F; Giuranna, M; Maturilli, A; Palomba, E; Colangeli, L; Formisano, V. Albedo and photometric study of Mars with the Planetary Fourier Spectrometer on-board the Mars Express mission, *Icarus*, 2007, 186 (2), 527-546.

[36] Fedorova, AA.; Korablev, OI.; Perrier, S; Bertaux, JL; Lefevre, F; Rodin, A; Kiselev, A. Exploration of Mars in the SPICAM-IR experiment onboard the Mars-Express spacecraft: 2. Nadir observations: Simultaneous observations of water vapor and O-2 glow in the Martian atmosphere, *Cosmic Research*, 2006, 44 (4), 294-304.

[37] Rauschenbach, I; Lazic, V; Jessberger, EK; Gentner, A. A miniaturized LIBS/Raman instrument for the comprehensive in situ analysis of the Martian surface. *Meteoritics & Planetary Science*, 2006, 41 (8) A146-A146 Suppl. S.

[38] Wang, A; Haskin, LA; Lane. AL; Wdowiak. TJ; Squyres, SW; Wilson, RJ; Hovland. LE; Manatt. KS; Raouf, N; Smith. CD. Development of the Mars microbeam Raman spectrometer (MMRS): *Journal of Geophysical Research-Planets*, 2003, 108 (E1), Art. No. 5005.

[39] Silverman, S; Peralta, R; Christensen, P; Mehall, G. Miniature thermal emission spectrometer for the Mars Exploration Rover. *Acta Astronautica*, 2006, 59 (8-11), 990-999.

[40] Miller, Dp; Varsi, G. Microtechnology For Planetary Exploration And Education *Acta Astronautica*, 1993, 29 (7), 561-567.

[41] Ellery, A; Ball, AJ; Cockell, C; Dickensheets, D; Edwards, H; Kolb, C; Lammer, H; Patel, M; Richter, L. Vanguard - a European robotic astrobiology-focussed Mars sub-surface mission proposal Source, *Acta Astronautica*, 2005, 56 (3), 397-407.

[42] Dickensheets, DL; Wynn-Williams, DD; Edwards, HGM; Schoen, C; Crowder, C; Newton. EM. A novel miniature confocal microscope/Raman spectrometer system for biomolecular analysis on future Mars missions after Antarctic trials, *Journal of Raman Spectroscopy*, 2000, 31 (7), 633-635.

[43] Korablev, O; Bertaux, JL; Fedorova, A; Fonteyn, D; Stepanov, A; Kalinnikov, Y; Kiselev, A; Grigoriev, A: Jegoulev, V; Perrier, S; Dimarellis, E; Dubois, JP; Reberac, A; Van Ransbeeck, E; Gondet, B; Montmessin, F; Rodin, A. SPICAM IR acousto-optic spectrometer experiment on Mars Express, *Journal of Geophysical Research-Planets*, 2006, 111 (E9), Art. No. E09S03

[44] Formisano, V; Angrilli, F; Arnold, G; Atreya, S; Bianchini, G; Biondi, D; Blanco, A; Blecka, MI; Coradini, A; Colangeli, L; Ekonomov, A; Esposito, F; Fonti, S; Giuranna, M; Grassi, D; Gnedykh, V; Grigoriev, A; Hansen, G; Hirsh, H; Khatuntsev, I; Kiselev, A; Ignatiev, N; Jurewicz, A; Lellouch, E; Moreno, JL; Marten, A; Mattana, A; Maturilli, A; Mencarelli, E; Michalska, M; Moroz, V; Moshkin, B; Nespoli, F; Nikolsky, Y; Orfei, R; Orleanski, P; Orofino, V; Palomba, E; Patsaev, D; Piccioni, G; Rataj, M; Rodrigo, R; Rodriguez, J; Rossi, M; Saggin, B; Titov, D; Zasova, L. The Planetary Fourier Spectrometer (PFS) on board the European Mars Express mission: *Planetary and Space Science*, 2005, 53 (10), 963-974.

[45] Klingelhofer, G. Mossbauer in situ studies of the surface of Mars, *Hyperfine Interactions*, 2004, 158 (1-4), 117-124.

[46] Lunine, JI; Chambers, J; Morbidelli, A; Leshin, LA. The origin of water on Mars, *Icarus*, 2003, 165 (1), 1-8.

[47] Pepin, RO. Atmospheres on the terrestrial planets: Clues to origin and evolution *Earth and Planetary Science Letters*, 2006, 252 (1-2), 1-14.

[48] Kasting, JF; Catling, D. Evolution of a habitable planet: Annual Review Of *Astronomy and Astrophysics*, 2003, 41, 429-463.
[49] Kasting, JF; Howard, MT. Atmospheric composition and climate on the early Earth, *Philosophical Transactions of the Royal Society B-Biological Sciences*, 2006, 361 (1474), 1733-1741.
[50] Jortner, J. Conditions for the emergence of life on the early Earth: summary and reflections, *Philosophical Transactions of the Royal Society B-Biological Sciences,* 2006, 361 (1474), 1877-1891.
[51] McKay, CP. The search for life on Mars, *Origins of Life and Evolution of the Biosphere*, 1997, 27 (1-3), 263-289.
[52] Clifford, SM. a model for the hydrologic and climatic behavior of water on Mars, *Journal of Geophysical Research-Planets*, 1993, 98 (E6), 10973-11016.
[53] Doute, S; Schmitt, B; Langevin, Y; Bibring, JP; Altieri, F; Bellucci, G; Gondet, B; Poulet, F. South Pole of Mars: Nature and composition of the icy terrains from Mars Express OMEGA observations, *Planetary and Space Science*, 2007, 55 (1-2), 113-133.
[54] Armstrong, J C; Nielson, S K; Titus, T N. Survey of TES high albedo events in Mars' northern polar craters, *Geophysical Research Letters*, 2007, 34 (1), Art. No. L01202.
[55] Fishbaugh, KE; Hvidberg, CS. Martian north polar layered deposits stratigraphy: implications for accumulation rates and flow, *Journal of Geophysical Research-Planets*, 2006, 111 (E6), Art. No. E06012.
[56] Christensen, PR. Water at the poles and in permafrost regions of Mars, *Elements*, 2006, 2 (3), 151-155.
[57] Ng, FSL; Zuber, MT. Patterning instability on the Mars polar ice caps, *Journal of Geophysical Research-Planets,* 2006, 111 (E2), Art. No. E02005.
[58] Litvak, ML; Mitrofanov, IG; Kozyrev, AS; Sanin, AB; Tretyakov, VI; Boynton, WV; Kelly, NJ; Hamara, D; Shinohara, C; Saunders, RS. Comparison between polar regions of Mars from HEND/Odyssey data, *Icarus*, 2006, 180 (1), 23-37.
[59] Cordoba-Jabonero, C; Zorzano, MP; Selsis, F; Patel, MR; Cockell, CS. Radiative habitable zones in martian polar environments, *Icarus*, 2005, 175 (2), 360-371.
[60] Horneck, G. The microbial world and the case for Mars, *Planetary and Space Science*, 2000, 48 (11), 1053-1063.
[61] Edwards, HGM; Cockell, CS; Newton, EM; Wynn-Williams, DD. Protective Pigmentation In Uvb-Screened Antarctic Lichens Studied By Fourier Transform Raman Spectroscopy: An Extremophile Bioresponse To Radiation Stress, *Journal of Raman Spectroscopy*, 2004, 35 (6), 463-469.
[62] Diaz, B; Schulze-Makuch, D. Microbial survival rates of Escherichia coli and Deinococcus radiodurans under low temperature, low pressure, and UV-Irradiation conditions, and their relevance to possible martian life, *Astrobiology*, 2006, 6 (2), 332-347.
[63] Rothschild, LJ; Mancinelli, RL. Life in extreme environments, *Nature*, 2001, 409 (6823), 1092-1101.
[64] Rainey, FA; Oren, A. Extremophile microorganisms and the methods to handle them, *Methods in Microbiology*, 2006, 35, 1-25.
[65] Pakchung, AAH; Simpson, PJL; Codd, R. Life on earth. Extremophiles continue to move the goal posts, Environmental Chemistry, 2006, 3 (2): 77-93.

[66] Javaux, EJ. Extreme life on Earth - past, present and possibly beyond, *Research in Microbiology*, 2006, 157 (1), 37-48.

[67] Alpert P: Sharing the secrets of life without water, *Integrative and Comparative Biology* 45 (5), 2005, 683-684.

[68] Charlier, D; Droogmans, L. Microbial Life at high temperature, the challenges, the strategies, *Cellular and Molecular Life Sciences*, 2005, 62 (24), 2974-2984.

[69] Marshall, CP; Carter, EA; Leuko, S; Javaux, EJ. Vibrational spectroscopy of extant and fossil microbes: Relevance for the astrobiological exploration of Mars, *Vibrational Spectroscopy*, 2006, 41 (2), 182-189 Sp. Iss. SI.

[70] Frosch, T; Tarcea, N; Schmitt, M; Thiele, H; Langenhorst, F; Popp, J. UV Raman imaging - A promising tool for astrobiology: Comparative Raman studies with different excitation wavelengths on SNC martian meteorites, *Analytical Chemistry*, 2007, 79 (3), 1101-1108.

[71] Ellery, A; Wynn-Williams, D; Parnell. J; Edwards, HGM; Dickensheets, D. The role of Raman spectroscopy as an astrobiological tool in the exploration of Mars, *Journal of Raman Spectroscopy*, 2004, 35 (6), 441-457.

[72] Villar, SEJ; Edwards, HGM. Raman spectroscopy in astrobiology, *Analytical and Bioanalytical Chemistry*, 2006, 384 (1), 100-113.

[73] Ellery, A; Wynn-Williams, D. Why Raman Spectroscopy on mars? A case of the right tool for the right job, *Astrobiology*, 2003 3 (3), 565-579.

[74] Villar, SEJ; Edwards, HGM; Benning, LG. Raman spectroscopic and scanning electron microscopic analysis of a novel biological colonisation of volcanic rocks, *Icarus*, 2006, 184 (1), 158-169.

[75] Edwards, HGM; Moody, CD; Newton, EM; Villar, SEJ; Russell, MJ. Raman spectroscopic analysis of cyanobacterial colonization of hydromagnesite, a putative martian extremophile, *Icarus*, 2005, 175 (2), 372-381.

[76] Edwards, HGM; Moody, CD; Villar, SEJ; Wynn-Williams, DD. Raman spectroscopic detection of key biomarkers of cyanobacteria and lichen symbiosis in extreme Antarctic habitats: Evaluation for Mars Lander missions, *Icarus*, 2005, 174 (2), 560-571.

[77] Moody, CD; Jorge Villar, SE; Edwards, HGM; Hodgson, DA; Doran, PT; Bishop, JL. Biogeological Raman spectroscopic studies of Antarctic lacustrine sediments, *Spectrochimica Acta Part A-Molecular and Biomolecular Spectroscopy*, 2005, 61 (10), 2413-2417.

[78] Villar, SEJ; Edwards, HGM; Cockell, CS. Raman spectroscopy of endoliths from Antarctic cold desert environments, *Analyst*, 2005, 130 (2), 156-162.

[79] Villar, SEJ; Edwards, HGM. Near-infrared Raman spectra of terrestrial minerals: relevance for the remote analysis of Martian spectral signatures, *Vibrational Spectroscopy*, 2005, 39 (1): 88-94.

[80] Edwards, HGM; Villar, SEJ; Jehlicka, J; Munshi, T. FT-Raman spectroscopic study of calcium-rich and magnesium-rich carbonate minerals, *Spectrochimica Acta Part A-Molecular and Biomolecular Spectroscopy*, 2005, 61 (10), 2273-2280.

[81] Jorge Villar, S E; Edwards, HGM. Spectroscopic requirements for Raman instrumentation on a planetary lander: potential for the remote detection of biosignatures on Mars, *International Journal of Astrobiology*, 2005, 3,165-174.

[82] Villar, SEJ; Edwards, HGM; Worland, MR. Comparative evaluation of Raman spectroscopy at different wavelengths for extremophile exemplars, *Origins of Life And Evolution of the Biosphere,* 2005, 35 (5), 489-506.

[83] Mueller, DR; Vincent, WF; Bonilla, S; Laurion, I. Extremotrophs, extremophiles and broadband pigmentation strategies in a high arctic ice shelf ecosystem, *Fems Microbiology Ecology*, 2005, 53 (1), 73-87.

[84] Cockell, CS; Rettberg, P; Horneck, G; Scherer, K; Stokes, MD; Measurements of microbial protection from ultraviolet radiation in polar terrestrial microhabitats, *Polar Biology*, 2005, 26 (1), 62-69.

[85] Karl, O. Stetter: Extremophiles and their adaptation to hot environments, *FEBS Letters*, 1999, 452, 22-25.

[86] Poli, A; Esposito, E; Orlando, P; Lama, L; Giordano, A; Appolonia, F; Nicolaus, B; Gambacorta, A. Halomonas alkaliantarctica sp. nov., isolated from saline lake Cape Russell in Antarctica, an alkalophilic moderately halophilic, exopolysaccharide-producing bacterium, *Systematic and Applied Microbiology*, 2007, 30, 31-38.

[87] Edwards, HGM; Newton, EM; Wynn-Williams, DD; Coombes, SR. Molecular spectroscopic studies of lichen substances 1: parietin and emodin, *Journal of Molecular Structure*, 2003, 648 (1-2), 49-59.

[88] Edwards, HGM; Newton, EM; Wynn-Williams, DD. Molecular structural studies of lichen substances II: atranorin, gyrophoric acid, fumarprotocetraric acid, rhizocarpic acid, calycin, pulvinic dilactone and usnic acid, *Journal of Molecular Structure*, 2003, 651, 27-37.

[89] Villar, SEJ; Edwards, HGM; Seaward, MRD. Raman spectroscopy of hot desert, high altitude epilithic lichens. *Analyst*, 2005, 130 (5): 730-737.

In: Planet Mars Research Focus
Editor: L. A. Costas, pp. 243-283
ISBN: 978-1-60021-826-2

Chapter 9

MARTIAN TECTONICS, WATER AND LIFE

E. I. Alves [(1)*]***, M. T. Barata*** [(1)]***, D. A. Vaz*** [(1)]***, L. C. G. Pereira*** [(1)]***,***
A. R. Baptista [(1, 2)] ***and M. J. Chorro*** [(3)]

(1) Centro de Geofisica da Universidade de Coimbra, Coimbra, Portugal
(2) Institut de Physique du Globe de Paris
(3) Departamento de Física, Universidade de Aveiro

ABSTRACT

In this chapter, we contend that Mars has had global-scale plate tectonics.

Consequences can be seen today in the form of traces of continental rifting with well developed shear zones that are only compatible with successive, differently oriented, horizontal stress fields.

Plate-tectonic rifting also explains the banded magnetic anomalies and their observed local offsetting.

As on Earth, plate tectonics could have had a majour role in the development and sustainability of life on Mars. This would happen by providing carbon and water recycling and magnetic sheltering from space radiation.

1. INTRODUCTION

There is ample evidence that Mars has had large-scale tectonic activity and, moreover, that plate tectonics (as opposed to isostatic tectonics) was once active in the planet.

One such evidence is the widespread presence, mainly in the Southern hemisphere, of parallel thermoremanent magnetic anomalies of alternating polarities. Connerney *et al.* (2005) have shown that there is a strong correlation of those anomalies with previously identified faulting and inferred the presence of majour tectonic lineaments from offset magnetic contours.

---

* Corresponding author can be contacted at Tel. +351 239 793 420; Fax. +351 239 793 428

A tool that was previously developed by us (Vaz *et al.*, 2006b) allowed the identification and geometric characterisation of those lineaments and associated fault systems on MOLA (Mars Orbiter Laser Altimeter aboard the Mars Global Surveyor satellite) digital elevation models (DEM). We apply this automated and quantitative methodology to the Tempe Fossae region in order to extract the fracturation pattern. This pattern analysis is complemented by the analysis of MOC visual imagery. Based on the results we propose an oblique rift model to the Tempe Fossae area, which is analogous to the Clifton and Schlische (2003) model for the Icelandic rift on Earth. The rift could be related with one of the transform faults identified by Connerney *et al.* (2005) based on the offset of magnetic anomalies. The obtained results are consistent with a plate tectonic regime on Mars.

Such tectonic regime would have had two consequences. First, shear stresses and strains must have influenced the hydrogeologic properties of crustal materials, constraining the Hanna and Phillips (2004) model and helping to explain the presence of large outflow channels emerging from active tectonic zones. Second, the model implies a large angle between present and past planetary rotation axes, with global climatic consequences namely on the stability of surficial liquid water. Both these aspects are verified in the Mangala Vallis area.

The existence of liquid water is essential to the onsetting of life as we know it. Persistence and development of life, even of a most rudimentary form, on the other hand, supposes the existence of shelter from space radiation. We present the first global map of radiation incidence on Mars, based on data from the MARIE (Martian Radiation Environment) experiment aboard the 2001 Mars Odyssey spacecraft. This map shows that even the weak remanent magnetic fields can shelter the surface from space radiation.

On Mars, as on Earth, global plate tectonics could have provided a suitable environment for life.

2. PALEOMAGNETISM

2. 1. Magnetic Field Determination and its Crustal Origin

On Mars the MAG-ER instrument (Magnetometer-Electron Reflectometer, aboard Mars Global Surveyor – MGS) took magnetic field measurements between 150 and 400 Km altitude (Acuña *et al.*, 1999). Purucker *et al.* (2000) developed a technique of reduction and analysis of terrestrial magnetic field data obtained by satellite, in order to better understand the distribution of the magnetic field for Mars obtained with MAG-ER.

The radial magnetic field component values (Br) are inverted, obtaining a distribution of the equivalent magnetization source, which was used to build a map (figure 1), normalized for an altitude of 200 km, and assuming a 40 Km thick magnetized layer.

Mars does not have, in the present, an internal or global scale magnetic field (Acuña *et al.,* 1999). However, in the absence of a field generated by an internal core, there are crustal remnant magnetic anomalies, (Purucker *et al.*, 2000). The remnant crustal magnetization must have been formed at a time when Mars had global a magnetic field of internal origin (Acuña *et al.*, 1999). Therefore, the observations on figure 1 are represented by a potential function of crustal origin, consistent with a remanent origin of the magnetization verified in Mars.

Moreover, Purucker *et al.* (2000) refer that the bands, which are truncated in some regions, reveal a tectonic origin, like those from Valles Marineris and Ganges Chasma, suggesting higher crustal properties alteration linked to faulting.

Even more, crustal magnetism, like most other geomorphologic features, shows a dichotomized distribution on Mars. The Southern hemisphere and near the equator are most strongly magnetized, namely in the oldest terrains of Noachian and Hesperian age.

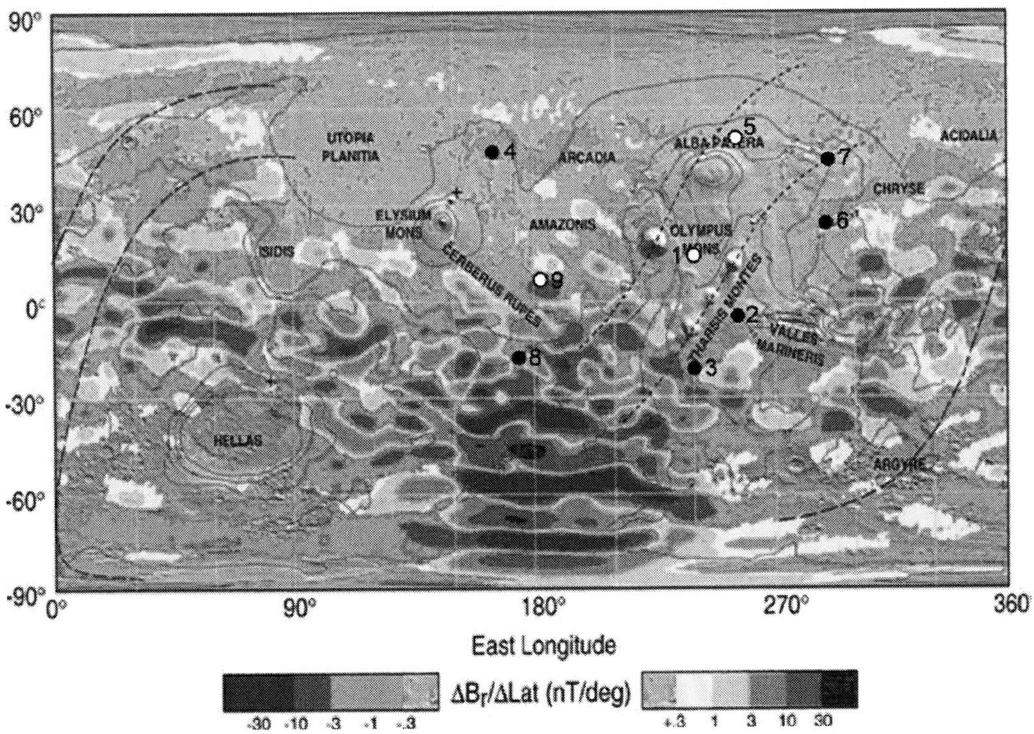

Figure 1: Radial magnetic anomaly, from Connerney *et al.*, 2005. 1-9: Locations of the paleomagnetic poles (solid circles are North poles and open circles are South poles, from Wieczorek *et al.*, 2005). [24-bit colour images are available on http://www.uc.pt/iguc/planets.htm]

The region of Terra Sirenum, between latitudes 40°S and 70°S and longitudes 180°W and 210°W, is where the strongest magnetic anomalies are observed. The crust of the Terra Sirenum region is interpreted as being of Noachian age (>3.6 Ga), is characterized by high impact cratering rate (up to 25 craters wider than 16 Km per 10^6 Km^2) and complex tectonic features, so far interpreted as relating to the Tharsis upwelling (*e. g.* Alves and Baptista, 2004).

The northern hemisphere does not show traces of remnant magnetization because it has suffered several resurfacing episodes by volcanic lavas. So, these surfaces formed in a time when the dynamo was already ceased and do not show magnetization (Stevenson, 2001).

Furthermore, in Hellas, southern hemisphere, the values of magnetization are quite low (or inexistent). The crater, in depth, might have reached zones close to the mantle; in the impact zone the scavenging would provoke reduction of the crustal thickness and consequent ascension of the mantelic surface by isostatic effect. Knowing that these demagnetized craters are about 4000 My (million years) old, the planetary dynamo would have ceased at about that time, according to Acuña (1999).

2.2. Location of the Magnetic Anomalies and Paleopole Positions

On Mars, the crustal anomalies can be determined and isolated to locate the magnetic paleopoles (figure 1). Using magnetic field orbital data, and assuming that an isolated magnetic anomaly formed on a cooling region of the crust when a dipolar internal field existed, the crustal magnetization direction can be used to infer the magnetic pole location, at a time when the anomaly has formed. (e. g. Arkani-Hamed and Boutin, 2004).

Modelling of each magnetic anomaly was done using each magnetic field value point for the first seven anomalies identified by Arkani-Hamed and Boutin in 2004. Then, Wieckzorek (2005) determined nine paleopole positions as shown on figure 1. The North and South paleopoles are located close to the martian equator, mainly around the Tharsis province.

3. Tectonics

3.1. Global-Scale Faulting

The Tharsis dome and the Elysium region concentrate must of the traces of tectonic and volcanic activity on Mars. The Tharsis bulge is surrounded by radiating sets of grabens and concentric sets of compressive wrinkle ridges. Several primary and secondary centers of tectonic activity on Tharsis have been identified by Anderson *et al.* (2001) relying on the statistical analysis of grabens' strikes and wrinkle ridges' normals to test a radial distribution of the underlying stresses. The separation of tectonic features by their possible ages allows the identification of five main stages of tectonic activity from the Noachian to the Amazonian.

The origin of this global radial pattern of extensive grabens is controversial. Regional isostatic strain, thermal plume uplift, giant dike swarms emplacement or a combination of all these factors, are some of the proposed explanations (Mège and Masson, 1996; Ernst *et al.*, 2001; Wilson and Head III, 2002).

Until recently, plate tectonic dynamics is not believed to have occurred at Mars (Solomon *et al.*, 2005). Despite that, identification of new transcurrent structures that are not compatible with isostatic and flexural stress models, supports a more mobilistic (not only isostatic/flexural) model for the formation of Mars tectonic structures (Anguita *et al.*, 2001). The same authors proposed an Early Hesperian period when the Mars crustal rheology would be favourable to a plate mobility regime. They proposed that buckling and thrust faults in the Coprates Rise arch originated by an East-West striking horizontal σ_1. The same stress field would be responsible for the Valles Marineris formation, which those authors interpret as a right-lateral trantensive megashear.

A posterior tectonic horizontal compressive episode at Amazonian times is responsible for wrinkle ridge and associated strike-slip faults. At this time the crust rheology would only allow thin-skin tectonics (Anguita *et al.*, 2001).

A more pervasive plate tectonic regime, involving seafloor spreading and subduction, has been proposed in order to explain the Mars dichotomy and the formation of the Northern lowlands (Sleep, 1994).

Evidence of recent tectonic activity on Mars has been found near Arsia Mons, where faults cut recent, possible glacial, aureole deposits (Marquez *et al.*, 2004).

We will propose an algorithm for the automatic identification of tectonic lineaments on the surface of Mars and present the results of its application; we will also present a tentative structural and kinematic interpretation for the fracturation pattern at Tempe Terra. The proposed model will be correlated with the crustal magnetic record.

3.2. Regional-Scale Faulting: Tempe Fossae

Planetary mapping is traditionally made by visual interpretation of remote sensing data. The interpretation consist on the identification of several geomorphologic structures (craters, valleys, volcanoes, or other geomorphologic features) mostly from imagery data.

Some attempts have been recently made for the automation of the process of geologic/geomorphologic mapping, especially on Mars, due to the ever crescent amount of available data.

Several methodologies have been proposed to perform automatic crater extraction from imagery (Kim *et al.*, 2005; Saraiva *et al.*, 2006) and from MOLA DEMs (Matsumoto *et al.*, 2005; Bue and Stepinski, 2007). Stepinski and Collier (2004) have used MOLA (Mars Orbiter Laser Altimeter) MEGDRs (Mission Experiment Gridded Data Record) to automatically extract and characterize Martian drainage networks and Bue and Stepinski (2005) propose a numerical method for classification and characterization of landforms on Mars.

Tectonic mapping on Mars was made using visible wavelength imagery; MOLA data has allowed a better characterization of tectonic structures on Mars and topographic profiles are now widely used on the study of compressive wrinkle ridges (Tate *et al.*, 2001) and extensional structures (Borraccini *et al.*, 2005).

In this chapter we present an automatic method for tectonic characterization of planetary surfaces affected by extensional regimes. We will apply this method to the Mars region of Tempe Fossae and discuss the results. This method relies on a mathematical morphological post-processing of wavelet transformed MOLA DEMs. It allows not only the identification of the grabens' scarps but also their characterization (Vaz *et al.*, 2006a).

3.2.1. Mathematical Morphology – An Introduction

Mathematical Morphology is an image analysis theory created in the middle 1960s by George Matheron and Jean Serra in the *École des Mines de Paris*. Its initial purpose was related to an application in porous media to describe the geometric features of structures (Matheron, 1967). The further developments since then have permitted to construct a solid framework (Matheron, 1975; Serra, 1982, 1988) and have successfully reached new application areas (good overview in Soille, 2002) namely remote sensing.

Initially developed for binary images this theory was generalized for grey-scale images. The mathematical morphology operators can be used directly or applyed sequentially to obtain more elaborated morphologic transformations, for specific ends. Any operator or morphological transform implies the comparison of the features to analyze with a known object, the structuring element. The success of the application of any mathematical transform depends on the choice of the structuring element.

3.2.1.1. The Basic Transforms

The first morphological transforms defined by Matheron (1967), are the erosion and dilation. To grey scale images, the erosion ($\varepsilon_B(f)$) of an image f by a structuring element B of size λ, is the minimum of the translations of f by the vectors **–b** of B (Soille, 2002):

$$\varepsilon_B(f) = \min_{b \in B} f_{-b}$$

The dilation of an image ($\delta_B(f)$) of an image f by a structuring element B of size λ, is the maximum of the translations of f by the vectors **–b** of B:

$$\delta_B(f) = \max_{b \in B} f_{-b}$$

The final results for both operators are poor in details when compared with the initial image. If it is necessary to enhance dark areas in a grey image, the erosion shows good results and the dilation has the opposite effect, because white zones in an image are more easily detected.

3.2.1.2. The Morphological Gradient

The objective of the morphological gradient is to enhance and extract the contours of the homogeneous regions of grey levels in an image. From the two basic operators, erosion and dilation, the morphological gradient or *Beucher gradient* (1990) is defined as the arithmetic difference between the maximum and the minimum of the function f, divided by the size or diameter of the structuring element (λB):

$$\rho_B = \delta_{\lambda B} - \varepsilon_{\lambda B} / 2\lambda$$

One particular case of morphological gradient is the directional gradient, very useful to enhance features in a particular direction, and it is defined as:

$$\rho_{L\alpha} = \delta_{L\alpha} - \varepsilon_{L\alpha}$$

The isotropic structuring element is replaced by a line L in a give direction α.

3.2.1.3. Opening and Closing

The erosion and dilation can be combined to perform two important transforms: opening and closing. The opening (γ) consists on submitting an image f to an erosion followed by a dilation, both by a structuring element B of size λ. In the final image the opening cuts peaks and removes small object protuberances:

$$\gamma_B(f) = (\delta_B(f)[\varepsilon_B(f)]$$

In the same manner, the closing transform (ϕ) consists in applying a dilation to an image f followed by an erosion. This operation suppresses (or closes) all the valleys smaller than a certain dimension given by the size λ of the structuring element B:

$$\phi_B(f) = (\varepsilon_B(f)[\ \delta_B(f)]$$

The opening and closing are the basis of the majority of morphological filters. Although possible, the isolated application of each operator is not a very powerful filter. The alternate application of these operators (alternate filters (AF)) and the application of sequences of alternate filters allowed the construction of alternating sequential filters (ASF) (Serra, 1988). An alternating sequential filter is initiated with a small structuring element whose dimension increases successively in each following stage until reaching the desired size. For example, the sequence of an ASF of closing-opening type (starts by an opening), is given by the expression:

$$f_2 = \gamma_i\, \varphi_i \ldots \varphi_3 \gamma_3\, \gamma_2\, \varphi_2\, \varphi_1 \gamma_1\, (f)$$

3.2.1.4. Top-hat

The top-hat transform was introduced by Meyer (1979, 1986) based on combinations of openings and closings. The black or valley top-hat is computed in two steps: (i) by subtracting the initial image f to the closed one with the structuring element B of size λ and (ii) by thresholding T this image within adequate levels t_1 and t_2:

$$X=TH(f)=T[\varphi^{\lambda B}(f)-(f)]_{(t1,\,t2)}$$

The closing transform closes all the valleys smaller than a certain dimension. The subtraction of the initial image allows to recover these structures by putting them at the same level and also by simultaneously filtering the ones that were not modified.

The white top-hat is defined as:

$$X=WTH(f)=T\ [(f)-\ \phi^{\lambda B}(f)]_{(t1,\,t2)}$$

and consists on thresholds at levels t_1 and t_2, the grey level image resulting from the arithmetic difference between the original image f and the opened image with the structuring element λB. The white top-hat is used to extract peaks (white regions) in an image with respect to the background. The choice of the shape and the size of the structuring element for both top-hats, depends on the morphology of the structures to be extracted (Soille, 2002).

3.2.1.5. Hit-or-Miss Transforms and Skeletons

The hit-or-miss transforms are used to extract pixels with specific neighbouring configuration from an image. These operators uses two structuring element, B_1 and B_2, to find the foreground and the background of the specific configuration. The objective is to preserve all the points of the image that fit the neighbourhood configuration.

These kinds of transforms have many interesting applications, such as locating isolated foreground points, foreground endpoints, multiple foreground points or foreground contour points.

3.2.1.6. Thinning and Thickening

A thinning is the operation of removing foreground pixels matching a configuration given by a composite set of two structuring elements. The opposite transform, thickening, consists in adding, to the foreground pixels, background pixels having a specific configuration.

The expression below defines the thinning (O) of an image f by a composite set of two structuring elements B:

$$f\,O\,B = \text{THIN}_B\,(f) = f - \text{HMT}_B\,(f)$$

So, the thinning is defined by the arithmetic difference between the initial image f and the hit-or-miss transform of f, according a specific configuration B.

In the same way, the thickening ($\underline{O}$) is the addition of the initial image f and the hit-or-miss transform of f by B, and is given by the expression:

$$f\,\underline{O}\,B = \text{THICK}_B\,(f) = f + \text{HMT}_B\,(f)$$

The design of a neighbourhood configuration obeys specific criteria, defined by Golay in 1969. Later, in 1982, Serra has developed a set of configurations, "Golay's alphabet", where each configuration is assigned a letter.

3.2.1.7. Skeletons

The idea of a skeleton is to reduce the object to a line that preserves the original information. Several definitions of skeletons can be consulted in Soille (2002). A way to determine the skeleton of an object is through the thinning until idempotence, that is, to perform successive thinnings using all the possible configuration neighbourhoods for one letter of Golay's alphabet. For instance, the skeleton for a hexagonal grid X, with the L letter, is given by the expression:

$$\text{X O L} = ((((((\text{X O L}_1)\text{ O }(\text{X O L}_2)\text{O }(\text{X O L}_3)\text{O }(\text{X O L}_4)\text{O }(\text{X O L}_5)\text{O }(\text{X O L}_6)$$

At the end of the skeleton application, the object is reduced to one pixel. Every point of the skeleton has two points in the neighbourhood, except on the extremities, multiple points or isolated points. The identification of these particular points is very useful in many applications of image analysis, especially in shape recognition.

Depending on the configuration used, the skeleton produces more or less branches in the final image. Through the pruning transform those branches can be eliminated or selectively reduced (preserving the most interesting). This can be done by applying a sequential thinning ($\underline{O}$) by the E letter of Golay's alphabet:

$$\text{PRUNE}(X)^{(n)} = (X \underline{o} E)^{(n)}$$

where n is the number of the iterations.

3.2.1.8. Geodesic Transforms

Mathematical transforms can be defined either in Euclidean or Geodesic space. Geodesy was introduced in image analysis by Lantuéjoul and Beucher (1980). When applied in geodesic space the mathematical transforms are limited to a region or subgroup of the image and only part of the image is analysed. Generally the geodesic transforms are applied iteratively until stability and this is called *reconstruction*. This transform is a powerful tool in mathematical morphology and is used to filter images, find and fill local maxima or minima, to suppress minima and maxima lesser than a give size, to remove noise without affecting structures of interest and filling holes.

The geodesic transforms involve two images: the mask and the marker image, with the same definition, provided that the mask image is greater or equal than the marker image (Soille, 2002). The reconstruction by dilation ($R_g^{\delta}(f)$) of a mask image *g* from a marker image *f* is defined as the geodesic dilation of *f* with respect to *g*, iterated until stability (Soille, 2002):

$$R_g^{\delta}(f) = \delta^i(f)$$

Similarly, reconstruction by erosion $R_g^{\varepsilon}(f)$ of a mask *g* from a marker image *f* is the geodesic erosion of *f* with respect to *g*, iterated until stability, and is given by the expression:

$$R_g^{\varepsilon}(f) = \varepsilon^i(f)$$

The reconstruction by erosion is the mathematical transform used to eliminate and fill holes in an image. In a binary image, a hole is defined as the set of background components which are not connect to the image border (Soille, 2002). To eliminate the holes by the application of a reconstruction by erosion, the image marker is equal to the original image and the mask is an image with a constant value and with the same border as the original.

In grey images the concept of a hole is different and removes holes means eliminate all the minima which are not connected with the image border. To fill holes in a grey image is necessary to impose the set of minima which are connected to the border image. The image marker used in this transform is the set of maximal values on the original image, except along its border where the values of the initial image are kept (Soille, 2002). The reconstruction by erosion is given by the expression:

$$FILL(f) = R_f^{\varepsilon}(f_m)$$

Openings and closings are also used in reconstruction. Reconstruction by opening removes small structures (they are eliminated by the erosion) while keeping the larger structures intact (boundaries first destroyed by the erosion will grow to the original size by the dilation). In grey images, the reconstruction by opening tends to better contrast the bright structures in the image. An identical effect can be obtained for dark objects by reconstruction by closing, where the dark structures are better preserved.

3.3. MOLA DEM Gridding

Since the resolution of the available MEGDR's is 128 pixels/degree (~463 m/pixel), we chose to generate a higher resolution DEM using the precision experiment data records (PEDRs, http://pds-geosciences.wustl.edu/missions/mgs/pedr.html).

We have followed the general steps described by Okubo *et al.*(2004) in order to extract the topographic profiles and remove offset tracks. The natural neighbour interpolation method was used to interpolate points between tracks. This method, although computationally more intensive, produced DEMs with less visual interpolation artifacts and is quantitatively more accurate than other interpolation methods when used with MOLA data (Abramov and McEwen, 2004).

The study area is comprised between 90°W and 60°W longitudes, 48°N and 30°N latitudes. The interpolated DEM has a spatial resolution of 250 pixels/degree (~237 m/pixel) and encloses part of Tempe Fossae and Mareotis Fossae regions (figure 2).

Grabens are linear depressions delimited by steep parallel scarps. Due to the linear nature of grabens and since the MOLA tracks are oriented more or less perpendicular to the grabens on the selected region (figure 3), the natural neighbour interpolator tends to smooth the slope of the scarps that are between MOLA tracks data, creating systematic linear artifacts.

Those artifacts can be reduced by applying a combination of morphological alternated operators, using linear structuring elements (SE) with the same orientation of the scarps to be enhanced. We build four SEs assuming a main orientation for the grabens but also a secondary one. The SE with the orientation of the main set of grabens must be larger than the secondary one.

Two parallel alternated filtering (AF) opening-closing type operations, using different combination of SEs, are averaged on a final matrix. The same alternated filters are then applied using a disk with a small radius. The use of a disk will smooth the DEM, removing the linear artifacts that were produced by the previous operation.

This set of operations improves the scarps' spatial continuity between MOLA tracks. Continuity is controlled by the size and orientation of the used SEs.

The procedure tends to eliminate smaller non-linear structures, for example smaller craters. It corrects some of the artifacts produced by interpolation, but also suppresses some finer details that are present in the initial interpolated DEM. Figures 4 and 5 compare the original DEM with the results obtained by the mathematical operators.

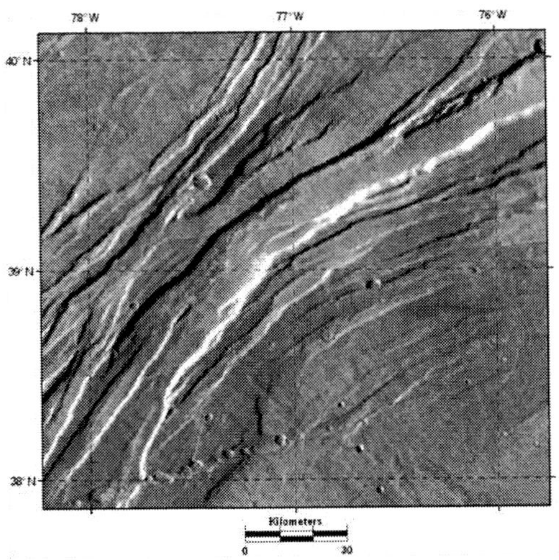

Figure 2: Viking mosaic image (MDIM 2.1) of the region chosen to illustrate the DEM enhancement methodology

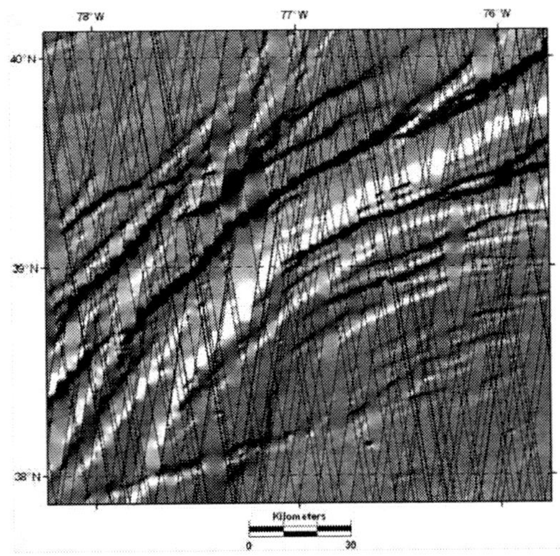

Figure 3: Shaded relief of the original interpolated DEM, with the MOLA track points superimposed. Illumination is from the Northwest.

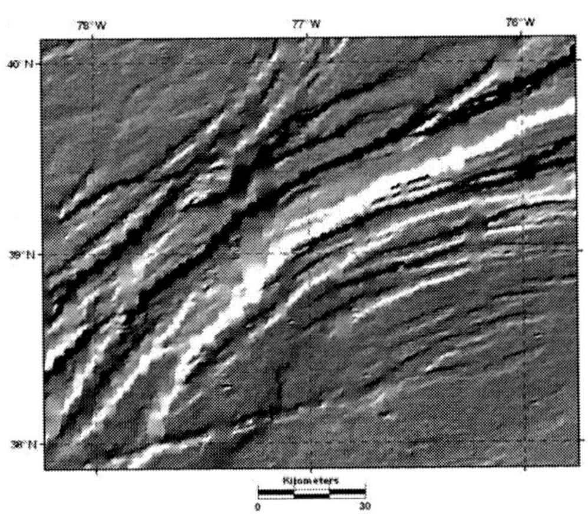

Figure 4: Shaded relief of the original interpolated DEM.

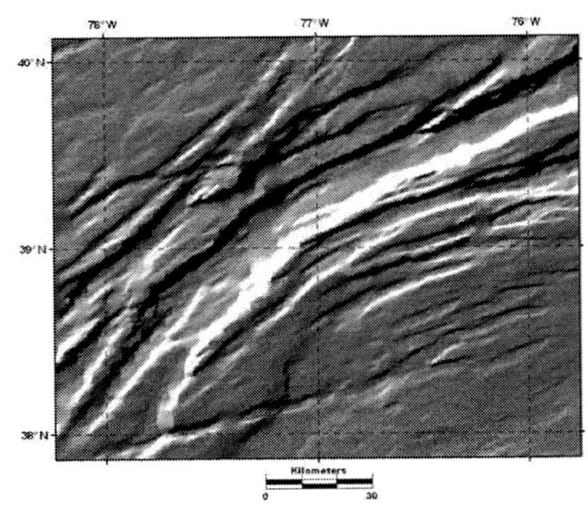

Figure 5: Shaded relief of the enhanced DEM using two linear structuring elements with size 12 and 3, and clockwise strike 70 and 40 degrees respectively.

The quality of the results depends on the orientation and size chosen for the SE's. Those parameters should match the orientation of the scarps and the spacing between MOLA data tracks. The described method reduces the artifacts that are visible on the original interpolated DEMs; nevertheless, it changes the digital values of the pixels. A quantitative test and comparison with other enhancing methods is out of the scope of this work and will be presented in a near future.

This method for MOLA-based DEM enhancement, taking into account the morphology of a desired structure, seems promising, and could constitutes the basis for a better method of interpolation for MOLA data (figures 6 and 7).

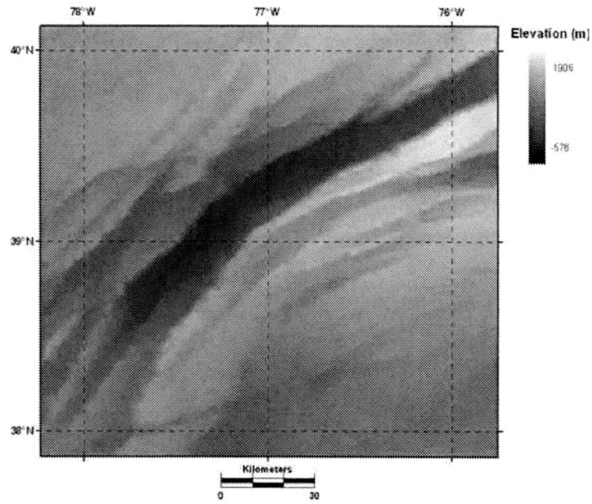

Figure 6: Original interpolated DEM.

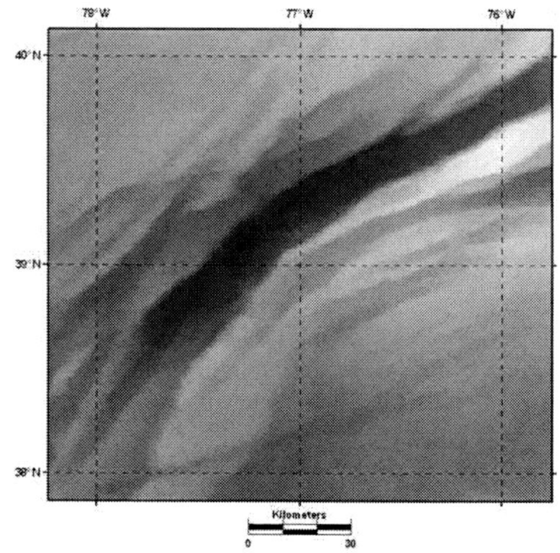

Figure 7: Enhanced DEM by applying the proposed method.

3.3.1. Automatic Lineament Extraction

In previous work we outlined a new method for extracting tectonic lineaments from Mars DEMs (Vaz *et al.*, 2006a; Vaz *et al.*, 2006b). We applied the wavelet transform for extracting the scarps from a DEM, and then morphological operators to extract the lineaments.

Wavelet analysis works by decomposing a signal using simpler blocks (wavelets) at different scales and positions (Mallat, 1989). The advantage of using wavelets is the capability to study features of a signal locally, with a detail that matches their scale (Kumar and Foufoula-Georgiou, 1997).

Dilations and translations of the analyzing wavelet $\Phi(x)$ define an orthogonal basis, the wavelet basis, defined recursively as

$$\Phi_{s,l}(x) = 2^{-\frac{s}{2}} \Phi(2^{-s} x - l)$$

where s and l are integers that scale and dilate the mother function Φ. The scale index s indicates the wavelet's width, and the location index l gives its position.

We follow the approach presented by Little (1994), who suggests the use of wavelet edge analysis to enhance and extract fault scarps from bathimetric data. We also use the first derivative of the cubic B-spline function as mother wavelet. The use of this anti-symmetric function allows not only the identification of the topographic discontinuities (that correspond to the scarps of the normal faults that bound the graben depressions), but also allows the segmentation of the scarps by dip direction.

The cubic B-spline function, $N_4(x)$ can be defined as:

$$x < 0 \Rightarrow N_4(x) = 0$$
$$0 \leq x < 1 \Rightarrow N_4(x) = x^3/6$$
$$1 \leq x < 2 \Rightarrow N_4(x) = -(1/3) + x/2 + (x-1)^2/2 - (x-1)^3/2$$
$$2 \leq x < 3 \Rightarrow N_4(x) = (2/3) - (x-2)^2 + (x-2)^3/2$$

$3 \leq x < 4 \Rightarrow N_4(x) = (5/3) - x/2 + (x-3)^2 - (x-3)^3/6$

$x \geq 4 \Rightarrow N_4(x) = 0$

We build two orthogonal filters by concatenating the 1-D first derivative of the cubic B-spline and then tapering the filters with a Hanning window. The length of the 1-D wavelet that is used determines the width of the fault to be isolated, and the number of identical filters concatenated to form the 2-D convolution matrix determines the fault length over which to average (or smooth) (Little, 1994; Vaz *et al.*, 2006c).

The wavelet decomposition framework permits the analysis of a signal at several scales. This characteristic is very suitable, since tectonic structures and deformation occurs at a large range of scales, from microscopic to planetary. In our case we are limited by the resolution of the DEM.

The size of the mother wavelet employed will condition the scale of the structures to be identified and the analysis of the tectonic pattern at several scales can this way be performed. Nevertheless in this case we will only analyse and illustrate the application of the method to one chosen scale. In the future a fully multiscale analysis will be done, and the tectonic/geologic meaning of the scale variation of the identified structures must be evaluated.

The two directional outputs obtained by convolution with the created filters are then binarized, segmenting the positive from the negative wavelet coefficients and using a hysteresis threshold (*DBLT*) (Soille, 2002), given by the expression:

$$DBLT_{[t1 \leq t2 \leq t3 \leq t4]}(f) = R^{\delta}_{T[t1,t4]}(f)\left[T_{[t2,t3]}(f)\right]$$

Several attempts have been made in order to automatically pick ideal threshold values. We have tried to use a combination of statistical parameters (mean, standard deviation, etc.) of the digital numbers of the matrices to estimate an ideal threshold value. This simple approach does not appear to be successful for all cases, and the threshold of the wavelet coefficients still constitutes one of the more awkward steps to the full automation of the lineament mapping process. Confronted with this problem we decided to manually define the threshold values used in this chapter.

By applying the threshold we obtain four binary images, each one presenting the scarps that dip to one of the geographic quadrants (figure 8). One of the particularities of using two orthogonal filters is that diagonal features will be present on two of the resulting binary images, for example a scarp that strikes NE-SW and dip SE will be present on the South and East dipping binary images which duplicates the number of identified structures. We have handled this problem by using a decision tree to choose which dip to attribute having as decision criterion the area of the objects.

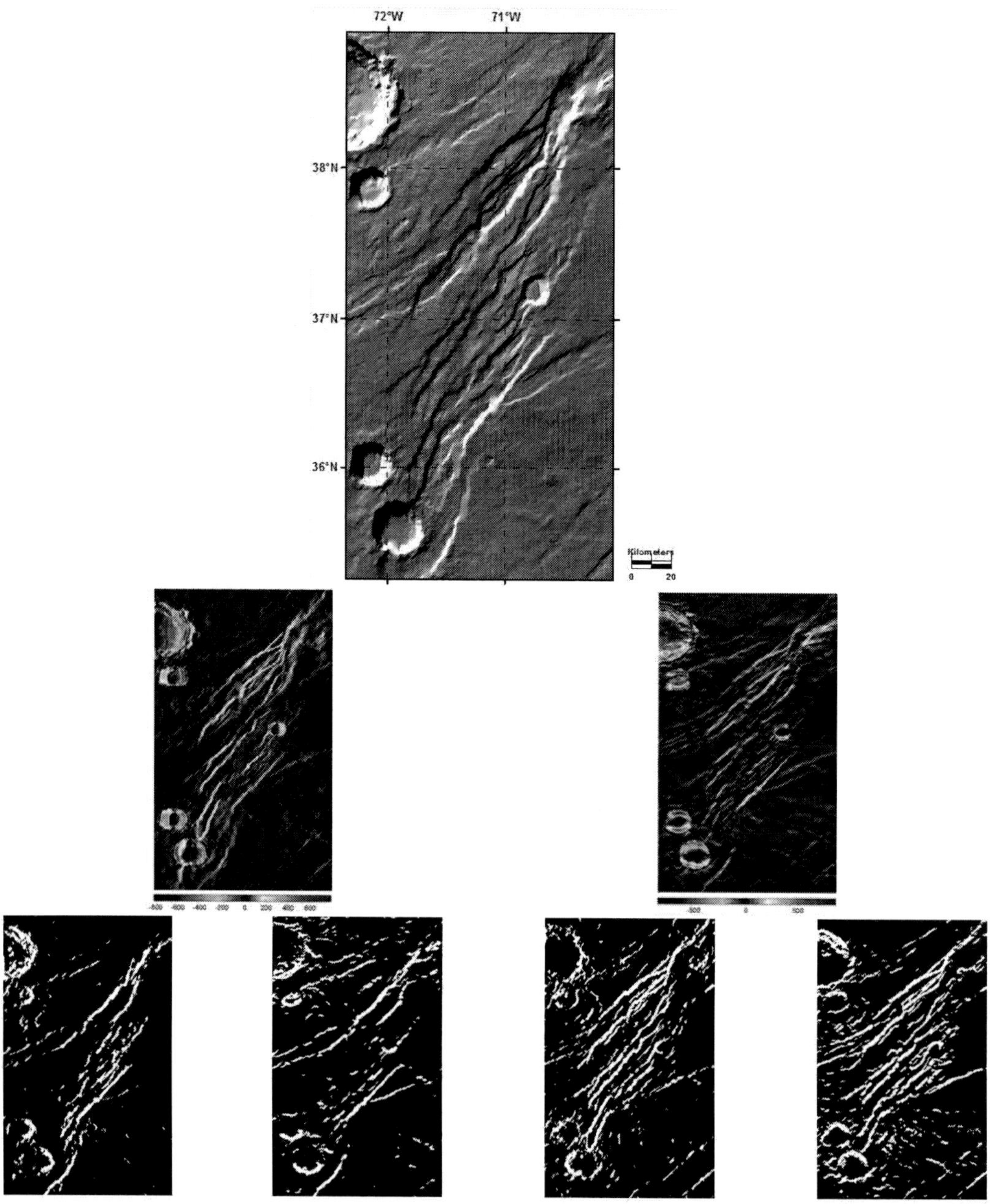

Figure 8: Shaded relief of a faulted area in Tempe Terra (up). The wavelet coefficients for the East-West and North-South directions (middle). Results from the application of the threshold on the two wavelet coefficients matrices presenting the scarps that dip West, North, East and South. Notice that the crater scarps are also extracted. [24-bit colour images are available on http://www.uc.pt/iguc/planets.htm]

A classification matrix is first created by adding the directional binary images, after multiplying each one by different prime numbers (1, 3, 7 and 13 for the South, North, West and East dipping structures respectively). By this scheme we expand from four possible dip directions to eight, each one coded by a different digital number.

Starting with the wider objects, we perform the intersection with the classification matrix. The object will be assigned to the correspondent new binary image based on the preponderant digital number found at the classification matrix. Next the pixels on the classification matrix that belong to the object, are all reclassificated with the value/dip attributed to the object in analysis if the number of pixels to be reclassified does not exceed a threshold level (we use 50% of the total of object pixels). This will force the next, smaller, objects that intersect with the previous one to be put on the same directional binary image, preventing the duplication of identified scarps. Figure 9 shows a flow diagram of the followed procedure.

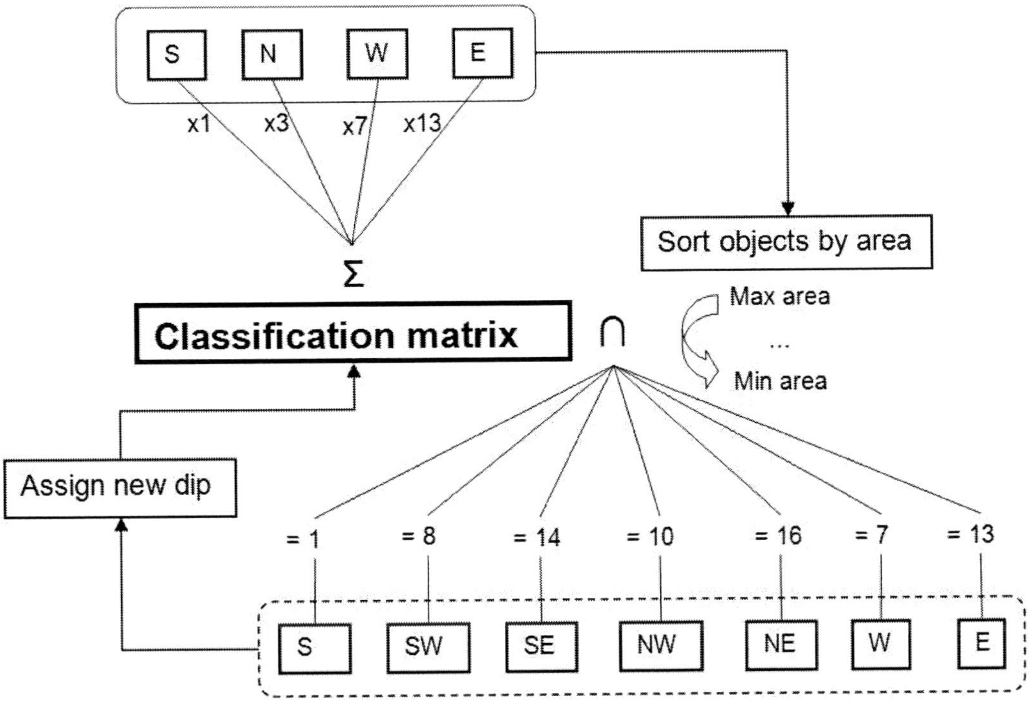

Figure 9: Flow diagram of the procedure to prevent duplication of identified objects in different binary images.

At the end we will have eight binary images (figure 10) that mark the horizontal projection of scarps on the surface of Mars, each one of those images presenting only those scarps that dip to one of the cardinal or intercardinal directions.

It is now necessary to reduce the identified structures to a one pixel-width line, and then vectorise those lineaments. To accomplish that we have used several morphological operators that are applied sequentially to all eight binary images.

First we remove the holes (figure 11a) that could be present on the objects; the holes will only be eliminated if they are non-linear with a length inferior to a threshold length. This is made by first identifying all holes present on the image, and then performing an opening operation using as SE a line with a defined length and an orientation similar to the strike of the objects. The surviving objects will be reconstructed by dilation and the complement is intersected with the non-hole image (figure 11b).

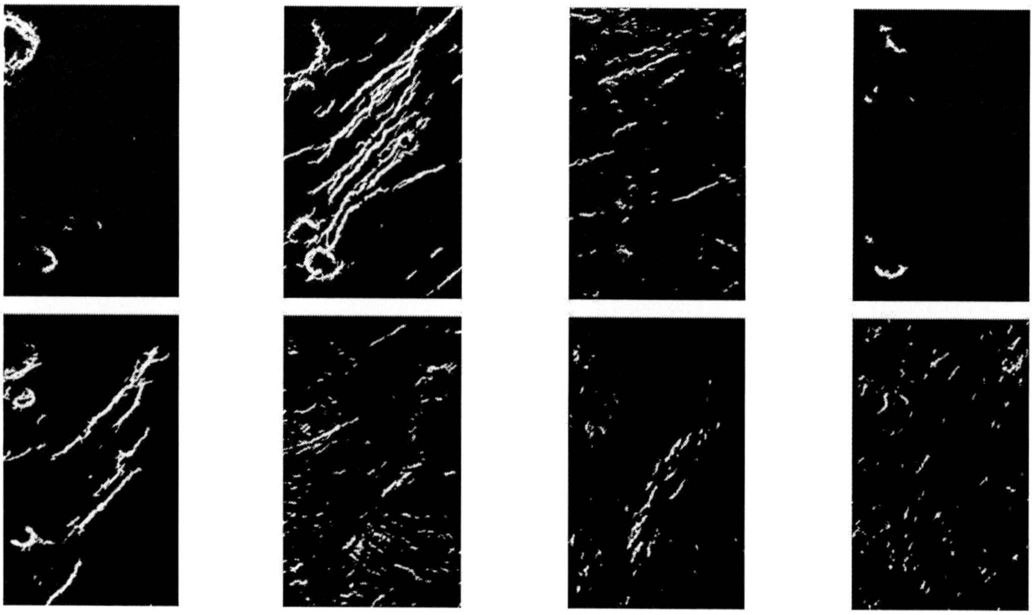

Figure 10: The eight binary images obtained by directional segmentation.

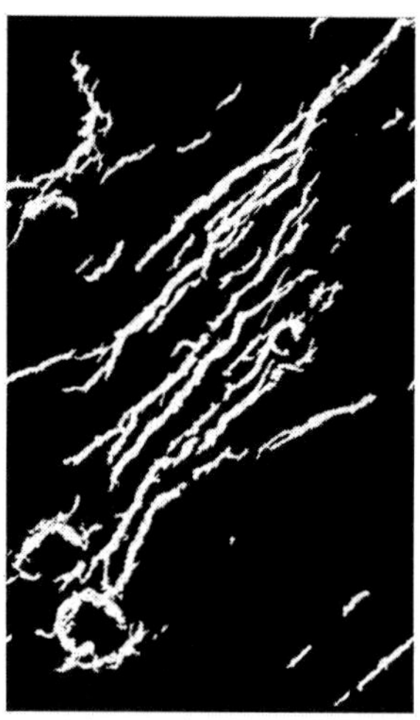

a

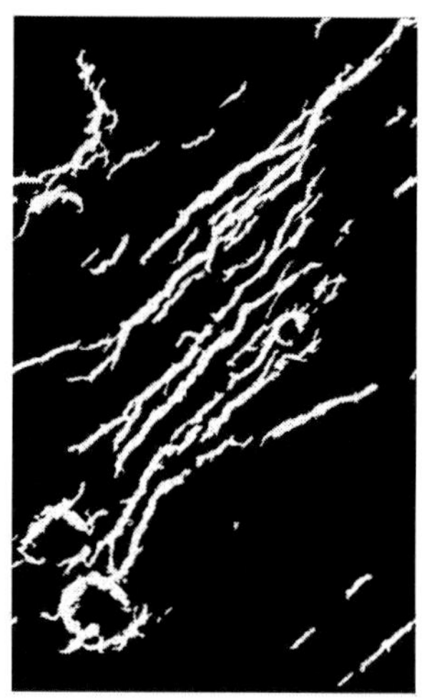

b

Figure 11: a) Image before the hole filling operation; and b) after the hole filling operation; notice that only small and non-linear holes are eliminated.

The next step concerns the reduction of the objects to one pixel width lines. This is accomplished by using a skeleton operation by the letter *L* of Golay's alphabet. The isolated pixels that result from the skeletonisation are eliminated by intersecting the skeleton image with the complement of a hit-or-miss operation with the appropriate SE (Soille, 2002).

The presence of a great amount of little branches can be easily observed on figure 12a.

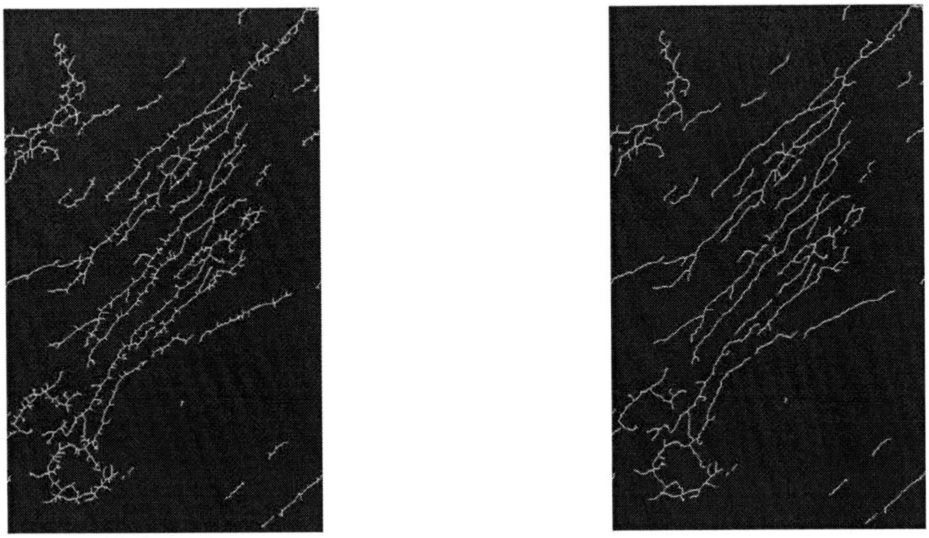

a b

Figure 12: a) Image after skeletonisation; b) final result of skeleton operation. Notice the number of little branches that are eliminated by the pruning operation.

Those branches can be eliminated by a pruning operation. This operation will remove *n* points from the extremities of the skeletons. On the other hand this operation will also reduce the initial length of the lines. This can be handled by adding the reconstruction of the eliminated branches, using the endpoints of the pruned skeleton as markers (Soille, 2002):

$$PRUNE(X)^{(n)}=((X\,\underline{o}\,E)^{(n)}\ldots\underline{\;\,}\underline{o}\,(X\,\underline{o}\,E)^{(1)})$$

We apply this skeleton filtering technique sequentially, using a decreasing value for *n*. This assures that little branches that are located near the extremities of a long line are also eliminated and that the length of the skeleton main line is preserved.

Smaller objects that have less than $2n$ pixels are transformed into points by the pruning operation. We use this fact to threshold objects by number of pixels, performing the reconstruction of the isolated pixels (if $2n > T$) or eliminating the isolated pixels (if $2n < T$) at each iteration. The final result can be observed in figure 12b.

The application of all these morphological operators produces one pixel-width lines although in some cases there are line segments which intersect each other. Those lines will then be vectorized using the Freeman chain coding technique (Freeman, 1961). Before vectorization, multiple points (those with more than two neighbours) are suppressed from the image so that vectorisation is made segment by segment. The several extracted segments are then subject to a merging operation. The mean strike of each segment and the distance between endpoints are used as arguments to this process.

The vectorized lineaments can now be overlaid on the DEM by importing them to a GIS platform. As can be seen (figure 13a) the method is accurate at identifying fault scarps yet the crater scarps and crater rims are also identified. To deal with this we decided to develop an algorithm to identify the craters. This algorithm will create a marker image that can be used to remove the lineaments that were previously identified and that correspond to crater rims scarps.

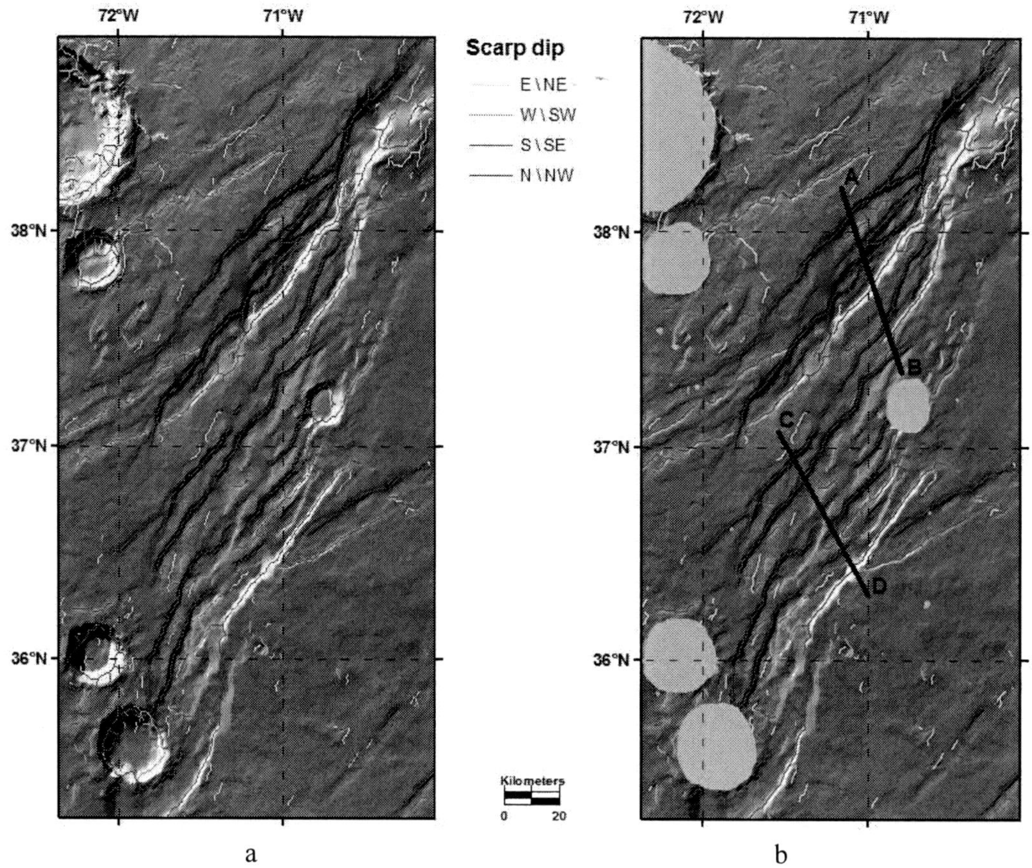

Figure 13: a) Identified lineaments superimposed on the shaded relief; b) lineaments and identified craters. The identified craters are used as markers to eliminated non tectonic lineaments. The dipping directions were grouped to improve visual interpretation by reducing the number of colors. The sections AB and CD are presented below, in figure 14. [24-bit colour images are available on http://www.uc.pt/iguc/planets.htm]

Our method consists on enhancing the craters by subtracting from the DEM the results of a black top-hat operation and the results of an external gradient operation (a different approach is presented by Bue and Stepinski (2007).The black top-hat is applied using a disk SE. The radius of the disk is the only input needed, and corresponds to the radius of the biggest crater present at the DEM. This operation is followed by a closing using a small SE in order to smooth the results of the black top-hat.

The external gradient of the DEM will allow the extraction of the inner scarp of the craters. A fill-hole operation on the gradient will enhance even more the craters. The next expression explains the adopted procedure:

$$Grad_{ext}(DEM) = DEM - \phi[TH(DEM) - FILL(DEM)]$$

After performing the described operations, the resulting matrix is subject to a level slicing and every object that appears on the successive slices will be circularity thresholded. Those objects whose circularity index is acceptable are added to an accumulation matrix. The linearization of the obtained accumulation matrix is dilated and used as crater marker (figure 13b). In the future this output could also be vectorized, for example by using the Hough transform (Matsumoto *et al.,* 2005). A preliminary qualitative evaluation of the results shows that the proposed method fails to detect heavily degraded craters. A detailed quantitative assessment of the algorithm remains to be done.

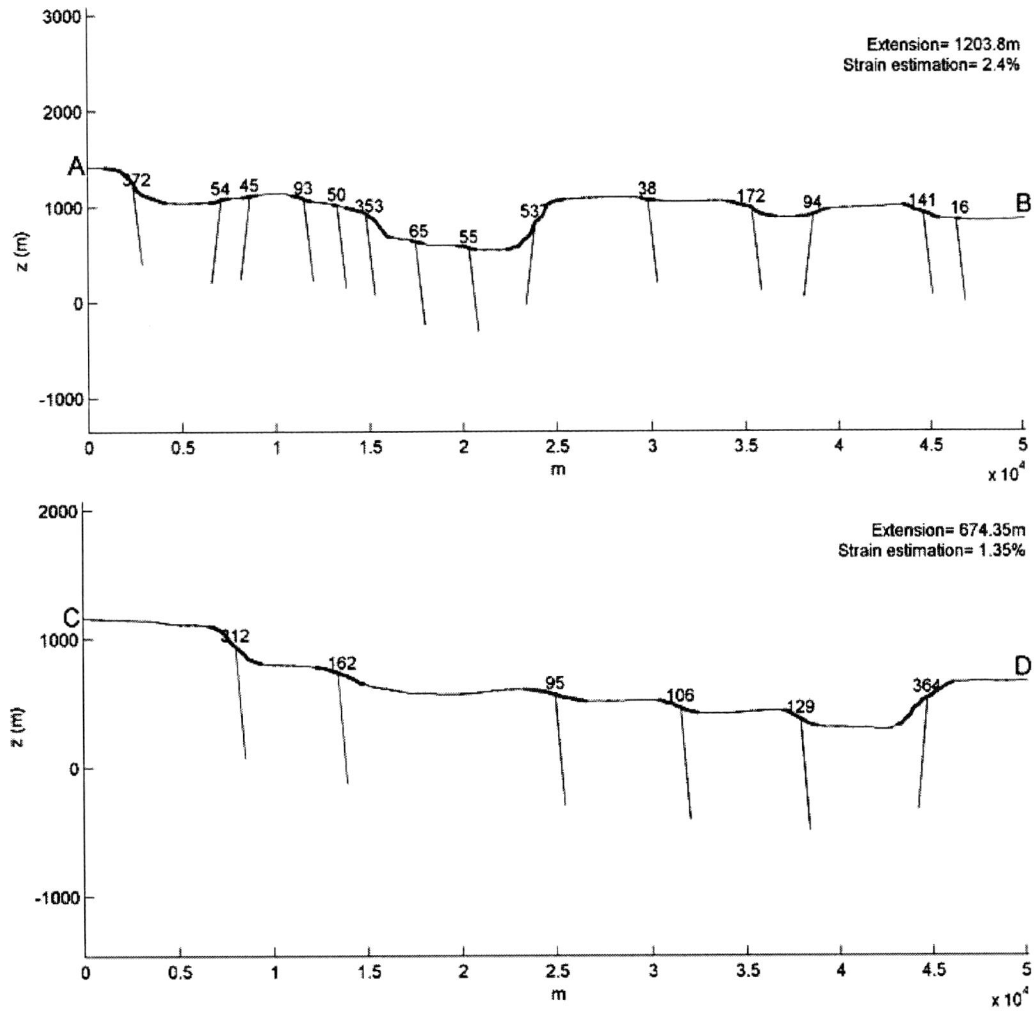

Figure 14: Example of structural sections that could be extracted automatically. In blue are the identified scarps that dip approximately South, and in red the scarps that dip North. The vertical throw of the scarps (in meters) is present above each one. For each scarp a line that marks the fault plane is drawn. [24-bit colour images are available on http://www.uc.pt/iguc/planets.htm]

The vectorized output can now easily be submitted to a statistical characterization. For example a simple azimuth rose diagram could be plotted. A by-product of all this processing is the capability of automatically constructing structural sections, intersecting the identified fault scarps with the altimetry profile, as in the example of the figure 14 (see Vaz *et al.*, (2006c) for details).

The vertical throw and the tectonic extension and strain for each fault can be automatically determined assuming a uniform fault dip of 60 degrees (Borraccini *et al.*, 2005). The computed throw should be considered as a minimum value because grabens' depressions might be covered by wind-blown sediments or volcanic deposits (Kronberg and Hauber, 1999).

A 3-D extension of the principle that is used on this 2-D sections could in the future be employed on the study of the geometry and strain of this kind of highly fractured terrains.

3.3.2. Automatic Detection from MOC Images

MOC imagery, previously corrected of the brightness variations, was used to build a mosaic of the surface of Mars with 256 pixels/degree resolution (more details on the web site http://www.msss.com). Figure 15 shows the original mosaic of the study area Tempe Fossae, used in this study.

The discontinuities of the surface (faults, lineaments) are high frequency phenomena. In remote sensing imagery they are characterized by different grey levels comparatively to other structures. The objective of this study is the automatic recognition of the principal discontinuities on the surface of Tempe Fossae, based on mathematical morphology.

The visual interpretation of the image of figure 15 allows the detection of the principal lineaments but this is an arduous work for minor structures. The principal cause of this difficulty is the poor image contrast; it can be enhanced by applying an alternating sequential filter (ASF) of opening-closing type (the sequence starts by an opening) on one dimension. Figure 16 shows the result of this operation.

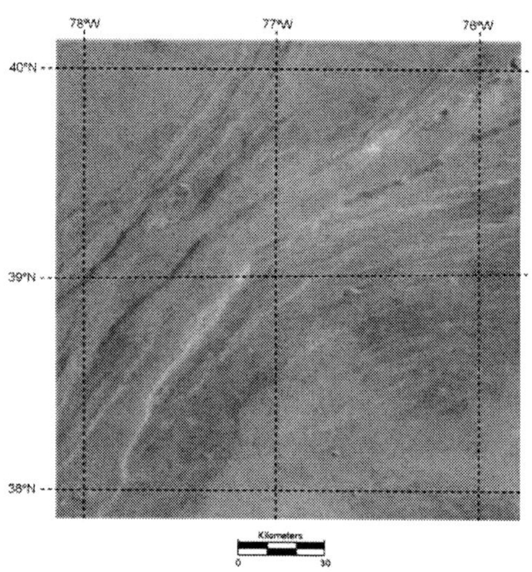

Figure 15: Mosaic obtained from MOC imagery: Tempe Fossae region.

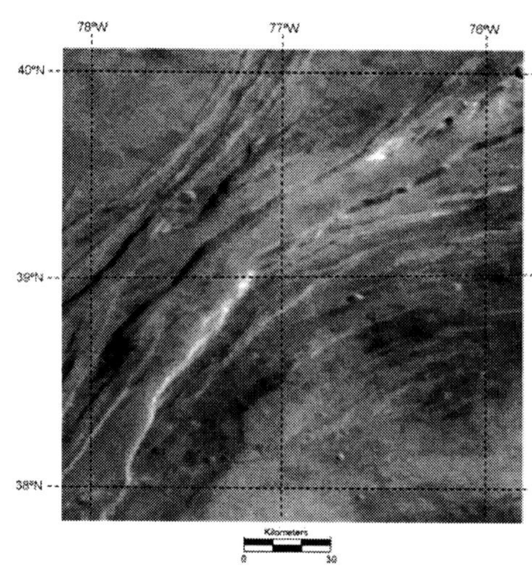

Figure 16: Alternating sequential filter of opening-closing type.

The distinction between high and low grey levels is now clearer and the discontinuities are more perceivable. The major lineaments present in the image have higher levels due to the solar incidence.

Trough a white top-hat transform the high values (peaks) of the image are extracted as is shown on figure 17. The great discontinuities of the Tempe Fossae region are now identified but the resulting image is very noisy. All the alignments have preferential directions, so they can be identified by applying a directional gradient. This transform was performed for all the principal directions of the hexagonal grid and the result for the direction 120°-330° is presented on figure 18. Although this is a visually more homogeneous image, noisy structures still remain. So, the image obtained by direction gradient was submitted to reconstruction by opening using a stucturing element of size two. The result of this transform is presented in figure 19. The structures smaller than the structuring element are removed by the erosion. Only the larger objects are preserved and detail is lost. Using a reconstruction by opening of size one the result is more effective, since the erosion has a milder effect (figure 20).

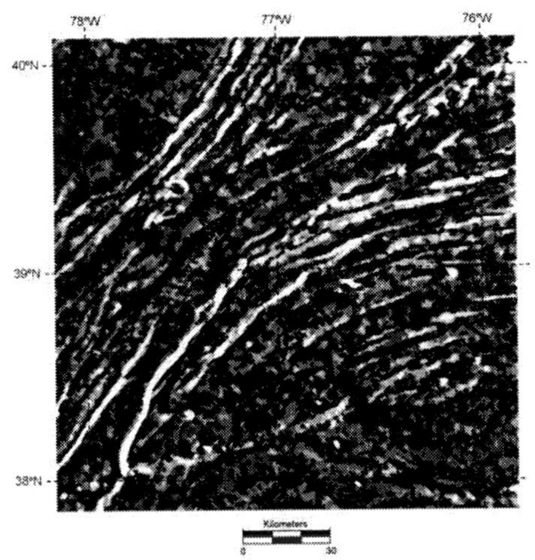

Figure 17: White top-hat applied to image of Figure (16).

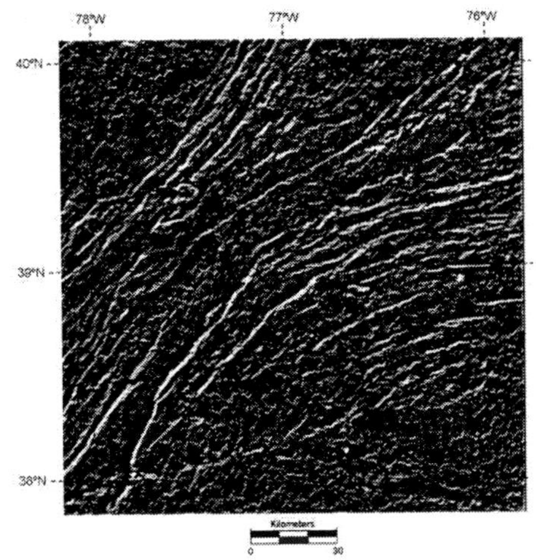

Figure 18: Directional gradient (120°-330°).

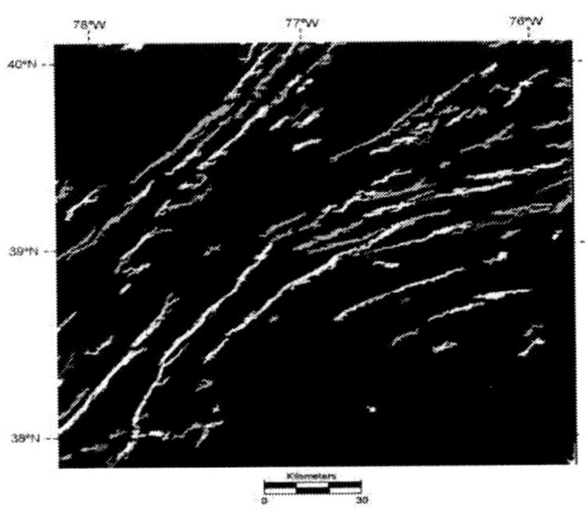

Figure 19: Reconstruction by opening with a structuring element of size two of the directional gradient (120° - 330°).

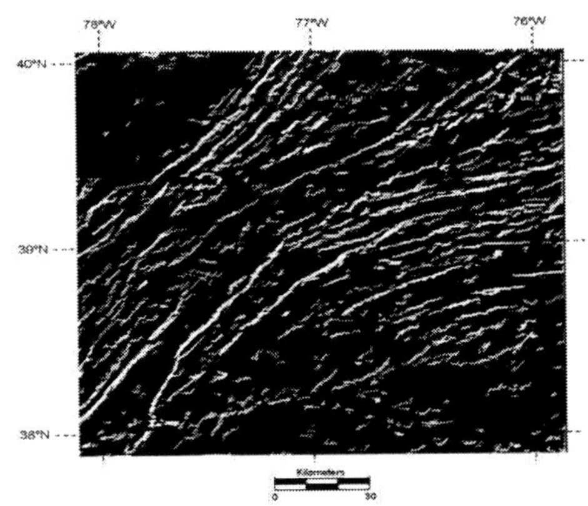

Figure 20: Reconstruction by opening with a structuring element of size one of the directional gradient (120°-330°).

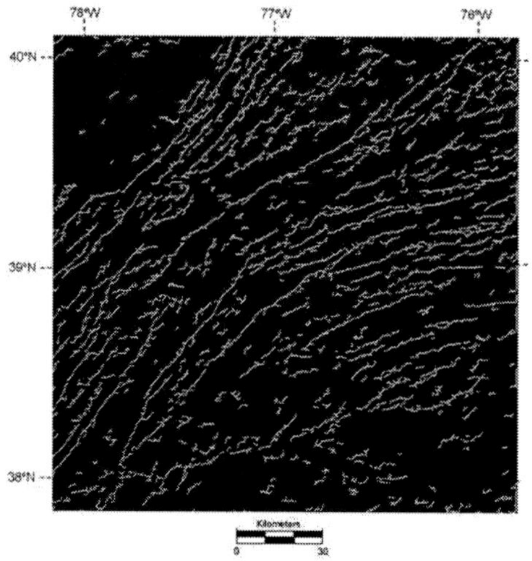

Figure 21: Skeleton by letter L of Golay'alphabet.

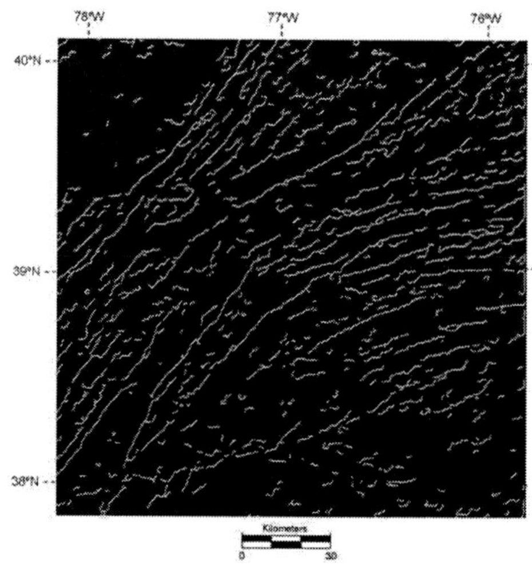

Figure 22: The pruning transform applied to skeleton image.

To reduce the dimension of the lineaments to one pixel dimension, it is enough to perform a skeleton transform by using all the possible neighbourhood configurations for the letter *L* of Golay'alphabet. This is shown on figure 21. As was mentioned above, the skeleton produces some undesirable branches and ramifications. This is easily eliminated by pruning, whose result is shown in figure 22. This algorithm was applied to all principal directions of the hexagonal grid: 0°-180°, 60°-240° and 120°-330° to define the principal directions of the lineaments.

The major problem of this algorithm is the identification of crater borders. They are also surface discontinuities and when we apply directional gradients in different directions they are also identified. To solve this problem, it will be necessary to further develop this algorithm by using isotropic sctruturing elements, such as a disk, to remove these structures. Another criterion to be incorporated is the size and the shape of the carter borders.

These results when compared with the results obtained for MOLA data are very similar. However there are minor structures not identified in MOLA data but they are detected in MOC images. On the other hand, the craters are very well removed in MOLA data contrarily to what happens on MOC images.

3.4. Interpretation of Tempe Terra Tectonics

Tempe Terra is a Noachian and Hesperian high plateau situated in the Northeastern flank of the Tharsis dome. Mareotis Fossae in the North, and Tempe Fossae in the South are zones that have undergone intense deformation that produced a complex array of tectonic structures. Figure 23 shows an image mosaic of Viking images, with MOLA DEM altimetry superimposed.

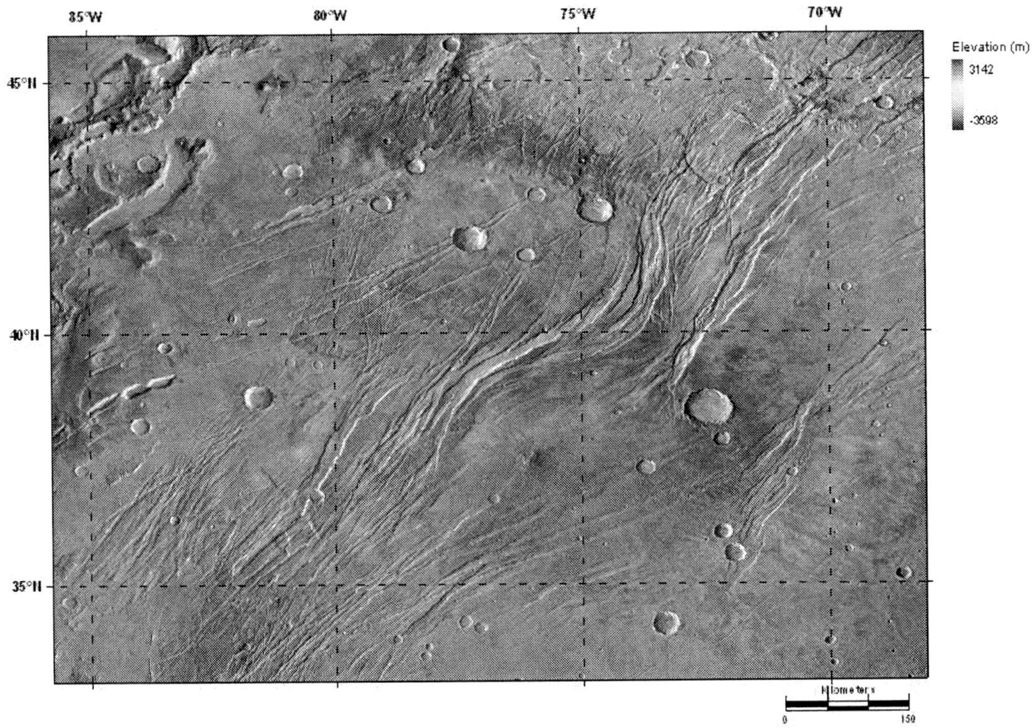

Figure 23: Tempe Terra Viking image mosaic with MOLA DEM altimetry. [24-bit colour images are available on http://www.uc.pt/iguc/planets.htm]

The main tectonic features present at Mareotis Fossae are extensive subparallel linear N65°E fractures interpreted as simple grabens. Individual grabens are 2 to 5 Km wide and in some cases hundreds of kilometres of length (Kronberg and Hauber 1999). A graben formation mechanism similar to volcanic tectonic models of terrestrial rifting on Earth has been proposed for the genesis of these "narrow grabens" (Mège *et al.*, 2003).

Tempe Fossae presents a very different kind of structures. Complex grabens, deeper and broader than those that prevail at Mareotis Fossae form a ~1500 Km length array of *en echelon* grabens and horsts. In agreement with other authors, we defend that the "narrow grabens" that are well exposed at Mareotis Fossae predate and are genetically different from those at Tempe Fossae, that have a different architecture and that intersect and obliterate the previous ones (Harrington *et al.*, 1999; Hauber and Kronberg, 1999; Hauber and Kronberg, 2001). Nevertheless we argue that those ancient N65°E structures have been reactivated and conditioned the formation of the following set of more recent N20-40°E fractures.

The elongated N20-40°E striking set of extensional features, the Tempe Fossae rift, has been interpreted before as a possible analogue to the Kenya Rift, based on a geomorphologic comparison (Hauber and Kronberg, 2001). We propose a different scheme for the tectonic evolution of this complex area. This scheme implies horizontal stress fields, the formation of shear bands, buckling and thick-skin tectonic on the Tempe area and has the advantage of not only explaining and integrating the fault patterns but also some regional topographic features.

As referred before, "narrow grabens" could represent the superficial expression of giant dikes (Mège and Masson, 1996; Ernst *et al.*, 2001). Those linear features when later subject to a new oblique regional horizontal stress field could have been reactivated as strike-slip faults, setting the limits for the development of Riedel shear zones (Atmaoui, 2006) (figure 24).

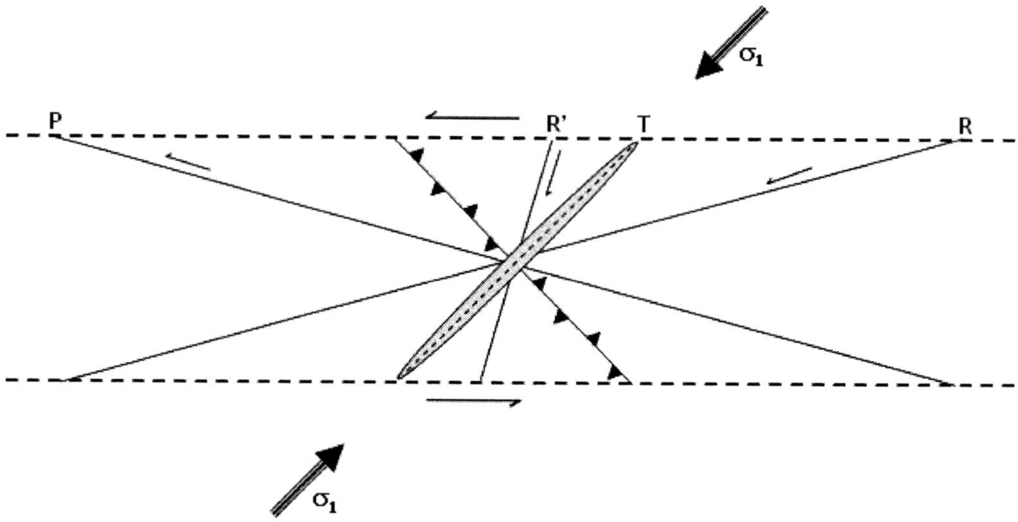

Figure 24: Idealized Riedel shear zone geometry.

The development of several shear zones, as predicted by the Riedel model, would lead to the formation of several tension cracks (T) that strike parallel to the maximum principal compressive stress (σ_1). Those fractures would originate the Tempe Rift grabens. Simultaneously, strike-slip conjugated Riedel (R and R') fractures would be developing and compressive structures striking perpendicular to the grabens would be formed. Another strike-slip set of fractures could also appear, making an angle of 15° with the main shear zone boundary. All these fractures would be forming and interacting at the same time and at several scales (Tchalenko, 1970; Katza *et al.,* 2004). This model is different from the "one strike one stress field" used by other authors to explain the tectonic features around Tharsis (Anderson *et al.*, 2001; Hauber and Kronberg, 2001) and has already be applied to explain fracturation patterns on Venus (Tuckwell and Ghail, 2003).

From the analysis of the lineament map that was obtained in Tempe Terra region (figure 25) it can be observed that many *en echelon* structures are present which appear to be controlled by the N65°E lineaments (this N65°E strike direction is only visible at some locations, in others it is concealed by the action of other fractures).

We interpret this as an *en echelon* pattern developed inside sinistral shear zones on a transtensional regime. Wilkins and Schultz (2001) have noted the similarity between this pattern and the patterns generated at the Earth on an oblique rift context (Dauteuil *et al.*, 2001; Schlische *et al.*, 2002; Clifton and Schlische, 2003).

The proposed shear zones, as can be seen in figure 26, should produce compressive structures that strike perpendicular to σ_1 and consequently to the grabens. Analysing the topography a great rise located north of the main rift, striking approximately NW-SE, is clearly evident (44°N, 76°W). This mount rises more than 1 Km above the surrounding terrains. In this case the rise is asymmetric and the transition to the lowlands at the SW is more abrupt than the NE scarp.

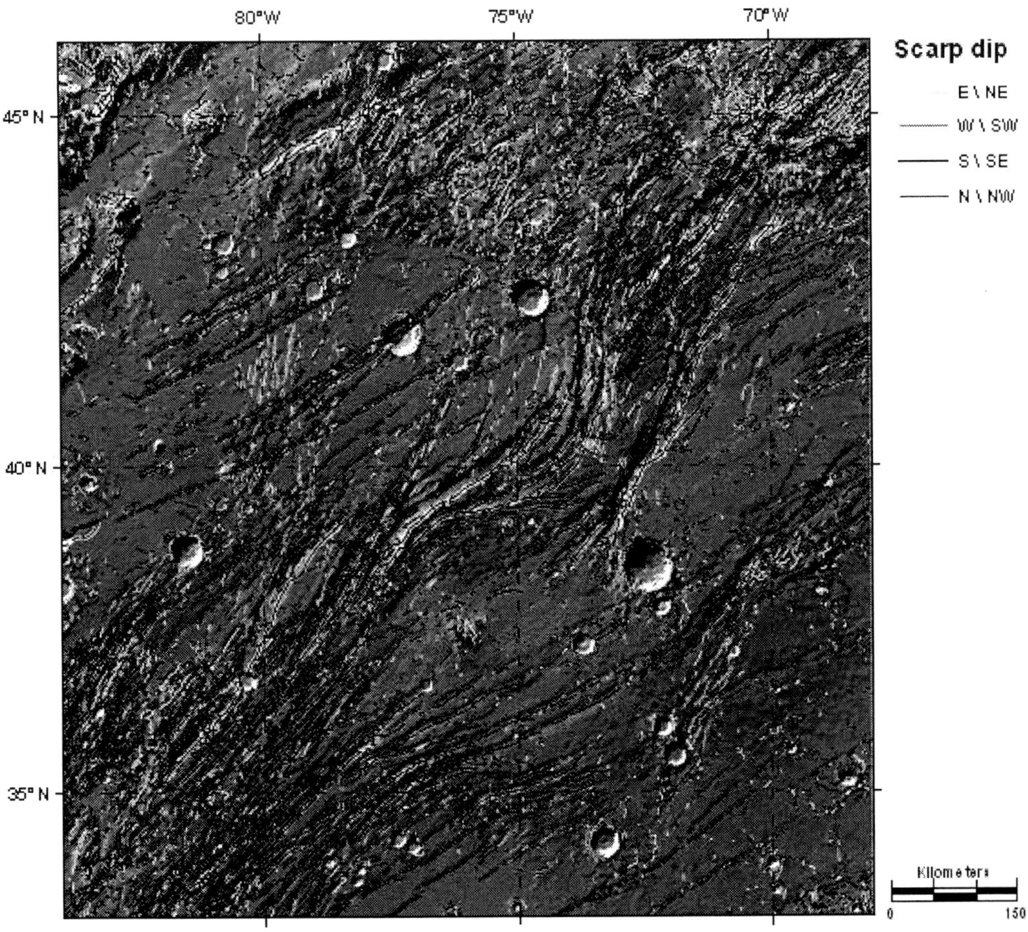

Figure 25: Lineament map obtained by the proposed methodology for Tempe Terra Region. The colours codify the dip direction of the identified scarps. [24-bit colour images are available on http://www.uc.pt/iguc/planets.htm]

The fractures in the NE side are well visible but they are less abundant on the SW side, maybe due to the sedimentary deposition and volcanic activity that is present more to the North. We interpret this 500 Km NW-SE striking linear mount as a compressive buckling similar to that proposed to model Coprates Rise on the Southern Tharsis Province (Anguita *et al.*, 2001). The asymmetric nature of the relief and the apparent obliteration of faults on the SW flank suggest a thrust-fault dipping to the NE. A careful analysis of the topography allows the detection of other linear bulges that are parallel to the giant rise described before (37°N, 81°W and 39°5', 80°W).

In the South of the main rift another gigantic linear bulge is visible (39°N, 73°W). In this case a flank symmetry is more evident which suggests the non-existence of a thrust fault, or at least a less active thrust fault. Again, the SW flank is associated with intense volcanism that masks the previous fracturation pattern. The grabens that cross the two high relief compressive structures are those that present the most curvilinear trend (Hauber and Kronberg, 2001). This characteristic could be related with a differential accumulation of compressive strain between the North and the South buckle structures.

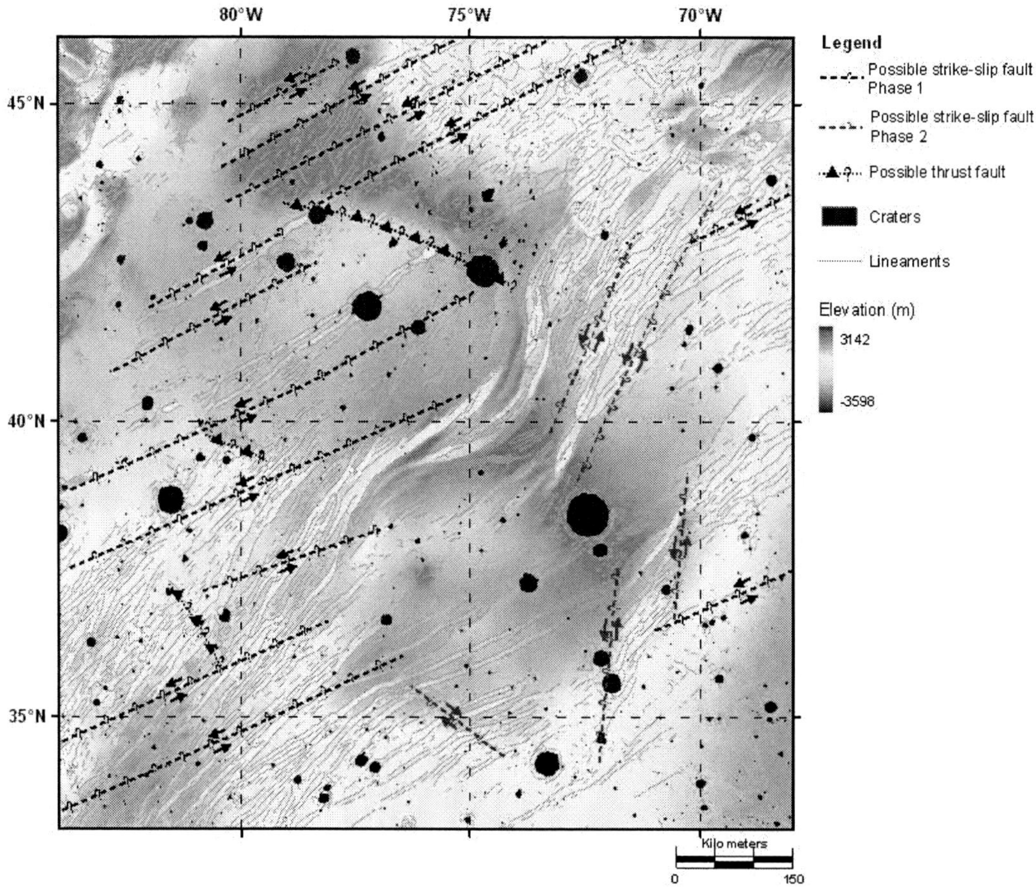

Figure 26: Tectonic interpretation for Tempe Terra. [24-bit colour images are available on http://www.uc.pt/iguc/planets.htm]

The described tectonic features are compatible with the action of shear zones striking N65°E originated by a NNE-SSW principal compressive stress. However another stress field (phase 2) is needed to explain the deflection of the previously formed structures (phase 1). This deflection is well visible in the Southeast part of the presented map, figure 26 (35°N, 75°N; 37°N, 72°W). This subsequent stress remobilised the previously formed structures. A sinistral movement on the North strike-slip faults and a right-lateral movement at the N60°W striking faults requires a NW-SE strike for the principal compressive stress of this second stress field. This second tectonic phase should have produced a tectonic inversion of the grabens that are perpendicular to the new σ_1 (Mège *et al.*, 2005).

Photogeological evidence for the proposed model must be sought. Volcanic activity in the area should help to constrain and date the two proposed tectonic phases since there are volcanic edifices that are clearly affected by the tectonic activity (45°N70°W), and others that are younger and cover the tectonic features (37°N, 76°W) (Hauber *et al.*, 2002).

The origin of the horizontal stresses that this scheme implies must also be sought. The lateral spreading of Tharsis could explain the thick-skinned thrust faults/folds that we identify at Tempe (Okubo and Schultz 2003). The phase 1 stress that strikes NNE-SSW, and the several shear zones that we propose to explain the tectonic structures and morphology of this area, are also consistent with the transform fault deduced from the magnetic data at Terra Sirenum by Connerney *et al.* (2005). This possible transform fault that disrupts the magnetic anomalies at Terra Sirenum passes under the three Tharsis Montes and across Tempe Terra, where the displacement should be left-lateral as deduced from our kinematic analysis.

4. Mars Climate: Past and Present

4.1. Energy in the Climatic System: Solar Radiation

Beyond the internal processes to the climate (the existence of oceans, the planetary circulation of the atmosphere, topography) that lead to a variability of climatic states there are external factors (solar radiation, volcanic activity) that force and shape climatic behaviour. It is enough to notice that the solar radiation conditions, immediately, the variations diurnal and seasonal of the climate. The solar radiation is the most important external conditioning factor. It is, in turn, conditioned by the solar constant, the eccentricity of the Earth's orbit, the obliquity of the ecliptic and the longitude of the perihelion.

The variations of these astronomical factors (Milankovitch parameters) are related with long-term variations of the planetary climates.

Table 1 – The orbital elements of Mars and the Earth and their variability

Parameter	Present Mars	Martian variability Range, Cycle (years)	Present Earth	Terrestrial variability Range, Cycle (years)
Obliquity (°)	25.2	14-48 (120,000)	23.5	22-24 (41,000)
Eccentricity	0.093	0-0.12 (120,000)	0.017	0-0.06 (100,000)
Longitude of perihelion (°)	250	0-360 (51,000)	285	0-360 (21,000)

Of all the planets of the solar system, Mars is the one which shows a more similar climate to the Earth.

Over the past 10 My, the obliquity of the Earth has ranged from 22° to 24,5°, while for Mars the range was 14° to 48°(Touma and Wisdom,1993). Today, the obliquity of Mars is 25.2°, similar to the Earth's, making daily and seasonal changes identical, but varied a lot during last the 10 My. In a similar way the eccentricity of the Earth's orbit varied from 0 to 0.06, wile for Mars the range was 0 to 0.12.

The consequent changes to insolation and seasonality at middle to high latitudes on Mars have inevitably caused significant changes in the seasonal cycles of carbon dioxide, water and dust. Thus Mars must have experienced the most significant quasi-periodic variations in its climate over the past 10 My of any planet in the Solar System.

Analysis of data from spacecraft missions before Mars Global Surveyor identified laminations and layers in both polar caps whose origin was attributed to quasi-periodic climate change. Data from the most recent Mars missions show multiple lines of evidence for surface deposits that formed as a result of recent quasi-periodic climate variability on Mars.

The observations span a large range of scales (metres to kilometres to hundreds of kilometres), are of diverse nature (morphology, topography and chemistry), and are remarkably consistent with models of current and past ground-ice stability. These results all point to the presence of a succession of metres-thick, latitude-dependent surface deposits that are young, ice-rich when formed, and whose deposition and removal is driven by climate change that is induced by spin-axis tilting.

The most likely cause to explain the deposition and removal of the mixtures of dust and water ice is climatic variation resulting from quasi-periodic variations in the orbital parameters (Head III *et al.*, 2003).

During periods of higher obliquity (>30°), the increased insolation in the summer, in the polar regions, results in the release of more water vapour of the polar caps, increasing the atmospheric water vapour content and thus the humidity (Jakosky *et al.*, 1995). Higher humidity moves the latitude of subsurface ice stability boundary closer to the equator and, within the soil profile, closer to the surface.

Climate models predict that ice removed from the polar caps is deposited at lower latitudes where it persists throughout the year (Richardson and Wilson, 2002). The Mars GCM predicts that the existing strong winds raise dust and redistribute it (Haberle *et al.*, 2003). This dust is thus incorporated in the thickening ice deposits. These factors, increased atmospheric water vapour content, increased atmospheric dust content, and equatorwards displacement of the surface ice zone are predicted to act in concert at high obliquity to produce the latitude-dependent deposits that we observe. Owing to the higher year-average and summer-average polar temperatures and denser atmosphere, the transport of water at high obliquities is predicted to be rapid. The periods most favourable to these processes are of the order of 40 Kyr, long enough to provide deposition of at least several metres of ice-rich material (Mischna *et al.*, 2003).

The model explains that the deposited layers should be geologically recent, latitude-dependent, dust–ice mixtures, superposed on underlying terrain, and varying in their character as a function of latitude.

When the obliquity decreases, the surface ice stability zones shrink toward the polar caps and ice sublimes from the uppermost layers, creating a protective ice-poor crust analogous to the sublimation of ice from loess on Earth. With a lower obliquity the atmospheric humidity also decreases. A dryer atmosphere induces poleward shrinking of the water-ice-rich cover and widespread desiccation between 30° and 60°. For latitudes >60° the ice is stable at the surface with obliquity of 20° and thus over the past 10 Myr dissection would be most important in the mid-latitudes (30°–60°). The periods of low obliquity would have been too short to desiccate and remove the ice mantle completely at all latitudes.

Evidence for the present desiccation and removal of the ice mantle layers includes the dissected terrains in the 30°–60° North and South latitude zones, the decrease of sub-kilometre-scale topographic concavity in this same latitude range, and the correspondence of this zone with the latitudes predicted to be at present desiccated in the near surface. In Antarctic dry valleys, a cold polar desert that represents a terrestrial analogue for regions with buried ice on Mars, such landforms are related to sublimation and thermal cycling of ice-rich sediments (Marchant *et al.*,2002).

Many of the surface textures of the ice mantle are consistent with morphologies formed over several glacial cycles in the Mars Arctic and Antarctic periglacial environments, suggesting that changing environmental conditions over extended time periods are required.

Despite the characteristics of the ice mantle suggesting that the deposits are geologically very young, it is composed of numerous sublayers and with characteristics that require multiple cycles of deposition.

As was said above, the obliquity, the eccentricity and the aerocentric longitude of the Sun at perihelion all affect water-ice migration on Mars, but with varying timescales and degree. 300 Kyr ago, the high eccentricity made that differences in seasonality between the hemispheres (the southern hemisphere today endures shorter, warmer summers and longer winters than the northern hemisphere), while retarding longitude of perihelion causes the seasonality difference to reverse every 25 Kyr.

When obliquity exceeds 30°, water ice is stable in the near surface down to the lower mid-latitudes, and climate models predict that water ice is efficiently moved from polar cap reservoirs by sublimation and deposited in the mid-latitudes (Mischna *et al.*,2003). When obliquity changes to values <30°, ice is sublimed from the mid-latitude deposits, creating a surface dust lag analogous to that seen in Earth environments.

Five My ago the obliquity exceeded 45° and the average obliquity during this time was >35°, while 4 My ago the average was 26° (Laskar, J. *et al.*, 2002). With obliquity 45° the polar sublimation rate would be very high and ice is predicted to persist at the equator. One concludes that the emplacement of the ice- and dust-rich mantle extending from polar regions down to low mid-latitudes represents the equivalent of an ice age on Mars. Poleward migration of the equatorward limit of the ice-dust-rich mantle, partial desiccation of the mantle in the latitudinal belt between 30° and 60°, and modification via sublimation and thermal cycling of long-lived ice-rich deposits poleward of 60°, represent some of the changes that occur during interglacial times.

Martian ice ages differ from the Earth's: the oceans are important heat reservoirs and distribution systems, and spendthrifts for the dust. The major reservoir response during climate cycles on Earth is the amount of water stored in the polar caps. The major atmospheric gas components are not thought to change significantly. Orbital forcing factors thought to be the pacemaker for the timing of Earth's recent ice ages are much less extreme than on Mars.

The ice ages on Earth have been characterized by colder temperatures in the poles on average, while on Mars the opposite happened. In cold polar deserts on Earth, evidence has been cited for the preservation of glacial ice > 8 Myr old at depths just below the surface. An extremely cold polar desert environment characterizes Mars, and the record of recent climate change on Mars should be similarly preserved in the shallow subsurface over wide latitude extensions (Head III *et al.*, 2003).

4.2. Polar Wander and the Stability of Liquid Water

The magnetic field data of Mars, namely the location of the magnetic paleopoles (Figure 1), give us important clues about the re-orientations, during the geological time, of the martian rotation axis, once it coincides with the principal magnetic field axis. Namely, it supports a polar wander hypothesis, which significantly constrains global climate changes during the geological history of Mars.

The alteration of the moment of inertia of a planet is the result of mass redistribution and can occur due to mantle convection, magmatic processes or giant impact events (Wieczorek *et al.*, 2005). If the angular momentum is conserved, and if there is any mechanism to dissipate the internal energy, then the planet tends to re-orient itself, during the geological timescale, until the maximum principal moment of inertia coincides with the spin axis (Goldreich and Toomre, 1969). On Earth, this re-orientation is called Inertial Interchange True Polar Wander (Evans, 2003) and it may also have occurred on Mars. For the Earth, there are different methods to determine magnetic paleopole positions. When a silicate magma is cooling in the presence of a magnetic field there is a preferential alignment of magnetic minerals. The dipole spin axis may have been coincident with the spin axis of the planet rotation when the rock cooled through its Curie temperature. For Mars, using magnetic field orbital data and assuming that an isolated magnetic anomaly formed on a cooling region of the crust when a dipolar internal field existed, the crust magnetization direction can be used to infer the magnetic pole location, at a time that the anomaly has formed (e.g. Arkani-Hamed *et al*, 2004).

A global magnetic field aligned close to the martian equator shows that the obliquity of the martian rotation axis changed during the martian geologic time, putting in evidence important global climate changes since the formation of this planet, where a warmer and wetter climate has alternated with a dryer and colder one.

Moreover, Mars presents several evidence of its wet past (e. g. Jakosky and Philips, 2001), exhibiting traces of liquid water visible at its surface and undoubtedly has great quantities of water stored in its subsurface.

On Earth, the water is preserved in great quantities at the surface. The presence of liquid water on the surface is important for planetary geodynamics because it allows the hydration of the crust which makes possible the conduction of water to the mantle during crust recycling (e.g. Iwamori, 2004). This phenomenon reduces the viscosity of the mantle that facilitates the setting up and the perpetuation of a regime of plate tectonics. Unlike the Earth, Mars is today in a stangnant lid regime.

The presence of a planetary core dynamo, as is observed on Earth and deduced for the Mars geologic past, requires an important heat flow out of the core. It was shown that an active regime of plate tectonics can make possible this constancy of high flow fluxes, whereas a mode of stagnant lid is less likely to reach it (Nimmo and Stevenson, 2000). This would tend to indicate the importance of the water on the determination of the convection regime.

Also, Mars has several enormous volcanic edifices. This suggests episodes of cataclysmic activity implying the release of large quantities of volatiles which would have had a considerable influence on the composition of the atmosphere.

The evolution of the planetary atmospheres, the presence of liquid water on the surface, and the maintenance of lenient atmospheric conditions are clearly essential for the generation, the survival and evolution of forms of life.

4.3. Liquid Water on Mars Today

For some time now, there is consensus that seepage of fluids occurred through Martian layers, with formation of channels and gullies that are much similar to those on Earth (Malin and Edgett, 2000). However, until recently, the time frame of these events was undefined.

This issue was first addressed by Alves *et al.* (2003) by comparing images that were acquired in two different missions: Mars Global Surveyor (MGS) and Mars Odyssey (MO). A comparison of image V01138003, obtained by the THEMIS camera onboard the MO spacecraft on March 18, 2002 and image MOC2-240, obtained by the Mars Orbiter Camera onboard the MGS spacecraft on July 12, 1999, shows a new gully on the northern scarp of Nirgal Vallis, near 29.7°S, 321.4°E, among previously existing gullies, with discernible formation of an incipient fan-shaped apron (figure 27). This showed that fluid outflow is happening in the present.

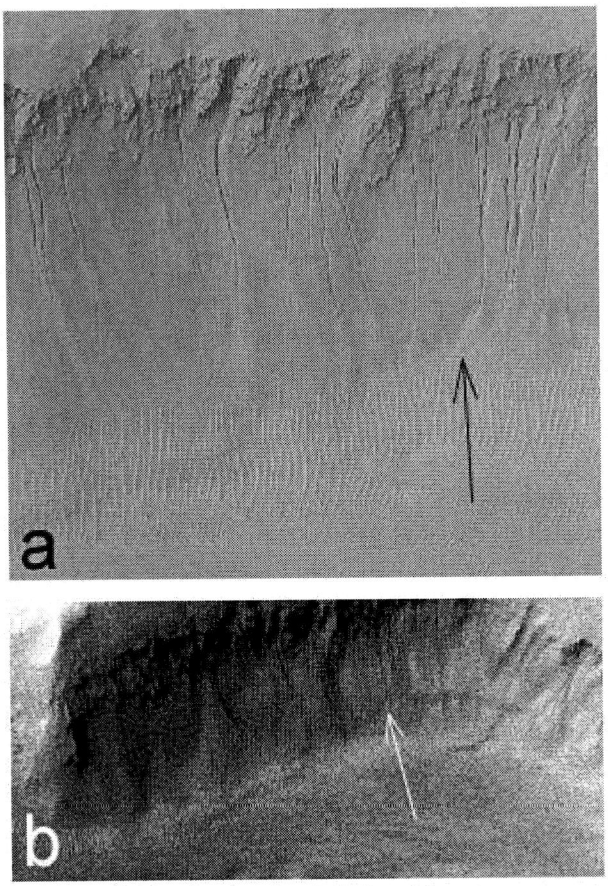

Figure 27: Comparison of images MOC2-240 (a – 7.12.1999) and V01138003 (b – 3.18.2002). In the space marked by the black arrow on *a* a new gully is visible on *b* (white arrow).

The fact that the THEMIS image was of medium-resolution (18 m/pixel) did not allow seeing many details on the new gully.

This time frame was confirmed by Malin *et al.* (2006) who used high-resolution MOC images and identified new deposits, formed between 1999 and 2006, which "are light toned and exhibit attributes expected from emplacement aided by a fluid with the properties of liquid water" on 20 locations.

5. Conditions for the Existence of Life

Since the work of Friedmann *et al.* (1993) on the possibility of terraforming Mars, several research groups have tried to find suitable candidate organisms on Earth which would first survive, then increase the atmospheric pressure and O_2 partial pressure and initiate the biochemical carbon cycle. Of course, if there are today or there have been organisms in the past on Mars, only the first condition must be met: to survive.

Thomas *et al.* (2006) list the characteristics that Earth organisms must have to survive on Mars. These include autotrophy, psychrophily, CO_2 tolerance, hypoxia tolerance, osmotic tolerance and dessication resistance, ultraviolet and ionizing radiation resistance and hypobaric tolerance. This last factor is the one that raises most difficulties, as studies indicate that 25 mbar is a lower limit for the development of bacteria on Earth (Schuerger *et al.*, 2006). Nevertheless, microorganisms collected in stratospheric regions where atmospheric pressure-temperature conditions are not unlike those of Mars ($P \approx -10$ °C; $T \approx 0.3$ kPa) have been successfully cultured (Wainwright *et al.*, 2003).

Thomas *et al.* (2006) experimentally tested the survavability of several Earth extremophiles under Martian conditions. The simulated conditions included daily thermal cycles between 26°C and -80°C and a similar atmospheric composition, albeit at 100 mb pressure. All tested bacteria survived 14 day trials.

As was laid out in Alves and Madeira (2003) the most restrictive conditions for the existence of life, anywhere, are (1) the presence of liquid water, (2) the availability of energy sources and (3) protection from radiation.

5.1. The Presence of Liquid Water

Liquid water is mandatory for the existence of life as we know it. If liquid water was absent, we could forget about the existence of life on Mars in the present. Other important condition is a suitable range of temperature and temperature stability. The proposed range is based on the knowledge of Earth life, i. e. from 0 to ca. 100 ˚C.

Whatever the temperature, water must be liquid. Obviously, the pressure range is very important since it conditions the state of water together with temperature. Therefore, the search for Martian life requires a very complete information on the space- and time-distribution of atmospheric temperature and pressure. The crustal surficial thermal and baric gradients must also be more fully understood and modelled since the most probable niches are soil pores.

Another obvious way to have liquid water available on, or near, the surface of Mars is by the presence of solutes which would change the phase conditions. Strong candidates are the SO_4^{2-}, Cl^-, Ca^{2+} and Na^+ ions. Figure 28 shows that liquid salt water has a broad stability range on the surface of Mars, which is consistent with present P-T conditions.

A high concentration of salts in Mars groundwater would also explain the recent formation of the light-toned deposits which were identified by Malin *et al.* (2006).

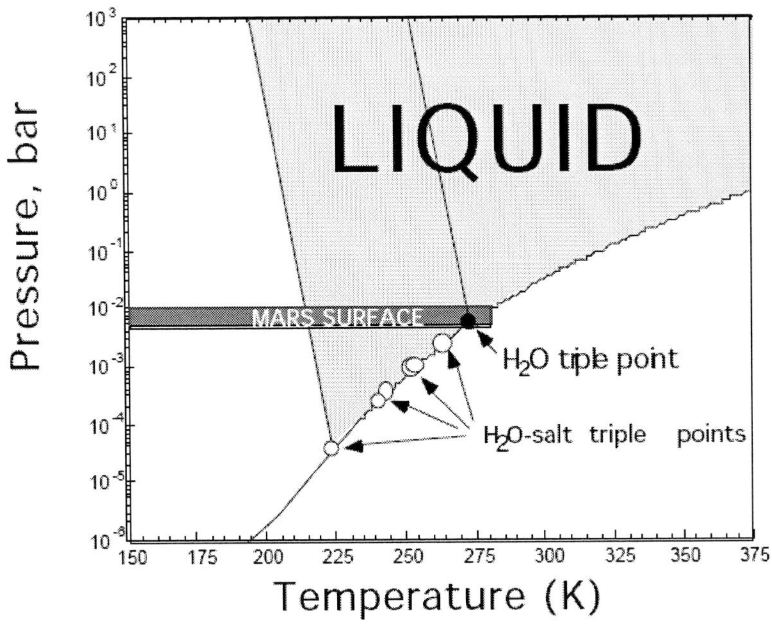

Figure 28: P-T diagram showing the stability field for liquid brines with solute NaCl (Bodnar, 2001).

5.2. The Availability of Energy

Another requirement for life is the availability of energy. Visible light cannot operate as an energy source in buried domains as it operates photosynthesis on Earth. Nevertheless, there are photosynthetic centers energized by the near infrared of ca. 1000 nm that are collected by special bacteriochlorophyls. Suitable modifications of the porphyrin may result in an effective collection of far infrared photons extending to 1000 cm^{-1} (10000 nm) (Allen and Williams, 1998). This warm radiation may penetrate deeper and reach microdomains where living communities might settle.

Although visible light is the major form of energy that supports life on Earth, "energetic" inorganics also support restricted living communities in the vicinity of marine hydrothermal vents (black smokers). The most important chemical is H_2S, efficiently oxidized by chemolithotrophs to yield useful energy for life and biomass accumulation. The chained organisms in the community are subsidiaries of the lithotrophic primary producers. Other communities living at considerable depths (found in mines) depend energetically on H_2S biochemistry.

Less expressive lithotrophism depends on the use of H_2 in methanogenic organisms which generate methane by reduction of CO_2. Recent data acquired by the Planetary Fourier Spectrometer (PFS – Mars Express mission) on the atmospheric abundance of CH_4 on Mars (Formisano, 2004) could be explained by the action of such organisms. Moreover, the same author found a significant overlap of water-rich and methane-rich areas on the surface. Still other microorganisms process lithotrophic biochemistry through the oxidation of pyrites (lixiviation to ferrous or ferric sulphate), but have little expression in terms of life support on Earth (Madigan *et al.*, 1999). Nevertheless, these can not be excluded as possible forms of life on Mars, especially since the PFS instrument also detected HS^- traces on the same areas where methane was detected.

5.3. Shelter from Space Weather

If conditions for the availability of liquid water and energy are found on Mars, one of the main foreseeable hazards to the possible life-forms are cosmic rays and solar wind radiations.

These radiations are extremely harmful to most known forms of life. A known exception is *Deinococcus radiodurans* which can be found striving in the coolant fluid of nuclear reactors. The genetic repair mechanisms that allow it to survive seem to be also responsible for its exceptional resistance to dessication (Mattimore and Battista, 1996). Biomolecules, particularly polymers and macromolecules, are rapidly destroyed by energetic radiations. Generation of free radicals is very likely to occur. In most cases, radicals undergo chain reactions turning into cyclic autocatalysis and autoamplification. This is a major issue of the effects of radicals towards biomolecules and living structures (Shea and Smart, 2001).

Although the presence of regional magnetic fields is putatively required to allow life activity, it is not absolutely mandatory. The magnetic field(s) may be present in the organisms themselves, similarly to what is found in magnetotactic bacteria (Blakemore, 1975) which contain inclusions, called magnetosomes, arranged in rows along the cell to generate magnetic fields at the microscopic scale. On Earth, magnetotactic bacteria are very scarce. It is our planetary permanent dipolar magnetic field that protects Earth's surface, and life as a whole, from radiations of cosmic and solar origin.

Besides knowing if there are radiologic conditions for sustainement of indigenous life on Mars, radiation incidence data are crucial if any projects are to be made for the human exploration of the planet. Landing on a naturally sheltered area may mean that radiation shielding requiements are less stringent, hence reducing mission weight and cost.

One question that arises naturally is whether the crustal magnetic fields on Mars can provide some form of protection from space radiation. If that is the case, some relation should be found between a map of magnetic anomalies and a map of the distribution of space radiation incidence. However, generation of such a map is not trivial.

Completing a full circle, we are back where we began: planetary magnetism.

5.3.1. The MARIE Experiment

The two main sources of space radiation are galactic cosmic rays (GCR) and solar energetic particle (SEP) events. Concerning GCR, the lower the solar activity, the higher their energy. On the other hand, SEP are directly correlated with solar activity.

The continuous energy flux of GCR is responsible for long term effects in living tissue, namely cancer. SEP are highly energetic particles which can lead, in the limit, to death in a short time. The MARIE instruments were designed to measure those radiations. Its detectors were sensitive to cosmic radiations between 15 and 500 MeV per nucleon (Badhwar, 2004).

The MARIE radiation spectroscope was built as a telescope. Its basic geometry was defined by two 25.4 x 25.4 x 1 mm thick ion-implanted silicon solid-state detectors A1 and A2. In between A1 and A2 there were two 24 x 24 mm position sensitive detectors PSD1 and PSD2, each with a 24 x 24 mm wire grid, to define the incident direction of charged particle with respect to the axis of the telescope. The A2 detector was followed by a set of 5 mm thick lithium-drifted silicon solid-state detectors (B1, B2, B3, and B4), and a high refractive index Schot-glass Cherenkov (C) detector (Badhwar, 2004). The PSD and Cerenkov detectors never worked nominally.

We intended to map the total energy that was detected by six of the MARIE sensors during the Mars Odyssey mapping phase, which began on March 13, 2002, until instrument failure at October 27, 2003.

5.3.2. MARIE Data Processing

Several difficulties arise when trying to use MARIE data, which were obtained from http://pds-geosciences.wustl.edu/missions/odyssey/marie.html.

Some of the data records have been marked "bad" by the experiment team; these records were obviously not used by us.

Then, since GCR are assumed to be almost isotropic and SEP originate in the Sun, MARIE was not designed to spatially locate its measurements. So, the acquired data (particle counts) have only a time stamp and are not geo-referenced.

There were consistency problems between instrument and ship clocks. Each data sample thus contains two time tags: corrected and uncorrected. Due to uncertainty arising in the data time tag – leading to uncertainty in location – only those data that had less than 0.01% difference between uncorrected and corrected time tags were used.

Our next task was to geo-reference the filtered MARIE calibrated event data. This was accomplished by using NASA's Navigation and Ancillary Information Facility (NAIF) SPICE toolkit routines, available on http://naif.jpl.nasa.gov/naif/toolkit.html.

The sums of the energies recorded by all six MARIE A and B detectors over 10°x10° bins were then mapped in geographic co-ordinates at the ship's nadir (figure 29).

5.3.3. Radiation Incidence on Mars

A comparison of the radiation incidence map with the most recent maps of magnetic anomaly (figure 1) suggests that both processes are not independent. The dominant direction of ridges and valleys on the radiation surface is N-S, orthogonal to that of the magnetic anomalies (E-W) as would be expected. Also, the radiation incidence is less where there are the strongest magnetic anomalies. The planetary dichotomy is clearly seen here.

It must be noted that this radiation map is not plotted onto the surface of Mars. The map is represented at the flight altitude of the 2001 Mars Odyssey spacecraft. Future work shall take the filtering effect of the atmosphere into account and estimate radiation doses to be expected on the surface.

The fact that the strongest magnetic anomalies are found in the Terra Sirenum region indicates that, apart from being one of the geologically most interesting areas on Mars, for radiation safety reasons this might be a suitable landing site for the first human explorers of the planet.

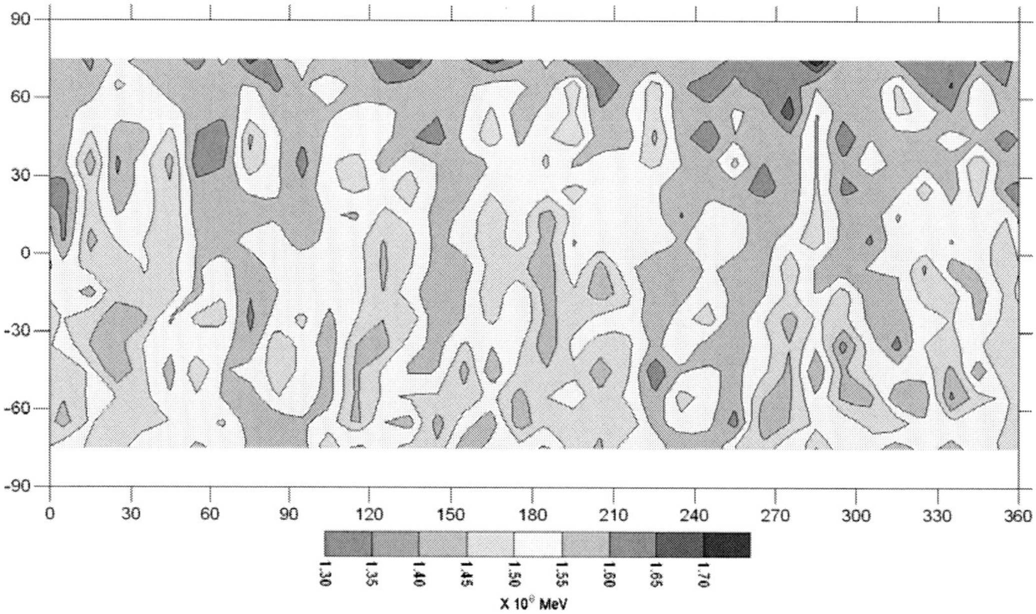

Figure 29: Radiation incidence on Mars. [24-bit colour images are available on http://www.uc.pt/iguc/planets.htm]

6. Conclusion

Planetary history – Geohistory, on Earth – is a succession of events which are correlated in time and space.

We have tried to show here that plate tectonics, the most integrated theory of Earth, is also the theory that can best explain some recent, otherwise puzzling, observations of Mars such as:

1. Parallel, successively reversed, magnetic anomalies;
2. Well developed shear zones;
3. Liquid water-carved gullies.

If climatic, atmospheric and geochemical conditions were, at some time in the history of Mars, adequate for the onset of life, this can be largely attributed to plate tectonics.

Acknowledgements

This work was financed by project *MAGIC* with EC (FEDER) and Portuguese (FCT) funds.

REFERENCES

Abramov O.; McEwen A. (2004). An evaluation of interpolation methods for Mars Orbiter Laser Altimeter (MOLA) data. *Int. Jour. Remote Sensing* 25, 3: 669 - 676.

Acuña M. H.; Connerney J. E. P.; Wasilewski, P. J.; Ness, N. F.; Reme H.; Mazelle C.; Vignes D.; Lin, R. P.; Mitchell D. L.; Cloutier P. A. (1999). Global distribution of crustal magnetism discovered by the Mars Global Surveyor MAG-ER Experiment. *Science* 284, 790-793.

Allen J. P.; Williams J. C. (1998). Photosynthetic reaction centers. *FEBS Letters* 438, 5-9.

Alves E. I.; Madeira V. M. C. (2003). Rationale for the deployment of a magnetic gradiometer on Mars. In C. A. P. Martins, ed., *Proc. Sixth Int. Conf. Mars*, 3014.

Alves E. I.; Baptista A. R. (2004). Rock Magnetic Fields Shield the Surface of Mars from Harmful Radiation. *XXXV LPSC,* 1689.

Alves E.I; Baptista A. R.; Conde L.; Azevedo J. M. (2003). Identification of present fluid seepage in Nirgal Vallis, by comparing Mars Global Surveyor and Mars Odyssey images. *Geophys. Res. Abs.* 5, 12651.

Anderson, R. C., Dohm, J. M., Golombek, M. P., Haldemann, A. F. C., Franklin, B. J., Tanaka, K. L., Lias, J.; Peer, B. (2001). Primary centers and secondary concentrations of tectonic activity through time in the western hemisphere of Mars, *J. Geophys. Res. 106*(E9), 20,563–20,586.

Anguita F.; Farelo A.; López V.; Márquez A.; Mas C.; Muñoz M.J.; Ruiz J. (2001). The Tharsis dome, Mars: new evidence for Noachian-Hesperian thick-skin and Amazonian thin-skin tectonics. *J. Geophys. Res.* 106-E4, 7577-7590.

Arkani-Hamed; Boutin C. (2004). Correction to "Paleomagnetic poles of Mars: Revisited". *J. Geophys. Res.* 109, E03001.

Atmaoui, N.; Kukowski, N.; Stöckhert, B.; König D. (2006). Initiation and development of pull-apart basins with Riedel shear mechanism: insights from scaled clay experiments. *Int. Jour. Earth Sci.* 95, 2, 1437-3254.

Badhwar, G. D. (2004). Martian Radiation Environment Experiment (MARIE). *Space Sci. Rev.*110, 1-2, 131-142.

Beucher S. (1990). Segmentation d'images et morphologie mathématique, *Thèse de Doctorat*, ENSMP, Paris.

Blakemore, R. (1975). Magnetotactic bacteria. *Science*, 190, 377-379.

Bockheim, J. G.; Hall K. J. (2202). Permafrost, active-layer dynamics and periglacial environments of continental Antarctica. *S. Afr. J. Sci.* 82-90.

Bodnar, R. J. (2001). PTX phase equilibria in the H2O-CO2-salt system at Mars near-surface conditions. *XXXII LPSC*, 1689.

Borraccini, F.; Lanci, L.; Wezel, F. C.; Baioni D. (2005). Crustal extension in the Ceraunius Fossae, northern Tharsis region, Mars. *J.. Geophys. Res.* 110, EO6006(doi: 10.1029/2004JE002373).

Bue, B. D.; Stepinski, T. F. (2005). Automated classification of landforms on Mars. *Comp. & Geosci.* 32(5), 604-614.

Bue, B. D.; Stepinski, T. F. (2007). Machine Detection of Martian Impact Craters From Digital Topography Data. *IEEE Trans. Geosci. Remote Sensing*, 45 (1), 265-274

Clifton, A. E.; Schlische, R. W. (2003). Fracture populations on the Reykjanes Peninsula, Iceland: comparison with experimental clay models of oblique rifting. *J. Geophys. Res.* 108 (B2), 2074.

Connerney, J. E. P.; Acuña, M. H.; Ness, N. F.; Kletetschka, G.; Mitchel, D. L.; Lin, R. P.; Reme, H. (2005). Tectonic implications of Mars crustal magnetism. *Proc. Nat. Acad. Sci.*102, 42, 14970.

Dauteuil, O.; Huchon, P.; Quemeneur, F.; Souriot, T. (2001). Propagation of an oblique spreading centre: the western Gulf of Aden. T*ectnophysics* 332, 423-442.

Ernst, R.; Grosfils, E.; Mège, D. (2001). Giant dike swarms: Earth, Venus, and Mars. *Ann. Rev. Earth Planet. Sci.* 29, 489-534.

Evans, D. A. D. (2003). True polar wander and supercontinents, *Tectonophysics*, 362, 303-320.

Formisano, V. (2004). PFS Observation of Methane, Formaldehide and HS: Extant life on Mars?. *International Mars Conference*, Ischia, September 19-23.

Freeman, H. (1961). On the Encoding of Arbitrary Geometric Configurations. *IEEE Trans. Elec. Comp.* EC-10, 260-268.

Friedmann, E. I.; Hua, M.; Ocampo-Friedmann, R. (1993). Terraforming Mars: dissolution of carbonate rocks by cyanobacteria. *Jour. British Interplanet. Soc.* 46, 291-292.

Goldreich, P.; Toomre, A. (1969). Some Remarks on Polar Wandering. *J. Geophys. Res.* 74, 2555.

Haberle, R.; Murphy, J.; Schaeffer, J. (2003). Orbital change experiments with a Mars general circulation model. *Icarus* 161(1) 66-89.

Hanna, J.C.; Phillips, R.J. (2005). Hydrological modeling of the Martian crust with application to the pressurization of aquifers. *J. Geophys. Res.* 110 (E1), 1004.

Harrington, B. W.; Phillips, R. J.; Golombek, M. P. (1999). Extension Across Tempe Terra, Mars from MOLA Topographic Measurements. *Fifth Int. Conf. Mars*, 6130.

Hauber, E.; Kronberg, P. (1999). Differences in style and age of extensional faulting – Examples from the Northern Tharsis Province, Mars. *XXX LPSC,* 1568.

Hauber, E., Kronberg, P. (2001).Tempe Fossae, Mars: A planetary analogon to a terrestrial continental rift?. *J. Geophys. Res.,* 106(E9), 20587-20602.

Hauber, E.; Kronberg, P.; Wolf, U.; Neukum, G. (2002). Tempe volcanic province revisited: new evidence for mid-late amazonian plains volcanism on Mars, and a hotspot beneath SW-Tempe Terra?. *XXXIII LPSC,* 1516..

Head III, J.W., Mustard, J.F., Kreslavsky, M.A., Miliken, R.E.; Marchant, D.R. (2003). Geological evidence for recent ice ages on Mars. *EOS Trans. Am. Geophys. U.* 84(46), P32B-02.

Iwamori, H. (2004). Phase relations of peridotites under H2O-saturated conditions and ability of subducting plates for transportation of H2O. *Earth Planet. Sci. Lett.* 227, 1-2: 57-71.

Jakosky, B. M.; Carr, M. H. (1985). Possible precipitation of ice at low latitudes of Mars during periods of high obliquity. *Nature* 315, 559-561.

Jakosky, B. M.; Henderson, B. G.; Mellon, M. T. (1995). Chaotic obliquity and the nature of the martian climate. *J. Geophys Res.* 100, (E1), 1579-1584.

Jakosky B. M.; Phillips R. J. (2001). Mars Volatile and Climate Evolution: Water the Real Constraints? *XXXII LPSC,* 1147.

Kim, J. R.; Muller, J.; Gasselt, S.; Morley, J. G.; Neukum, G.; Team, H. C. (2005). Automated Crater Detection, A New Tool for Mars Cartography and Chronology. *Photogram. Eng. Rem. Sensing.* 71(10), 1205-1217.

Kronberg, P.; Hauber, E. (1999). Extensional tectonics and new crater statistics of the Northern Tharsis province, Mars. *Fifth Int. Conf. Mars*, Pasadena.

Kumar, P.; Foufoula-Georgiou, E. (1997). Wavelet analysis for Geophysical applications. *Rev. Geophys.* 35, 4, 385-412.

Lantuéjoul, Ch.; Beucher, S. (1980). On the use of the geodesic metric in image analysis, *Jour. Microscopy*, 121, 29-49.

Laskar, J., Correia, A. C. M., Gastineau, M., Joutel, F., Levrard, B., Robutel P. (2002). Long term evolution and chaotic diffusion of the insolation quantities of Mars. *Icarus*, 2004, 170, 2, 343-364

Little, S. A. (1994). Wavelet Analysis of Seafloor Bathymetry: An Example. In *Wavelets in Geophysics*, Efi Foufoula-Georgiou and Praveen Kumar (eds.), 167-182. Academic Press.

Madigan, M.; Martinko, J.; Parker, J. (1999). *Biology of Microorganisms*, Prentice-Hall.

Malin, M. C.; Edgett, K. S. (2000). Evidence for Recent Groundwater Seepage and Surface Runoff on Mars. *Science*. 288, 2330-2335.

Malin, M. C.; Edgett, K. S.; Posiolova, L. V.; McColley, S. M.; Dobrea, E. Z. N. (2006). Present-Day Impact Cratering Rate and Contemporary Gully Activity on Mars. *Science* 314, 1573-1577.

Mallat, S. G. (1989). A theory for multiresolution signal decomposition: the wavelet representation. *IEEE Trans. Patt. Anal. Mach. Intel.* 11(7), 674-693.

Marchant, D.R.; Lewis, A.R.; Phillips, W. M.; Moore, E.J.; Souchez, R.A.; Denton, G.H.; Sugden, D.E.; Potter Jr., N.; Landis, G.P. (2002). Formation of patterned ground and sublimation till over Miocene glacier ice in Beacon Valley, southern Victoria Land, Antarctica. *Geol. Soc. Am. Bull.* 114, 718–730.

Marquez, A.; Fernandez, C.; Anguita, F.; Farelo, A.; Anguita, J.; Casa, M. (2004). New evidence for a volcanically, tectonically, and climatically active Mars. *Icarus* 172(2), 573-581.

Matheron, G. (1967). *Éléments pour une théorie des milieux poreux*, Masson, Paris.

Matheron, G. (1975). *Random sets and integral geometry*, John Wiley & Sons, New York.

Matsumoto, N.; Asada, N.; Demura, H. (2005). Automatic crater recognition on digital terrain model. *XXXVI LPSC*, 1995.

Mattimore, V.; Battista, J. R. (1996). Radioresistance of *Deinococcus radiodurans*: functions necessary to survive ionizing radiation are also necessary to survive prolonged desiccation. *Jour. Bact.* 178 (3), 633–637.

Mège, D.; Masson, P. (1996). A plume tectonic for the Tharsis province, Mars. *Plan. Space Sc.* 44, 1499 – 1566.

Mège, D.; Cook, A.; Garel, E.; Lagabrielle, Y.; Cormier, M. (2003). Volcanic rifting at Martian grabens. *J. Geophys. Res.* 108 (E5), 5044, doi:10.1029/2002JE001852.

Mège, D.; Peulvast, J.; Masson, P. (2005). A planetary example of tectonic inversion: folding and thrusting in the Valles Marineris graben system on Mars. *XXXVI LPSC*, 1772.

Meyer, F. (1979). *Cytologie quantitative et morphologie mathématique*, Thèse de Doctorat, ENSMP, Paris.

Meyer, F. (1986). Automatic screening of cytological specimes. *Comp. Vis. Graph. Im. Proc.* 35, 356-369.

Milliken, R.E.; Mustard, J.F.; Goldsby, D.L. (2003). Viscous flow features on the surface of Mars: Observations from high-resolution Mars Orbiter Camera (MOC) images. *J. Geophys. Res.* 108, E6, 5057, doi:10.1029/2002JE002005.

Mischna, M.; Richardson, M. I.; Wilson, R. J.; McCleese, D. J. (2003). On the orbital forcing of Martian water and CO2 cycles: A general circulation model study with simplified volatile schemes. *J. Geophys. Res.* 108 (E6), 5062, DOI 10.1029/2003JE002051

Mustard, J. F. (2003). Geological evidence for recent Martian ice ages. *Sixth Int. Conf. Mars*, 3250 - 3251.

Mustard, J. F.; Cooper, C.D.; Rifkin, M. K. (2001). Evidence for recent climate change on Mars from the identification of youthful near-surface ground ice. *Nature.* 412, 411-414.

Nimmo, F.; Stevenson, D. J. (2000). Influence of early plate tectonics on the thermal evolution and magnetic field of Mars. *J. Geophys. Res.* 105, 11969-11980.

Okubo, C. H; Schultz, R. A. (2003). Thrust fault vergence directions on Mars: A foundation for investigating global-scale Tharsis-driven tectonics. *Geophys. Res. Lett.* 30 (22), doi: 10.1029/2003GL018664.

Okubo, C. H.; Shultz, R. A.; Stefanelli, G. S. (2004). Gridding Mars Orbiter Laser Altimeter data with GMT: effects of pixel size and interpolation methods on DEM integrity. *Comp Geosci.* 30, 59-72

Peixoto, J.; Kettani, M. (1973). The Control of the Water Cycle. *Scientific American* 228, 46-61.

Purucker, M.; Ravat D.; Frey, H.; Voorhies, C.; Sabaka, C.; Acuña, M. (2000). An altitude-normalized magnetic map of Mars and its interpretation. *Geophys. Res. Lett.* 27, 2449-2452.

Richardson, M. I.; Wilson, R. J. (2002). Investigation of the nature and stability of the Martian seasonal water cycle with a general circulation model. *J. Geophys. Res.* 107, doi:10.1029/2001JE001536.

Saraiva J.; Bandeira L. P. C.; Pina, P. (2006). A structured approach to automated crater detection. *XXXVII LPSC,* 1142.

Schlische, R.; Withjack, M.; Eisenstadt, G. (2002). An experimental study of the secondary deformation produced by oblique-slip normal faulting. *AAPG Bulletin* 86 (5), 885-906.

Schuerger, A. C.; Berry, B.; Nicholson, W. L. (2006). Terrestrial bacteria typically recovered from Mars spacecraft do not appear able to grow under simulated martian conditions. *XXXVII LPSC,* 1397.

Serra J. (1982). *Image Analysis and Mathematical Morphology*, Academic Press, London.

Serra J. (1986). *Éléments de théorie pour l'optique morphologique*, Thèse de Doctorat d'État, Université Paris VI, Paris.

Shea M.A.; Smart D.F. (2001). Vertical cutoff rigidities for cosmic ray stations since 1955 *Proceedings of 27th International Cosmic Ray Conference.* 10, 4063-4066.

Sleep, N. H. (1994). Martian plate tectonics, *J. Geophys. Res.* 99(E3), 5639-5656.

Soille, P. (2002). *Morphological Image Analysis - Principles and Applications*, 2nd Ed. Berlin, Springer-Verlag.

Solomon, S.; Aharonson, O.; Aurnou, J.; Banerdt, W.; Carr, M.; Dombard, A.; Frey, H.; Golombek, M.; Hauck II, S.; Head III, J.; Jakosky, B.; Johnson, C.; McGovern, P.; Neumann, G.; Phillips, R.; Smith, D.; Zuber, M. (2005). New Perspectives on Ancient. *Mars Sci.* 307, 5713, 1214-1220.

Stepinski, T. F.; Collier, M. L. (2004). Extraction of Martian valley networks from digital topography. *J. Geophys. Res.* 109, E11005, doi:10.1029/2004JE002269.

Stevenson, D. (2001). Mars core and magnetism. *Nature*. 412, 214-219.

Tchalenko, J.S. (1970). Similarities between shear zones of different magnitudes. *Bull. Geol. Soc. Amer.* 8, 1625-1640.

Tate, T.; Golombek, M. P.; Mueller, K. J. (2001). Morphology of wrinkle ridges on Lunae and Solis Plana, Mars from MOLA topography: implications for their kinematic development. *XXXII LPSC*, 1444.

Thomas, D. J.; Boling J.; Boston, P. J.; Campbell, K. A.; McSpadden, T. M.; McWilliams, L.; Todd P. (2006). Extremophiles for ecopoiesis: desirable traits for and survivability of pioneer Martian organisms. *Gravitat. Sp. Biol.*19(2), 91-104.

Touma, J.; Wisdom, J. (1993). The chaotic obliquity of Mars. *Science* 259, 1294–1297.

Tuckwell, G. W.; Ghail, R. C. (2003). A 400-km-scale strike-slip zone near the boundary of Thetis Regio, Venus. *Earth Planet. Sci. Lett.*, 211, 1-2, 45-55.

Vaz, D. A.; Alves, E. I.; Barata, M. T. (2006a). Application of mathematical morphology to the enhancement of wavelet-detected fault lines on Mars. *Geophys. Res. Abs.* 8, 2611.

Vaz, D. A.; Barata M. T.; Alves E. I. (2006b). Automatic detection and classification of fault scarps on MOLA data. *XXXVII LPSC*, 1571.

Vaz, D. A.; Barata, M. T.; Alves, E. I. (2006c). Automatic tectonic and geomorphologic characterisation of grabens on Mars. *Proc. XIth Int. Congress Int. Assoc. Math. Geol.*

Wainwright, M.; Wickramasinghe, N. C.; Narlikar, J. V.; Rajaratnam, P. (2003). Microorganisms cultured from stratospheric air samples obtained at 41 km. *FEMS Microbiol. Lett.* 218, 161-165.

Ward, W. R. (1992). Long-term orbital and spin dynamics of Mars. In: *Mars*. H. H. Kieffer, B. M.. Jakosky, C. W. Snyder, and M. S. Matthews (eds.) Univ. Ariz. Press, Tucson, AZ, 298-320.

Wieczorek, M.; Greff-Lefftz, M.; Labrosse, S.; van Thienen, P.; Rouby, H.; Besse, J.; Feldman, W. C. (2005). The Case For a Martian Inertial Interchange True Polar Wander Event. *XXXVI LPSC,*1679

Wilkins, S.J.; Schultz, R.A. (2001). Structural Style and Mode of Extension in the Northern Tempe Rift. *XXXII LPSC*, 1253.

Wilson, L.; J. W. Head III (2002). Tharsis-radial graben systems as the surface manifestation of plume-related dike intrusion complexes: Models and implications, *J. Geophys. Res.* 107, E8, 5057, doi:10.1029/2001JE001593.

INDEX

A

B

C

D

E

F

I

N

O

P

Q

R

S

T

U

V

W

X

Y

Z